Amphibian Conservation

Amphibian Conservation

Global evidence for the effects of interventions

Rebecca K. Smith and William J. Sutherland

Synopses of Conservation Evidence, Volume 4

Pelagic Publishing | www.pelagicpublishing.com

Published by Pelagic Publishing
www.pelagicpublishing.com
PO Box 725, Exeter EX1 9QU

Amphibian Conservation
Global evidence for the effects of interventions
Synopses of Conservation Evidence, Volume 4
www.conservationevidence.com

ISBN 978-1-907807-85-5 (Pbk)
ISBN 978-1-907807-86-2 (Hbk)
ISBN 978-1-907807-87-9 (ePub)
ISBN 978-1-907807-88-6 (Kindle)

Series Editor: William J. Sutherland

British Library Cataloguing in Publication Data
A catalogue record for this book is available from the British Library.

Cover image: Masked treefrog *Smilisca phaeota* from the Choco forests of Colombia. Photograph by Robin Moore.

Contents

Advisory board

We thank the following people for advising on the scope and content of this synopsis:

Associate Professor Phil Bishop, University of Otago, New Zealand
Dr Jaime García Moreno, Amphibian Survival Alliance, the Netherlands
Professor Richard Griffiths, Durrell Institute of Conservation and Ecology, UK
Professor Tim Halliday, Open University, UK
Dr Tibor Hartel, Sapientia University, Cluj-Napoca, Romania
Professor Hamish McCallum, Griffith School of Environment, Australia
Dr Joe Mendelson, Zoo Atlanta, USA
Dr Robin Moore, IUCN SSC Amphibian Specialist Group, USA
Dr Kevin Zippel, IUCN SSC Amphibian Ark, USA

About the authors

Rebecca K. Smith is a Research Associate in the Department of Zoology, University of Cambridge, UK.

William J. Sutherland is the Miriam Rothschild Professor of Conservation Biology at the University of Cambridge, UK.

Acknowledgements

This synopsis was funded by Synchronicity Earth and Arcadia.

We would like to thank Stephanie Prior and Lynn Dicks for providing support throughout the project. We also thank all the people who provied help and advice, including Brian Gratwicke and Helen Meredith, and those who allowed us to access their research.

About this book

The purpose of Conservation Evidence synopses

Conservation Evidence synopses **do**	Conservation Evidence synopses **do not**
• Bring together scientific evidence captured by the Conservation Evidence project (over 4,000 studies so far) on the effects of interventions to conserve biodiversity	• Include evidence on the basic ecology of species or habitats, or threats to them
• List all realistic interventions for the species group or habitat in question, regardless of how much evidence for their effects is available	• Make any attempt to weight or prioritize interventions according to their importance or the size of their effects
• Describe each piece of evidence, including methods, as clearly as possible, allowing readers to assess the quality of evidence	• Weight or numerically evaluate the evidence according to its quality
• Work in partnership with conservation practitioners, policymakers and scientists to develop the list of interventions and ensure we have covered the most important literature	• Provide recommendations for conservation problems, but instead provide scientific information to help with decision-making

Who is this synopsis for?

If you are reading this, we hope you are someone who has to make decisions about how best to support or conserve biodiversity. You might be a land manager, a conservationist in the public or private sector, a farmer, a campaigner, an advisor or consultant, a policymaker, a researcher or someone taking action to protect your own local wildlife. Our synopses summarize scientific evidence relevant to your conservation objectives and the actions you could take to achieve them.

We do not aim to make your decisions for you, but to support your decision-making by telling you what evidence there is (or isn't) about the effects that your planned actions could have.

When decisions have to be made with particularly important consequences, we recommend carrying out a systematic review, as the latter is likely to be more comprehensive than the summary of evidence presented here. Guidance on how to carry out systematic reviews can be found from the Centre for Evidence-Based Conservation at the University of Bangor (www.cebc.bangor.ac.uk).

The Conservation Evidence project

The Conservation Evidence project has three parts:
• An online, **open access journal** *Conservation Evidence* that publishes new pieces of research on the effects of conservation management interventions. All our papers are written by,

or in conjunction with, those who carried out the conservation work and include some monitoring of its effects.

- An ever-expanding **database of summaries** of previously published scientific papers, reports, reviews or systematic reviews that document the effects of interventions.
- **Synopses** of the evidence captured in parts one and two on particular species groups or habitats. Synopses bring together the evidence for each possible intervention. They are freely available online and available to purchase in printed book form.

These resources currently comprise over 4,000 pieces of evidence, all available in a searchable database on the website www.conservationevidence.com.

Alongside this project, the Centre for Evidence-Based Conservation (www.cebc.bangor.ac.uk) and the Collaboration for Environmental Evidence (www.environmentalevidence.org) carry out and compile systematic reviews of evidence on the effectiveness of particular conservation interventions. These systematic reviews are included on the Conservation Evidence database.

Of the 107 amphibian conservation interventions identified in this synopsis, none are the subject of a specific systematic review. One systematic review has been undertaken on the effectiveness of a combination of mitigation actions for great crested newts:

Lewis B. (2012) Systematic evidence review of the effectiveness of mitigation actions for great crested newts. In Lewis B. (2012) *An Evaluation of Mitigation Actions for Great Crested Newts at Development Sites*. PhD thesis. The Durrell Institute of Conservation and Ecology, University of Kent. pp. 61–87.

The systematic review above has been included in three interventions:
- Create ponds
- Restore ponds
- Translocate amphibians

The following interventions we feel would benefit significantly from systematic reviews:
- Translocation of amphibians
- Release of captive-bred or head-started amphibians

In addition, Schmidt and Zumbach (2008) suggested that a systematic review should be undertaken to assess the effectiveness of underpasses and related methods to reduce road deaths.

Schmidt B.R. & Zumbach S. (2008) Amphibian road mortality and how to prevent it: a review. In J.C. Mitchell, R.E. Jung Brown & B. Bartolomew (eds) *Herpetological Conservation*, 3, 157–167.

Scope of the Amphibian Conservation synopsis

This synopsis covers evidence for the effects of conservation interventions for native wild amphibians.

Evidence from all around the world is included. Any apparent bias towards evidence from some regions reflects the current biases in published research papers available to Conservation Evidence.

Husbandry vs conservation of species

This synopsis does not include evidence from the substantial literature on husbandry of pet or zoo amphibians. However, where these interventions are relevant to the conservation of native wild species, they are included (e.g. 'Breed amphibians in captivity', 'Use hormone treatment to induce sperm and egg release during captive breeding', 'Use artificial fertilization in captive breeding' and 'Freeze sperm or eggs for future use').

How we decided which conservation interventions to include

A list of interventions was developed and agreed in partnership with an Advisory Board made up of international conservationists and academics with expertise in amphibian conservation. We have tried to include all actions that have been carried out or advised to support populations or communities of wild amphibians.

The list of interventions was organized into categories based on the International Union for the Conservation of Nature (IUCN) classifications of direct threats and conservation actions.

How we reviewed the literature

In addition to evidence already captured by the Conservation Evidence project, we have searched the following sources for evidence relating to amphibian conservation:

- Eighteen specialist amphibian journals, from their first publication to the end of 2012 (*Acta Herpetologica, African Journal of Herpetology, Amphibian and Reptile Conservation, Amphibia-Reptilia, Applied Herpetology, Australasian Journal of Herpetology, Bulletin of the Herpetological Society of Japan, Contemporary Herpetology, Copeia, Current Herpetology, Herpetologica, Herpetological Bulletin, Herpetological Conservation and Biology, Herpetological Journal, Herpetological Monographs, Journal of Herpetology, Russian Journal of Herpetology* and *South American Journal of Herpetology*).
- Thirty general conservation journals over the same time period.
- Where we knew of an intervention which we had not captured evidence for, we performed keyword searches on ISI Web of Science and www.scholar.google.com for this intervention.
- Evidence published in other languages was included when it was identified.

The criteria for inclusion of studies in the Conservation Evidence database are as follows:

- There must have been an intervention carried out that conservationists would do.
- The effects of the intervention must have been monitored quantitatively.

These criteria exclude studies examining the effects of specific interventions without actually doing them. For example, predictive modelling studies and studies looking at species distributions in areas with long-standing management histories (correlative studies) were excluded. Such studies can suggest that an intervention could be effective, but do not provide direct evidence of a causal relationship between the intervention and the observed biodiversity pattern.

Altogether 416 studies were allocated to interventions they tested. Additional studies published or completed in 2012 or before were added if recommended by the advisory board or identified within the literature during the summarizing process.

How the evidence is summarized

Conservation interventions are grouped primarily according to the relevant direct threats, as defined in the IUCN Unified Classification of Direct Threats (http://www.iucnredlist.org/technical-documents/classification-schemes/threats-classification-scheme). In most cases, it is clear which main threat a particular intervention is meant to alleviate or counteract.

Not all IUCN threat types are included, only those that threaten amphibians, and for which realistic conservation interventions have been suggested.

Some important interventions can be used in response to many different threats, and it would not make sense to split studies up depending on the specific threat they were studying. We have therefore separated out these interventions, following the IUCN's Classification of Conservation Actions (http://www.iucnredlist.org/technical-documents/classification-schemes/conservation-actions-classification-scheme-ver2). The actions we have separated out are: 'Habitat protection', 'Habitat restoration and creation', 'Species management' and 'Education and awareness raising'. These respectively match the following IUCN categories: 'Land/water protection', 'Land/water management – Habitat and natural process restoration', 'Species Management' and 'Education and awareness'.

Normally, no intervention or piece of evidence is listed in more than one place, and when there is ambiguity about where a particular intervention should fall there is clear cross-referencing. Some studies describe the effects of multiple interventions. Where a study has not separated out the effects of different interventions, the study is included in the section on each intervention, but the fact that several interventions were used is made clear.

In the text of each section, studies are presented in chronological order, so the most recent evidence is presented at the end. The summary text at the start of each section groups studies according to their findings.

At the start of each chapter, a series of **key messages** provides a rapid overview of the evidence. These messages are condensed from the summary text for each intervention.

Background information is provided where we feel recent knowledge is required to interpret the evidence. This is presented separately and relevant references are included in the reference list at the end of each background section.

Some of the references containing evidence for the effects of interventions are summarized in more detail on the Conservation Evidence website (www.conservationevidence.com). In the online synopsis, these are hyperlinked from the references within each intervention. They can also be found by searching for the reference details or species name, using the website's search facility.

The information in this synopsis is available in three ways:
- As a book, printed by Pelagic Publishing and for sale from www.pelagicpublishing.com
- As a pdf to download from www.conservationevidence.com
- As text for individual interventions on the searchable database at www.conservationevidence.com.

Terminology used to describe evidence

Unlike systematic reviews of particular conservation questions, we do not quantitatively assess the evidence or weight it according to quality. However, to allow you to interpret evidence, we make the size and design of each trial we report clear. The table below defines the terms that we have used to do this.

The strongest evidence comes from randomized, replicated, controlled trials with paired-sites and before and after monitoring.

Term	Meaning
Site comparison	A study that considers the effects of interventions by comparing sites that have historically had different interventions or levels of intervention.
Replicated	The intervention was repeated on more than one individual or site. In conservation and ecology, the number of replicates is much smaller than it would be for medical trials (when thousands of individuals are often tested). If the replicates are sites, pragmatism dictates that between five and ten replicates is a reasonable amount of replication, although more would be preferable. We provide the number of replicates wherever possible, and describe a replicated trial as 'small' if the number of replicates is small relative to similar studies of its kind. In the case of translocations or release of animals, replicates should be sites, not individuals.

Controlled	Individuals or sites treated with the intervention are compared with control individuals or sites not treated with the intervention.
Paired sites	Sites are considered in pairs, when one was treated with the intervention and the other was not. Pairs of sites are selected with similar environmental conditions, such as soil type or surrounding landscape. This approach aims to reduce environmental variation and make it easier to detect a true effect of the intervention.
Randomized	The intervention was allocated randomly to individuals or sites. This means that the initial condition of those given the intervention is less likely to bias the outcome.
Before-and-after trial	Monitoring of effects was carried out before and after the intervention was imposed.
Review	A conventional review of literature. Generally, these have not used an agreed search protocol or quantitative assessments of the evidence.
Systematic review	A systematic review follows an agreed set of methods for identifying studies and carrying out a formal 'meta-analysis'. It will weight or evaluate studies according to the strength of evidence they offer, based on the size of each study and the rigour of its design. All environmental systematic reviews are available at: www.environmentalevidence.org/index.html
Study	If none of the above apply, for example a study looking at the number of people that were engaged in an awareness raising project.

Taxonomy

Taxonomy has not been updated but has followed that used in the original paper. Where possible, common names and Latin names are both given the first time each species is mentioned within each synopsis.

Where interventions have a large literature associated with them we have sometimes divided studies along taxonomic lines. These do not follow strict taxonomic divisions, but instead are designed to maximize their utility. For example, salamanders and newts have been included together as they may respond to the specific interventions in similar ways.

Habitats

Where interventions have a large literature associated with them and effects could vary between habitats, we have divided the literature using broad habitat types.

Significant results

Throughout the synopsis we have quoted results from papers. Unless specifically stated, these results reflect statistical tests performed on the results.

Multiple interventions

Some studies investigated several interventions at once. When the effects of different interventions are separated, then the results are discussed separately in the relevant sections. However, often the effects of multiple interventions cannot be separated. When this is the case, the study is included in the section on each intervention, but the fact that several interventions were used is made clear.

How you can help to change conservation practice.

If you know of evidence relating to amphibian conservation that is not included in this synopsis, we invite you to contact us, via our website www.conservationevidence.com. You can submit a published study by clicking 'Submit additional evidence' on the right hand side of an intervention page. If you have new, unpublished evidence, you can submit a paper to the Conservation Evidence journal. We particularly welcome papers submitted by conservation practitioners.

1 Threat: Residential and commercial development

The greatest three threats from development tend to be destruction of habitat, pollution, and impacts from 'transportation and service corridors'. Interventions in response to these threats are described in 'Habitat restoration and creation', 'Threat: Pollution' and 'Threat: Transportation and service corridors'. Three interventions that are more specific to development are discussed in this section.

Key messages

Protect brownfield or ex-industrial sites
We captured no evidence for the effects of protecting brownfield sites on amphibian populations.

Restrict herbicide, fungicide and pesticide use on and around ponds on golf courses
We captured no evidence for the effects of restricting herbicide, fungicide or pesticide use on or around ponds on golf courses on amphibian populations.

Legal protection of species
Three reviews, including one systematic review, in the Netherlands and UK found that legal protection of amphibians was not effective at protecting populations during development. Two reviews found that the number of great crested newt mitigation licences issued in England and Wales increased over 10 years.

1.1 Protect brownfield or ex-industrial sites

- We found no evidence for the effects of protecting brownfield sites on amphibian populations.

Background
Brownfield sites include land that was once used for industrial or other human activity, but is then left disused or partially used, for example, disused quarries or mines, demolished or derelict factory sites, derelict railways or contaminated land. Natural recolonization of these sites can result in valuable habitats for wildlife and provide migration corridors in built-up or disturbed areas.

1.2   Restrict herbicide, fungicide and pesticide use on and around ponds on golf courses

- We found no evidence for the effects of restricting herbicide, fungicide or pesticide use on or around ponds on golf courses on amphibian populations.

Background
Studies investigating the effect of reducing chemical applications are discussed in 'Threat: Pollution – Reduce pesticide, herbicide or fertilizer use'.

1.3   Legal protection of species

- Three reviews (including one systematic review) in the Netherlands and UK[2-4] found that legal protection of amphibian species was not effective at protecting populations during development.

- Two reviews in the UK[1, 4] found that the number of great crested newt mitigation licences issued over 10 years increased to over 600 in England and Wales.

Background
Legal protection can be given to species on a national or international scale. Levels of protection vary for species and may include protection against killing, capturing, disturbing or trading, or damaging or destroying their breeding sites or resting places. Depending on the level of protection, activities such as development that are likely to affect protected species in these ways may be against the law and require licences from a government licensing authority.

Other studies that discuss legal protection of species are included in 'Threat: Biological resource use – Use legislative regulation to protect wild populations'.

A review from 1990 to 2001 of great crested newt *Triturus cristatus* mitigation licences in England, UK (1) found that the number issued had increased, from 3 in 1990 to 153 in 2000 and 97 in 2001. Of the 737 licences examined, only 45% contained reporting ('return') documents, a condition of the licence. Great crested newts are a European Protected Species. Licences are therefore issued for certain activities that involve mitigation and/or compensation for the impacts of activities such as development. Licensing information collected by the governmental licensing authorities (1990–2000: English Nature; 2000–2001: Department of the Environment, Food and Rural Affairs) was analysed.

A review of habitat compensation for amphibians in the Netherlands (2) found that legislation was not effective at protecting habitats and amphibians. Only 10% of 20 development projects had completed habitat compensation measures as set out within legal contracts. Some of the compensation required was provided by 55% of projects and none by 35% of projects. Three of the projects created a compensation habitat before destroying a habitat as required, three provided it after destruction and timing was unknown for seven projects. No monitoring data were available from any project. For 11 of 31 projects work had not yet started. In the Netherlands, amphibian species are protected and loss of habitat for these species must be compensated by creating new equivalent habitat.

Thirty-one projects required to undertake compensation were selected from government files. Projects were assessed on the implementation of proposed measures in the approved dispensation contracts and on monitoring data. Field visits were undertaken.

A review in 2011 of compliance with legislation during development projects in the Netherlands (3) found that evidence was not provided to suggest that legislation protected a population of moor frogs *Rana arvalis*. By 2011 only 42% of the compensation area that was required had been provided. Translocation of frogs started in 2007, but as the compensation area was not complete they were released into a potentially unsuitable adjacent habitat. Monitoring before and after translocation was insufficient to determine population numbers or to assess translocation success. The ecological function of the landscape was not preserved during development. In the Netherlands, the Flora and Fauna Act protects amphibians. The development project was required by law to provide a 48 ha compensation area for moor frogs and to translocate the species from the development site to that area.

A review from 2000 to 2010 of great crested newt *Triturus cristatus* mitigation licences issued in England and Wales, UK (4) found that the number issued had increased. Licences issued in England increased from 273 in 2000 to over 600 in 2009. In Wales numbers increased from 7 in 2001 to 26 in 2010. Of the licences examined, only 41% of English licences and 30% of Welsh licences contained reporting ('return') documents, a condition of the licence. Reporting had therefore decreased since 1990–2001 (45%; (1)). Of those that reported, only 9% provided post-development monitoring data; a further 7% suggested surveys were undertaken but no data were provided. The majority of English (71%) and Welsh (56%) licences were for small populations (<10 recorded). Just over half of projects were considered to be of 'low impact', a quarter 'medium impact' and 20% 'high impact' to newts. A review of the governmental licensing authorities (Natural England and Welsh Assembly Government) licence files was undertaken.

In a continuation of a study (4), a systematic review in 2011 of the effectiveness of mitigation actions for legally protected great crested newts *Triturus cristatus* in the UK (4) found that neither the 11 studies captured or monitoring data from licensed mitigation projects showed conclusive evidence that mitigation resulted in self-sustaining populations or connectivity to populations in the wider countryside. Only 5% of 460 licensed projects provided post-development monitoring data and of those, 16 reported that small populations, 3 medium and 1 large population, were sustained. Two reported a loss of populations. The review identified 11 published or unpublished studies and 309 Natural England and 151 Welsh Assembly Government (licensing authorities) mitigation licence files. Mitigation measures were undertaken to reduce the impact of the development and included habitat management, as well as actions to reduce mortality, including translocations.

(1) Edgar P. W., Griffiths R. A. & Foster J. P. (2005) Evaluation of translocation as a tool for mitigating development threats to great crested newts (*Triturus cristatus*) in England, 1990–2001. *Biological Conservation*, 122, 45–52.
(2) Bosman W., Schippers T., de Bruin A. & Glorius M. (2011) Compensatie voor amfibieën, reptielen en vissen in de praktijk. *RAVON*, 40, 45–49.
(3) Spitzen-van der Sluijs A., Bosman W. & De Bruin A. (2011) Is compensation for the loss of nature feasible for reptiles, amphibians and fish? *Pianura*, 27, 120–123.
(4) Lewis B. (2012) An evaluation of mitigation actions for great crested newts at development sites. PhD thesis. The Durrell Institute of Conservation and Ecology, University of Kent.

2 Threat: Agriculture

In Europe, much of the conservation effort is directed at reducing the impacts of agricultural intensification on biodiversity on farmland and in the wider countryside. A number of the interventions that we have captured reflect this. However, the two greatest threats from agriculture tend to be loss of habitat and pollution (e.g. from fertilizer and pesticide use). Interventions in response to these threats are described in 'Habitat restoration and creation', 'Threat: Natural system modifications' and 'Threat: Pollution'.

Key messages – engage farmers and other volunteers

Pay farmers to cover the costs of conservation measures
Four of five studies, including two replicated studies, in Denmark, Sweden and Taiwan found that payments to farmers increased amphibian populations, numbers of species or breeding habitat. One found that amphibian habitat was not maintained.

Engage landowners and other volunteers to manage land for amphibians
Three studies, including one replicated and one controlled study, in Estonia, Mexico and Taiwan found that engaging landowners and other volunteers in habitat management increased amphibian populations and axolotl weight. Six studies in Estonia, the USA and the UK found that up to 41,000 volunteers were engaged in habitat restoration programmes for amphibians and restored up to 1,023 ponds or 11,500 km^2 of habitat.

Key messages – terrestrial habitat management

Manage cutting regime
Studies investigating the effects of changing mowing regimes are discussed in 'Habitat restoration and creation – Change mowing regime'.

Manage grazing regime
Two studies, including one replicated, controlled study, in the UK and the USA found that grazed plots had lower numbers of toads than ungrazed plots and that grazing, along with burning, decreased numbers of amphibian species. Five studies, including four replicated studies, in Denmark, Estonia and the UK found that habitat management that included reintroduction of grazing maintained or increased toad populations.

Reduced tillage
We captured no evidence for the effects of reduced tillage on amphibian populations.

Maintain or restore hedges
We captured no evidence for the effects of maintaining or restoring hedges on amphibian populations.

Plant new hedges
We captured no evidence for the effects of planting new hedges on amphibian populations.

Manage silviculture practices in plantations
Studies investigating the effects of silviculture practices are discussed in 'Threat: Biological resource use – Logging and wood harvesting'.

Key messages – aquatic habitat management

Exclude domestic animals or wild hogs by fencing
Four replicated studies, including one randomized, controlled, before-and-after study, in the USA found that excluding livestock from streams or ponds did not increase overall numbers of amphibians, species, eggs or larval survival, but did increase larval and metamorph abundance. One before-and-after study in the UK found that pond restoration that included livestock exclusion increased pond use by breeding toads.

Manage ditches
One controlled, before-and-after study in the UK found that managing ditches increased toad numbers. One replicated, site comparison study in the Netherlands found that numbers of amphibians and species were higher in ditches managed under agri-environment schemes compared to those managed conventionally.

Engage farmers and other volunteers

2.1   Pay farmers to cover the costs of conservation measures

- Three studies (including one replicated study) in Denmark, Sweden and Taiwan found that payments to farmers created amphibian breeding habitat[1] or increased frog or toad populations[2, 4]. However, a second study in Taiwan[3] found that payments did not maintain green tree frog habitat.

- One replicated, site comparison study in the Netherlands[5] found that ditches managed under agri-environment schemes had higher numbers of amphibian species and higher abundance than those managed conventionally.

Background
Agri-environment schemes are government or inter-governmental schemes designed to compensate farmers financially for changing agricultural practice to be more favourable to biodiversity and landscape. In Europe, agri-environment schemes are an integral part of the European Common Agricultural Policy and Member States devise their own agri-environment prescriptions to suit their agricultural economies and environmental contexts.

Financial incentives to undertake specific management actions with the aim of increasing biodiversity on farmland may also be provided by governmental departments or non-governmental organisations.

Payments to farmers can be provided for many different specific interventions, and where a study's results can be clearly assigned to a specific intervention, they also appear in the appropriate section. This section includes evidence about the success of the actions for amphibian populations following payments.

A study in 1986–1993 of payments to landowners to create ponds on the island of Samsø, Denmark (1) found that landowners created 29 ponds following payments, of which 17 were colonized and 12 used for breeding by green toads *Bufo viridis*. Breeding was successful in 10 of the 12 ponds. Toads colonized the ponds over three years. Private landowners were offered payment by the county to build ponds. Twenty-nine ponds were created in 1989–1992. Fish, crayfish and ducks could not be introduced and a 10 m pesticide-free zone was required around each pond.

A replicated, before-and-after study in 1986–2004 of coastal meadows in Funen County, Denmark (2) found that green toad *Bufo viridis* and natterjack toad *Bufo calamita* populations increased significantly following habitat management supported by agri-environment schemes. On 10 islands with management, green toads increased from 1,132 to over 10,000 adults. In contrast, numbers remained stable on four islands without management. Pond occupancy increased from 27 to 61 ponds in 1997 and ponds with successful breeding from 11 to 22. Natterjacks increased from 3,106 to 4,892 adults in 1997. Ponds with successful breeding remained similar (28 increased to 34). In 2000–2004, numbers dropped and small populations were lost due to insufficient grazing. In 1987–1993, cattle grazing was reintroduced to 111 ha of coastal meadows on 6 islands and continued on a further 10. From 1990, farmers could get financial support from agri-environmental schemes. In addition, 31 ponds were created and 31 restored on 16 islands. Green toad eggs were translocated to one island. Four populations were monitored annually and others less frequently during two or three call, visual and dip-net surveys.

A before-and-after study in 2001–2006 of subsidising farmers to maintain bamboo bushes in Taiwan (3) found that following five years of subsidies, the area of green tree frog *Rhacophorus arvalis* habitat had decreased by approximately 50%. This was considered by the authors to be the result of aging farmers changing from growing bamboo to crops that were less physically demanding and the low price of bamboo. Before agreement finalization in 2006, farmers asked for double the subsidies otherwise they would change their crops. Some did change crops. Taipei Zoo, Taipei Zoological Foundation, the Wild Bird Society of Yunlin and the Farmers' Association of Gukeng Township raised funds for the conservation project. A 5-year agreement was drawn up with 21 farmers to maintain a 5 ha area of bamboo bush that they owned. Farmers were given approximately US $150 each year provided that original farming patterns were maintained, pesticide use was avoided, fallen leaves were left on the ground and bamboo bushes were watered.

A before-and-after study in 1999–2006 of a water lily paddy field in Taipei County, Taiwan (4) found that providing financial incentives resulted in a farmer adopting organic farming practices. Halting herbicide and pesticide use along with habitat management more than doubled a population of Taipei frogs *Rana taipehensis* (from 28 to 85). In 2002, a proportion of a farmer's crop was sold for him and additional expenses resulting from no longer using herbicides and pesticides were paid for. Habitat management, with participation from the local community, included cutting weeds in the field. Community-education programmes about wetland conservation were also carried out in the area.

A replicated, site comparison study of 42 ditches within pasture in the Western Peat District of the Netherlands (5) found that amphibian diversity and abundance were significantly higher in ditches managed under agri-environment schemes compared to conventional management. Adult green frog *Rana esculenta* numbers in conventional ditches declined with distance from reserves; this was not the case in agri-environment scheme ditches. Farmers managing ditches under agri-environment schemes are encouraged to reduce grazing/mowing intensity, reduce fertilizer inputs, and not to deposit mowing cuttings or sediments from ditch cleaning on the ditch banks. Relative amphibian abundance was measured in ditches in April–May and/or May–July 2008. Ditches were perpendicular to eight nature reserve borders and monitoring was just inside reserves and at four distances (0–700 m) from reserve borders. Three methods were used during each sampling period: 5 minute counts, 20 dip net samples and 2 overnight funnel traps.

(1) Amtkjær J. (1995) Increasing populations of the green toad *Bufo viridis* due to a pond project on the island of Samsø. *Memoranda Societatis pro Fauna et Flora Fennica*, 71, 77–81.
(2) Briggs L. (2004) Restoration of breeding sites for threatened toads on coastal meadows. In R. Rannap, L. Briggs, K. Lotman, I. Lepik & V. Rannap (eds) *Coastal Meadow Management – Best Practice Guidelines*, Ministry of the Environment of the Republic of Estonia, Tallinn. pp. 34–43.
(3) Chang J.C.-W., Tang H.-C., Chen S.-L. & Chen P.-C. (2008) How to lose a habitat in 5 years: trial and error in the conservation of the farmland green tree frog *Rhacophorus arvalis* in Taiwan. *International Zoo Yearbook*, 42, 109–115.
(4) Lin H.-C., Cheng L.-Y., Chen P.-C. & Chang M.-H. (2008) Involving local communities in amphibian conservation: Taipei frog *Rana taipehensis* as an example. *International Zoo Yearbook*, 42, 90–98.
(5) Maes J., Musters C.J.M. & De Snoo G.R. (2008) The effect of agri-environment schemes on amphibian diversity and abundance. *Biological Conservation*, 141, 635–645.

2.2   Engage landowners and other volunteers to manage land for amphibians

- Two before-and-after studies (including one replicated study) in Estonia and Taiwan found that habitat management with participation of volunteers increased natterjack toad[1] and Taipei frog[2] populations.

- One controlled study in Mexico[5] found that engaging landowners in aquatic habitat management increased axolotl weight.

- Six studies in Estonia[1], the USA[3, 4, 6, 7] and the UK[8] found that between 8 and 41,000 volunteers were engaged in aquatic and terrestrial habitat restoration programmes for amphibians. Individual programmes restored up to 1,023 ponds[8] or over 11,500 km^2 of habitat[3].

Background
Only 11.5% of the world's land surface is protected (Rodrigues *et al.* 2004). This means that it is vital to engage effectively with landowners so that they manage their land in ways that help to maintain amphibian populations. Volunteers can make a valuable contribution to the management of habitats for amphibians, on private and public land. In some cases the long-term success of habitat management can depend on the involvement of local people.

As well as the direct effects from habitat restoration, volunteer programmes help raise awareness about amphibians and the threats that they face. For example, a study found that participants with high levels of engagement in conservation projects learned more (Evely *et al.* 2011). For interventions that involve engaging volunteers to help manage or monitor amphibian populations see 'Threat: Transportation and service corridors – Use humans to assist migrating amphibians across roads' and 'Education and awareness raising – Engage volunteers to collect amphibian data'.

Evely A.C., Pinard M., Reed M.S. & Fazey L. (2011) High levels of participation in conservation projects enhance learning. *Conservation Letters*, 4, 116–126.
Rodrigues A.S.L., Andelman S.J., Bakarr M.I., Boitani L., Brooks T.M., Cowling R.M., Fishpool L.D.C., da Fonseca G.A.B., Gaston K. J., Hoffmann M., Long J.S., Marquet P.A., Pilgrim J.D., Pressey R.L., Schipper J., Sechrest W., Stuart S.N., Underhill L.G., Waller R.W., Watts M.E.J. & Yan X. (2004) Effectiveness of the global protected area network in representing species diversity. *Nature*, 428, 640–643.

A replicated, before-and-after study in 2001–2004 of three coastal meadows in Estonia (1) found that habitat restoration with participation from 200 volunteers resulted in increased numbers of natterjack toads *Bufo calamita* on 1 island and a halt in the decline of the species on the other 2 islands. In 2001–2004, habitats were restored with the help of 200 volunteers during 14 work camps. Restoration included reed and scrub removal, mowing (cuttings removed) and implementation of grazing where it had ceased. Sixty-six breeding ponds and natural depressions were cleaned, deepened and restored. The project also involved educational and informational activities.

A before-and-after study in 1999–2006 of a water lily paddy field in Taipei County, Taiwan (2) found that participation from the local community resulted in the doubling of a population of Taipei frogs *Rana taipehensis*. Habitat management by the community, along with the halting of herbicide and pesticide use by providing financial incentives to a farmer, resulted in a significant population increase (from 28 to 85). Habitat-improvement work including cutting weeds in the field was undertaken with participation from a local school and the Tse-Xing Organic Agriculture Foundation. Community-education programmes about wetland conservation were also carried out in the area.

A study in 2008 of a partnership programme in the USA (3) found that since establishment the Partners for Fish and Wildlife Program supported over 41,000 private landowners and developed partnerships with over 3,000 nationwide organizations to restore huge areas of habitat. Working together, partners have restored and enhanced 324,000 ha of wetlands, 800,000 ha of uplands and 10,500 km of stream habitat. Data were not provided to determine the effect on target species. The programme run by the US Fish and Wildlife Service was a voluntary habitat restoration programme. It provided technical and financial assistance to private landowners to support the habitat needs of species of conservation concern. Projects included creating and restoring ponds and wetlands for the Puerto Rican crested toad *Peltophryne lemur*, chiricahua leopard frog *Lithobates chiricahuensis* and the California red-legged frog *Rana draytonii*.

A study in 2008 of a pond restoration project within pasture in California, USA (4) found that eight livestock ponds had been restored by ranchers with more restorations planned. To encourage participation, regulatory agencies developed a coordinated permit for pond restorations. The new system enabled ranchers to go to one, rather than up to six, agencies to obtain permits and funding for pond and other management projects. The permit provided guidance on wildlife-friendly pond design and management. Ranchers who participated in the programme were given assurances that they would not encounter extra regulatory obligations under the Endangered Species Act if they restored and maintained ponds to benefit California red-legged frog *Rana draytonii* and California tiger salamander *Ambystoma californiense*.

A controlled study in 2009 of axolotls *Ambystoma mexicanum* in canals through agricultural land in Xochimilco, Mexico (5) found that filters to improve water quality and exclude competitive fish installed with participation of landowners resulted in increased weight gain of axolotls. Only 4 of 12 previously marked axolotls were recaptured; however, their weight had increased by 16%. Weight gain was greater than that of axolotls in control colonies over the same period. Farmers benefited from better-quality farm products as a result of improved water quality and from the protection of traditional agricultural practices. In 2009, with participation from farmers, a canal used as a refuge by axolotls was isolated from the main system using filters made of wood. Filters excluded fish and improved water quality.

A study in 2010 of landowner agreements to manage habitats for amphibians in California, USA (6) found that 8 ranchers and a Municipal Utility District enrolled in 30-year agreements. The eight ranchers managed over 4,000 ha and the Municipal Utility District 8,000 ha of habitat for 2 amphibians of conservation concern, the California red-legged frog *Rana draytonii* and the California tiger salamander *Ambystoma californiense*. Data were not provided to determine the effect on target species. Agreements were made between the US Fish and Wildlife Service and private landowners, with landowners agreeing to carry out management activities for the benefit of priority conservation species. Management included maintenance of stock ponds and surrounding uplands and bullfrog and fish removal. At the end of the agreement landowners were authorized to cease management and return their property to its original condition.

A study in 2012 of a Houston toad *Anaxyrus houstonensis* project in Texas, USA (7) found that landowners attended a workshop and became involved in habitat restoration and protection. Over 200 landowners attended a workshop on wildlife, woodlands and drought. At least 25 landowners (2,000 ha) expressed interest in the project and participated in some form of restoration and stewardship effort for toad habitat. In 2012, a workshop was hosted for landowners, who owned the majority of remaining habitat for the toads. Topics included forest resiliency, wildlife management, Houston toad ecology and landowner cost-share and assistance programmes.

A study in 2012 of the Million Ponds Project in England and Wales, UK (8) found that in 2008–2012 the project team worked with landowners and managers to create 1,023 ponds for rare and declining species. Over 60 organizations were involved and more than 1,016 people were trained in pond creation at 57 events. The aim of the 50-year initiative, started in 2008, was to change attitudes so that pond creation became a routine activity within land management. Pond creation and management training courses were provided to partner and non-partner organizations. Over 50 factsheets were produced as part of an online toolkit and funding for pond creation was also provided.

(1) Rannap R. (2004) Boreal Baltic coastal meadow management for *Bufo calamita*. In R. Rannap, L. Briggs, K. Lotman, I. Lepik & V. Rannap (eds) *Coastal Meadow Management – Best Practice Guidelines*, Ministry of the Environment of the Republic of Estonia, Tallinn. pp. 26–33.
(2) Lin H.-C., Cheng L.-Y., Chen P.-C. & Chang M.-H. (2008) Involving local communities in amphibian conservation: Taipei frog *Rana taipehensis* as an example. *International Zoo Yearbook*, 42, 90–98.
(3) Milmoe J. (2008) Partnerships to conserve amphibian habitat. *Endangered Species Bulletin*, 33, 36–37.
(4) Symonds K. (2008) Ranchers restore amphibian-friendly ponds. *Endangered Species Bulletin*, 33, 30–31.
(5) Valiente E., Tovar A., Gonzalez H., Eslava-Sandoval D. & Zambrano L. (2010) Creating refuges for the axolotl (*Ambystoma mexicanum*). *Ecological Restoration*, 28, 257–259.
(6) Kuyper R. (2011) The role of safe harbor agreements in the recovery of listed species in California. *Endangered Species Bulletin*, 36, 10–13.
(7) Crump P. (2012) The recovery program for the Houston Toad. *Amphibian Ark Newsletter*, 21, 13–14.
(8) Million Ponds Project (2012) Million Ponds Project pond conservation – year 4 report. Pond Conservation Report.

Terrestrial habitat management

2.3   Manage cutting regime

- Studies investigating the effects of changing mowing regimes are discussed in 'Habitat restoration and creation – Change mowing regime'.

Background

Many amphibians require damp terrestrial habitat once they move out of water. If vegetation surrounding water bodies is cut very short, it will not retain sufficient humidity and cover for amphibians during their terrestrial stages. Cutting can also disturb amphibians.

2.4   Manage grazing regime

- One replicated, controlled study in the UK[1] found that grazed plots did not have higher abundance of natterjack toads than ungrazed plots and had lower abundance of common toads. Five studies (including four replicated studies) in Denmark, Estonia and the UK found that habitat management that included reintroduction of grazing increased green toad populations[2, 3], maintained or increased natterjack toad populations[3–5, 7] and maintained common toad populations[4].

- One before-and-after study in the USA[6] found that the decline in amphibian species was similar under traditional season-long or intensive-early cattle stocking.

Background

Livestock grazing changes habitats in a number of ways such as reducing vegetation height, changing plant diversity, creating openings for seed growth and preventing reed and shrub growth. Such changes can have beneficial or detrimental effects on amphibian populations, depending on the amphibian species, grazing intensity and timing.

For an intervention that aims to reduce the detrimental effects of grazing see 'Exclude domestic animals or wild hogs by fencing'.

A replicated, controlled study in 1992–1995 of natterjack toad *Bufo calamita* terrestrial habitat in southern England, UK (1) found that natterjacks did not use grazed plots more than ungrazed plots. There was no significant difference between average numbers in grazed and ungrazed plots for toadlets (13 vs 13) or adults (5 vs 5). Total common toad *Bufo bufo* numbers were lower in grazed compared to ungrazed plots and surroundings (1 vs 11). Four plots (20 x 20 m) of each of four habitats were established: grassy clearfell, sandy clearfell, heath and moss habitat. Two plots of each were grazed in May–September by highland cattle (1 adult/3 ha). Captive-reared natterjack toadlets were released onto each square in summer, 75 in 1992 and 20 in 1993. Toads were monitored twice monthly in April–September 1992–1995.

A replicated, before-and-after study in 1989–1997 of coastal meadows and abandoned fields on two islands in Funen County, Denmark (2) found that the green toad *Bufo viridis* population increased significantly following reintroduction of grazing to fields, along with pond creation and restoration. The population increased from 92 to 2,568. Pond occupancy increased from 10 to 29 ponds and ponds with successful breeding from 4 to 7. In 1989–1997, cattle grazing was reintroduced to 48 ha of coastal meadows and abandoned fields. Four ponds were created and eight restored by removing plants and dredging. Populations were monitored annually in 1990–1997 during two or three call, visual and dip-net surveys. One population was also monitored in 1987–1989.

A replicated, before-and-after study in 1986–2004 of coastal meadows in Funen County, Denmark (3) found that green toad *Bufo viridis* and natterjack toad *Bufo calamita* populations increased significantly following reintroduction of grazing of fields, along with pond creation and restoration. On 10 islands, the total number of green toad adults increased from 1,132 to over 10,000 in 2004. Numbers remained stable on four islands without management. Pond occupancy increased from 27 in 1988 to 61 ponds in 1997 and ponds with successful breeding doubled from 11 to 22. Natterjacks increased from 3,106 in 1988 to 4,892 adults in 1997. Ponds with successful breeding remained similar (28–34). In 2000–2004, numbers dropped and small populations were lost due to insufficient grazing. In 1987–1993, cattle grazing was reintroduced to 111 ha of coastal meadows on 6 islands and continued on a further 10. From 1990, farmers could get financial support from EU agri-environmental schemes. In addition, 31 ponds were created and 31 restored by removing reeds on 16 islands. Green toad eggs were translocated to one island. Four populations were monitored annually and others less frequently during two or three call, visual and dip-net surveys.

A before-and-after study in 1994–2004 of a coastal meadow on a small island in Estonia (4) found that reintroduction of grazing along with aquatic and terrestrial habitat restoration resulted in a stable population of natterjack toads *Bufo calamita*. A total of 17 natterjacks were counted in 1992 and 7 in 2004, with numbers ranging from 1–17/year. The author considered that without management the population may have declined or become extinct. Common toad *Bufo bufo* counts were 8 in 1992 and 4 in 2004 and ranged from 3 to 40/year. Restoration on the 16 ha island involved implementation of sheep grazing, reed and scrub removal and mowing. Toads were counted along a 1 km transect.

A replicated, before-and-after study in 2001–2004 of three coastal meadows in Estonia (5) found that reintroduction of grazing along with aquatic and terrestrial habitat restoration increased the population of natterjack toads *Bufo calamita* on one island and halted the decline on the other two islands. In 2001–2004, habitats were restored where the species still occurred. Restoration included reintroduction of grazing where it had ceased, reed and scrub removal and mowing. Sixty-six breeding ponds and natural depressions were cleaned, deepened and restored.

A before-and-after study in 1989–2003 of tallgrass prairie in Kansas, USA (6) found that there was no significant difference in the decline in amphibian species richness during season-long cattle stocking compared to intensive-early stocking. Although not significant, species richness tended to decline faster during season-long stocking than during intensive-early stocking. Authors considered that strong conclusions could not be reached because of confounding effects of changes in both grazing and burning. From 1989 to 1998, the ranch was managed with traditional season-long stocking (0.6 cattle/ha) with burning in alternate years. From 1999, management changed to intensive-early stocking (1.0 cattle/ha) for three months from late spring combined with annual burning. Amphibians were surveyed in April each year along a 4 km transect.

A replicated, site comparison study in 1985–2006 of 20 sites in the UK (7) found that natterjack toad *Bufo calamita* populations increased with species specific habitat management including introduction of grazing to fields. Populations declined at unmanaged sites.

Individual types of habitat management (aquatic, terrestrial or common toad *Bufo bufo* management) did not significantly affect trends, but length of management did. Overall, 5 of the 20 sites showed positive population trends, 5 showed negative trends and 10 showed no significant trend. Data on populations (egg string counts) and management activities over 11–21 years were obtained from the Natterjack Toad Site Register. Habitat management was undertaken at seven sites. Management varied between sites, but included introduction of grazing, pond creation, adding lime to acidic ponds, maintaining water levels and vegetation clearance. Translocations were also undertaken at 7 of the 20 sites using wild-sourced (including head-started) or captive-bred toads.

(1) Denton J.S. & Beebee T.J.C. (1996) Habitat occupancy by juvenile natterjack toads (*Bufo calamita*) on grazed and ungrazed heathland. *Herpetological Journal*, 6, 49–52.
(2) Briggs L. (2003) Recovery of the green toad *Bufo viridis* Laurenti, 1768 on coastal meadows and small islands in Funen County, Denmark. *Deutsche Gesellschaft für Herpetologie und Terrarienkunde*, 14, 274–282.
(3) Briggs L. (2004) Restoration of breeding sites for threatened toads on coastal meadows. In R. Rannap, L. Briggs, K. Lotman, I. Lepik & V. Rannap (eds) *Coastal Meadow Management – Best Practice Guidelines*, Ministry of the Environment of the Republic of Estonia, Tallinn. pp. 34–43.
(4) Lepik I. (2004) Coastal meadow management on Kumari Islet, Matsalu Nature Reserve. In R. Rannap, L. Briggs, K. Lotman, I. Lepik & V. Rannap (eds) *Coastal Meadow Management – Best Practice Guidelines*, Ministry of the Environment of the Republic of Estonia, Tallinn. pp. 86–89.
(5) Rannap R. (2004) Boreal Baltic coastal meadow management for Bufo calamita. In R. Rannap, L. Briggs, K. Lotman, I. Lepik & V. Rannap (eds) *Coastal Meadow Management – Best Practice Guidelines*, Ministry of the Environment of the Republic of Estonia, Tallinn. pp. 26–33.
(6) Wilgers D.J., Horne E.A., Sandercock B.K. & Volkmann A.W. (2006) Effects of rangeland management on community dynamics of the herpetofauna of the tall grass prairie. *Herpetologica*, 62, 378–388.
(7) McGrath A.L. & Lorenzen K. (2010) Management history and climate as key factors driving natterjack toad population trends in Britain. *Animal Conservation*, 13, 483–494.

2.5   Reduce tillage

- We found no evidence for the effects of reduced tillage on amphibian populations.

Background

Conventional ploughing uses a mould-board plough, cultivating to a depth of around 20 cm. A number of methods can be used to reduce the depth or intensity of ploughing, such as layered cultivation, non-inversion tillage and conservation tillage. Such have been found to be beneficial for some farmland biodiversity (Holland & Luff 2000).

Holland J.M. & Luff M.L. (2000) The effects of agricultural practices on Carabidae in temperate agroecosystems. *Integrated Pest Management Reviews*, 5, 109–129.

2.6   Maintain or restore hedges

- We found no evidence for the effects of maintaining or restoring hedges on amphibian populations.

Background

Hedgerows can provide valuable migration corridors for wildlife, particularly in disturbed landscapes. For example, newts migrating away from breeding ponds were found to use hedgerows more than expected within a pastoral landscape (Jehle & Arntzen 2000).

Jehle, R. & Arntzen, J.W. (2000). Post-breeding migrations of newts (*Triturus cristatus* and *T. marmoratus*) with contrasting ecological requirements. *Journal of Zoology*, 251, 297–306.

2.7   Plant new hedges

• We found no evidence for the effects of planting hedges on amphibian populations.

Background
Hedgerows can be planted to provide migration corridors for amphibians and for resources for other wildlife.

2.8   Manage silviculture practices in plantations

• Studies investigating the effects of silviculture practices are discussed in 'Threat: Biological resource use – Logging and wood harvesting'.

Background
Forestry practices, particularly clear-cutting all trees and vegetation, can have significant effects on amphibian populations. There are a number of silviculture management practices that can be carried out to try to reduce the effect of timber harvest on wildlife. These include retaining some scattered or groups of trees, which ensures that some canopy cover remains and therefore that forest floor or stream conditions are maintained in some areas.

Aquatic habitat management

2.9   Exclude domestic animals or wild hogs by fencing

• Three replicated, site comparison studies in the USA[1, 3, 5] found that excluding livestock from streams or ponds did not increase numbers of amphibian species or overall abundance, but did increase larval abundance[1, 3] and abundance of green frog metamorphs[5]. Two studies found that the abundance of green frogs and/or American toads was higher with grazing[1, 5].

• One randomized, replicated, controlled, before-and-after study in the USA[4] found that excluding cattle from ponds did not increase numbers of eggs or larval survival of Columbia spotted frogs. One before-and-after study in the UK[2] found that pond restoration that included livestock exclusion increased pond use by breeding natterjack toads.

Background
Livestock grazing can have significant effects on aquatic habitats through disturbance, trampling, erosion, reduced water quality and changes in vegetation structure and composition. Such changes may have detrimental effects on amphibian populations. However, grazing can also have beneficial effects on amphibians and their habitats. For example, a study found that 3 years after excluding livestock grazing by fencing, ungrazed temporary pools dried 50 days earlier than grazed pools (Pyke & Marty 2005).

Other studies investigating the effects of grazing on amphibians are discussed in 'Manage grazing regime'.

Pyke C.R. & Marty J. (2005) Cattle grazing mediates climate change impacts on ephemeral wetlands. *Conservation Biology*, 19, 1619–1625.

A replicated, site comparison study in 1998–1999 of streams in pasture in Pennsylvania, USA (1) found that excluding livestock from stream banks did not increase amphibian species richness or abundance overall, but did increase tadpole numbers. There was no significant difference in overall species richness, abundance or biomass, or in the abundance of salamanders, bullfrogs *Rana catesbeiana* or wood frog *Rana sylvatica* between fenced and unfenced streams. However, tadpole captures were higher in fenced compared to unfenced areas (20 vs 6). In comparison, captures were higher in unfenced compared to fenced areas for green frogs *Rana clamitans* (8 vs 5/site) and American toads *Bufo americanus* (2.4 vs 1.5). Ten grazed and 10 recently fenced (1–2 yrs) streams were selected over 20 farms. Sites were 100 m long by 10–15 m wide on both banks. Monitoring was undertaken using two drift-fences per site. Each fence had a pitfall trap, side-flap pail-trap and funnel trap that were checked 3–4 times/week in April–July.

A before-and-after study in 1991–1999 of 17 ponds in a reserve in Caerlaverock, Scotland, UK (2) found that pond restoration with livestock exclusion increased natterjack toad *Bufo calamita* use of ponds for breeding. Out of 12 ponds restored in 1995–1998, 11 were used for breeding every year until 1999, compared to just 4 before restoration. Toads started to breed in the additional ponds one or two years after restoration. Toads continued to breed in ponds used before restoration and there was little change in use of unmanaged ponds. Of the 11 ponds restored in 1995–1996, 10 were used for breeding every year until 1999. In 1995–1999, 17 ponds were restored by clearing aquatic vegetation and excavation. Electric fences were installed around ponds during the summer to exclude cattle and sheep. Fences were removed after toadlet emergence. Eggs, tadpoles and toadlets were counted at least four times in each pond in May–August 1991–1992 and 1994–1999.

A replicated, site comparison study in 2002–2003 of streams within pasture in southwestern Georgia, USA (3) found that excluding cattle did not result in increased amphibian species richness or abundance along stream banks, but did result in significantly higher numbers of in-stream larvae. There was no significant difference in amphibian species richness between buffered and unbuffered streams, although species richness tended to be higher where cattle were excluded. Abundance of adult salamanders and treefrogs *Hyla* spp. did not differ between sites. At 3 sites cattle grazed stream banks and at 2 other sites cattle had been excluded by fencing for over 25 years. Amphibians were monitored by walking a transect (100 x 4 m) along one side of each stream from March 2002 to March 2003. Bimonthly surveys under natural and artificial cover objects (30 tiles/site) and monthly surveys using tree pipes (10/site) and stream bottom samplers were undertaken.

A randomized, replicated, controlled, before-and-after study in 2002–2006 of 12 ponds in Oregon, USA (4) found that there was no effect of complete or partial cattle exclusion on Columbia spotted frog *Rana luteiventris* egg numbers, larval survival or size at metamorphosis. There was no significant difference between treatments for egg mass counts (exclusion: 8; partial exclusion: 4; access: 7); pre-treatment counts were 6–11. The same was true for larval survival index (exclusion: 25; partial exclusion: 52; access: 33; pre-treatment: 30–72) and size at metamorphosis (pre-treatment: 28–33 mm; post-treatment: 29–31). Fishless ponds within four blocks were randomly assigned to one of three treatments: complete cattle exclusion, exclusion from a section of pond (where most eggs were laid) or no exclusion. Fences were installed in 2003–2005 creating a 1–5 m buffer around ponds. Cattle were present in June–September (25–31 ha/cow-calf pair). Egg masses were counted and a sample of juveniles marked in 2002–2006.

A replicated, site comparison study in 2005–2006 of eight farm ponds in Tennessee, USA (5) found that the effects of excluding cattle from ponds depended on amphibian species. There was no significant difference in captures or egg mass abundance for 12 species. However, significantly higher numbers of green frog *Rana clamitans* metamorphs were captured at exclusion ponds compared to those with cattle grazing (0.06–0.10 vs 0.01–0.03 relative captures/day). The opposite was true for American toads *Bufo americanus* (0 vs 0.01–0.03). Length and/or mass were significantly greater at exclusion ponds for one and grazed ponds for four species. Four ponds had been exposed to grazing (132 cattle/pond ha/month) and 4 fenced to prevent grazing for 10 years. Ponds were 0.1–1.0 ha and within similar habitat. Amphibians were monitored using pitfall traps both sides of drift-fencing enclosing half of each pond. Traps were set for two days/week in March–August 2005–2006. Weekly egg mass counts were also undertaken along transects.

(1) Homyack J.D. & Giuliano W.M. (2002) Effect of streambank fencing on herpetofauna in pasture stream zones. *Wildlife Society Bulletin*, 30, 361–369.
(2) Phillips R.A., Patterson D. & Shimmings P. (2002) Increased use of ponds by breeding natterjack toads, Bufo calamita, following management. *Herpetological Journal*, 12, 75–78.
(3) Muenz T.K., Golladay S.W., Vellidis G. & Smith L.L. (2006) Stream buffer effectiveness in an agriculturally influenced area, southwestern Georgia: Responses of water quality, macroinvertebrates, and amphibians. *Journal of Environmental Quality*, 35, 1924–1938.
(4) Adams M. J., Pearl C. A., Mccreary B., Galvan K., Wessell S. J., Wente W. H., Anderson C. W. & Kuehl A. B. (2009) Short-term effect of cattle exclosures on Columbia spotted frog (*Rana luteiventris*) populations and habitat in northeastern Oregon. *Journal of Herpetology*, 43, 132–138.
(5) Burton E.C., Gray M.J., Schmutzer A.C. & Miller D.L. (2009) Differential responses of postmetamorphic amphibians to cattle grazing in wetlands. *Journal of Wildlife Management*, 73, 269–277.

2.10   Manage ditches

- One controlled, before-and-after study in the UK[2] found that managing ditches increased common toad numbers.

- One replicated, site comparison study in the Netherlands[1] found that numbers of amphibian species and abundance was significantly higher in ditches managed under agri-environment schemes compared to those managed conventionally.

Background

Intensification of agricultural and other land management can result in loss of ditch biodiversity through activities such as mowing, grazing and use of fertilizer and pesticides leading to water pollution. These can have significant effects on amphibian populations. Ditch management practices such as the frequency, season and technique used to clean or dredge ditches have also been found to affect the presence of amphibians (Twisk *et al.* 2000). Management practices that maintain and increase species diversity should therefore be encouraged.

Twisk W., Noordervliet M.A.W. & ter Keurs W.J. (2000) Effects of ditch management on caddisfly, dragonfly and amphibian larvae in intensively farmed peat areas. *Aquatic Ecology*, 34, 397–411.

A replicated, site comparison study of 42 managed ditches within pasture in the Western Peat District of the Netherlands (1) found that amphibian diversity and abundance was significantly higher in agri-environment schemes compared to conventionally managed ditches. Adult green frog *Rana esculenta* numbers in conventional ditches declined with distance from reserves; this was not the case in agri-environment scheme ditches. Farmers managing ditches under agri-environment schemes are encouraged to reduce grazing/ mowing intensity and reduce fertilizer inputs compared to conventional management, and not to deposit mowing cuttings or sediments from ditch cleaning on the ditch banks. Monitoring was undertaken along 18 agri-environment and 24 conventionally managed ditches in April–July 2008. Ditches were perpendicular to eight nature reserve borders and monitoring was just inside reserves and at four distances from reserve borders (0–700 m). Three methods were used during each sampling period: 5 minute counts, 20 dip net samples and 2 overnight funnel traps.

A controlled, before-and-after study in 1999–2012 of seven ditches in pasture in Suffolk, UK (2) found that common toad *Bufo bufo* numbers increased after restoring ditch management. Numbers of adults counted three to seven years after management (after 3–4 years toad maturation) were significantly higher than in the subsequent five years once management ceased (563 vs 245). The year after ditch clearance, large numbers of tadpoles were seen and toadlets increased from 10s–100s to 1,000s in one of the dredged ditches. In comparison, highly vegetated unmanaged ditches supported few or no tadpoles through to metamorphosis. Ditch management including dredging was undertaken in five of seven ditches in 1999. Monitoring was undertaken three times in March by egg counts, torchlight surveys, netting ditches and counting breeding adults.

(1)   Maes J., Musters C.J.M. & De Snoo G.R. (2008) The effect of agri-environment schemes on amphibian diversity and abundance. *Biological Conservation*, 141, 635–645.
(2)   Beebee T. (2012) Decline and flounder of a Sussex common toad (*Bufo bufo*) population. *Herpetological Bulletin*, 121, 6–16.

3 Threat: Energy production and mining

Energy production (renewable and non-renewable) and mining can have significant impacts on amphibian populations through the destruction and pollution of habitats. Interventions in response to these threats are discussed in 'Habitat restoration and creation' and 'Threat: Pollution – Industrial pollution'.

Key messages

Artificially mist habitat to keep it damp
One before-and-after study in Tanzania found that installing a sprinkler system to mitigate against a reduction of river flow did not maintain a population of Kihansi spray toads.

3.1   Artificially mist habitat to keep it damp

- One before-and-after study in Tanzania[1, 2] found that installing a sprinkler system to mitigate against a 90% reduction of river flow did not maintain a population of Kihansi spray toads.

Background
Reduction in river flow due to activities such as the implementation of hydropower projects can have significant effects on wetland habitats and the amphibians they support. In cases where alternative habitat is not available, intensive management may be undertaken to recreate natural habitats in an attempt to conserve particular species. For example, the wetland habitat in the study described below was the only known habitat for the Kihansi spray toad *Nectophrynoides asperginis*.

A before-and-after study in 1996–2004 of a sprinkler system to mitigate against a 90% reduction of river flow caused by a hydropower project along the Lower Kihansi River, Tanzania (1, 2) found that following a brief recovery, the Kihansi spray toad *Nectophrynoides asperginis* declined rapidly. Following the implementation of the sprinkler system, numbers increased to approximately 20,000 by June 2003 from 11,000 in 2000. However, the population then declined rapidly to 40 in August 2003 and 5 in January 2004. The authors suggest the cause for the sudden decline may have been the introduction of the chytrid fungus or pesticides. The population estimate for the toads had decreased from 50,000 in 1996 to 11,000 toads in 2000 once the river flow was reduced. The hydropower project was implemented in May 2000 resulting in a reduction of water flow, but the sprinkler system was not completed until February 2001. The system comprised a several kilometre-long gravity-fed pipe system that delivered mist from hundreds of spray nozzles onto a quarter of the suitable toad habitat.

(1) Krajick K. (2006) The lost world of the Kihansi toad. *Science*, 311, 1230–1232.
(2) Conservation Breeding Specialist Group (IUCN/SSC) (2007) Kihansi spray toad (*Nectophrynoides asperginis*) population and habitat viability assessment: briefing book. Conservation Breeding Specialist Group (IUCN/SSC) Report.

4 Threat: Transportation and service corridors

The greatest threats from transportation and service corridors tend to be from the destruction of habitat and pollution. Interventions in response to these threats are described in 'Habitat restoration and creation' and 'Threat: Pollution'.

Key messages

Install culverts or tunnels as road crossings
Thirty-two studies investigated the effectiveness of installing culverts or tunnels as road crossings for amphibians. Six of seven studies, including three replicated studies, in Canada, Europe and the USA found that installing culverts or tunnels decreased amphibian road deaths. One found no effect on road deaths. 15 of 24 studies, including 1 review, in Australia, Canada, Europe and the USA found that tunnels were used by amphibians. Four found mixed effects depending on species, site or culvert type. Five found that culverts were not used or were used by less than 10% of amphibians. Six studies, including one replicated, controlled study, in Canada, Europe and the USA investigated the use of culverts with flowing water. Two found that they were used by amphibians. Three found that they were rarely or not used. Certain culvert designs were found not to be suitable for amphibians.

Install barrier fencing along roads
Seven of eight studies, including one replicated and two controlled studies, in Germany, Canada and the USA found that barrier fencing with culverts decreased amphibian road deaths, in three cases depending on fence design. One study found that few amphibians were diverted by barriers.

Modify gully pots and kerbs
One before-and-after study in the UK found that moving gully pots 10 cm away from the kerb decreased the number of great crested newts that fell in by 80%.

Use signage to warn motorists
One study in the UK found that despite warning signs and human assistance across roads, some toads were still killed on roads.

Close roads during seasonal amphibian migration
Two studies, including one replicated study, in Germany found that road closure sites protected large numbers of amphibians from mortality during breeding migrations.

Use humans to assist migrating amphibians across roads
Three studies, including one replicated study, in Italy and the UK found that despite assisting toads across roads during breeding migrations, toads were still killed on roads and 64–70% of populations declined. Five studies in Germany, Italy and the UK found that large numbers of amphibians were moved across roads by up to 400 patrols.

4.1   Install culverts or tunnels as road crossings

- Thirty-two studies investigated the effectiveness of installing culverts or tunnels as road crossings for amphibians.

- Six of seven studies (including three replicated studies) in Canada, Germany, Italy, Hungary and the USA[7, 17, 18, 23, 29, 32, 33] found that installing culverts or tunnels significantly decreased amphibian road deaths; in one study this was the case only when barrier fencing was also installed. One found no effect on road deaths[32].

- Fifteen of 24 studies (including one review and 17 replicated studies) in Australia, Canada, Europe and the USA found that culverts/tunnels were used by amphibians[1, 9, 31], by 15–85% of amphibians[2, 6, 7, 10, 19, 32] or 3–15 species[15, 17, 21], or that 23–100% of culverts or tunnels were used by amphibians[11, 16] or used in 12 of 14 studies reviewed[28]. The majority of culverts/tunnels had barrier fencing to guide amphibians to entrances. Four found mixed effects depending on species[22], or for toads depending on the site or culvert type[4, 8, 18]. Five found that culverts were used by less than 10% of amphibians[3, 14, 30] or were not used[5, 13]. The use of culverts/tunnels was affected by diameter in three of six studies, with wider culverts used more[4, 11–13, 26, 27]; length in one of two studies, with long culverts avoided[26, 27]; lighting in all three studies, with mixed effects[10, 13, 26]; substrate in three of six studies, with natural substrates used more[12, 13, 19, 20, 25–27]; presence of water in two of three studies, with mixed effects[3, 11, 13]; entrance location in one[11]; and tunnel climate in one study[31].

- Six studies (including one replicated, controlled study) in Canada, Spain, the Netherlands and the USA investigated the use of culverts with flowing water and found that they were used by amphibians[12, 17], or rarely used by salamanders[20, 25] or not used[24], and were used more[11] or the same amount as dry culverts[13].

- Certain culvert designs were not suitable for amphibians; one-way tunnels with vertical entry chutes resulted in high mortality of common toads[4] and condensation deposits from steel culverts had very high metal concentrations[13]. One study found that thousands of amphibians were still killed on the road[1].

Background
Roads and traffic can have major impacts on amphibian populations. This is particularly the case if they cut across annual migration routes between hibernation and breeding habitats. Underpasses can be installed to try to reduce mortality on roads. Unlike methods such as toad patrols and road closures, which tend to target breeding adults, tunnels could help reduce deaths of dispersing juveniles. Tunnels may be designed specifically for amphibian migrations: wildlife pipes over land, wildlife culverts over water channels designed for small- to medium-sized animals or drainage culverts that were engineered for water passage, but that can be modified to encourage wildlife passage.

Culverts or tunnels are usually associated with barrier walls that prevent amphibians reaching the road and direct them towards tunnels. Studies that specifically investigated the effect of barrier fencing along roads are discussed in 'Install barrier fencing along roads'.

A study in 1983–1984 of a tunnel with guide fencing in Oberbergischer Kreis, Germany (1) found that 640 common toads *Bufo bufo* and 4 frogs migrated through the tunnel. Overall, 85% of adult and 90% of young migrating toads used the tunnel, a nearby brook pipe, footpath or bridge to get across the road. However, on one night thousands of young toads were killed on the road. The tunnel was 19 m long and 0.75 m high and was completed in March 1984. A fence was constructed to direct amphibians to the tunnel. Monitoring was undertaken using pitfall traps at the ends of the tunnel and by observing toads during the migration.

A study in 1982–1989 of a tunnel under a road through woodland in Schleswig Holstein province, Germany (2) found that 21% of amphibians recorded along the drift-fencing used the tunnel. In 1988, a total of 2,446 amphibians were recorded along the fence, of which 21% passed through the tunnel. Seven species were recorded using the tunnel. For the 4 species for which more than 10 individuals were recorded (136–1278/species) 12–45% passed through the tunnel. The tunnel was installed in 1987 (0.2 m diameter, 10 m long). Drift-fencing 360 m long and 0.4 m high already existed at the site. Amphibians were monitored using 28 pitfall traps along the fence and one at the tunnel exit.

A study in 1984–1985 of a tunnel with barrier fencing in Lower Saxony, Germany (3) found that only 15% of amphibians recorded entered the tunnel and few passed through the tunnel. It was considered that this may have been due to high water levels which resulted in a stream flowing through the tunnel. Fences 350 m long were installed on both sides of the road. The concrete tunnel was located in the centre of the fences. Common toads *Bufo bufo* and common frogs *Rana temporaria* were monitored in March–April. Toads were tagged.

A replicated study in 1987 of tunnels with guide fencing at 13 locations in West Germany (4) found that tunnel use by amphibians varied with site. Some tunnels were not used by amphibians while others were used by the majority of migrating amphibians. Large two-way tunnels (diameter: 1 m; length: 15 m) were used by a larger proportion of common toads *Bufo bufo* in the area than those with smaller diameters. However, even those with a diameter of 0.3 m were used by some toads. One-way tunnels with vertical entry chutes resulted in high mortality of amphibians. There were no deaths with angled chutes. Three types of tunnels were investigated: two-way systems or one-way systems with angled or vertical entry chutes. One-way tunnels were laid in pairs to allow migration in both directions. At one site, nine two-way systems of various dimensions were investigated. Guide fences were also used at sites.

A replicated study in 1980–1988 of eight tunnels with barrier fencing in Bavaria, West Germany (5) found that common toads *Bufo bufo* did not use the tunnels. Tunnels were 60 cm in diameter. Wire-netting fences 25 cm high were installed on both sides of the road. Fences were bent over at the top to prevent toads climbing over.

A small, replicated study in 1987 of two tunnels with barrier fencing in Henley-on-Thames, England, UK (6) found that approximately 2,750 common toads *Bufo bufo* used the tunnels during 18 migration nights. In the first 2 nights, only about 10% of 2,200 toads recorded behind the fence were estimated to have used the tunnels. This increased to a maximum of 43% of toads recorded in one night. Fencing was installed between the two tunnels creating a W-shaped catchment of 600 m. A trip counter was set 0.2 m into the entrance of the tunnels.

A small, replicated study in 1987–1988 of two amphibian tunnels with barrier fencing in the Mittelgebirge region of West Germany (7) found that once an effective fence was installed, 85% of amphibians recorded used the tunnels and road deaths decreased. Prior to the new fence, numbers killed were 109/night, compared to just 20 in 1987 and 30 in 1988. Between 2,432 and 2,050 individuals/year were captured at the fence and surroundings during the spring migration, of which 85% used the tunnels. Of 211 toads marked at the fence in 1987,

68% were recaptured at tunnel exits within 5 days. Two drain channels with metal grid roofs were installed in the road in 1981. A more effective fence of plastic fabric similar to wire mesh (1 m high) was installed at entrances and parallel to the road in 1987. Pitfall traps were set at each end of the fence and at tunnel exits.

A replicated study of five amphibian tunnels with barrier fencing in Overveen in the Netherlands (8) found that only 4% of the population of 2,000–3,000 common toads *Bufo bufo* used the tunnels. Ten percent of the population broke over the barrier fencing. The remaining toads walked along the fence, were captured in pitfall traps and were carried across the road. In an experiment, toads were placed at tunnel entrances and 43% passed through within 24 hours. The cast-iron tunnels had been installed nine years before the study. They were 12 m long, 0.3 m in diameter and were buried 0.7 m under a road between a wooded dune and stream. The road had permanent barrier fencing.

A replicated study in 1993 of 17 culverts in Madrid province, Spain (9) found that amphibians used the culverts. An average of 0.03 amphibian tracks/culvert/day (range: 0–0.19) were recorded. Two culverts were selected under a motorway, 10 under local roads and 5 under a railway line. Amphibian tracks were monitored within culverts using marble dust over the floor. Monitoring was undertaken over four to eight days each season.

A small, replicated study in 1988 of two amphibian tunnels under a road in Amherst, Massachusetts, USA (10) found that 76% of spotted salamanders *Ambystoma maculatum* that reached tunnel entrances successfully passed through ($n = 87$). Of the salamanders recorded along fences 68% ($n = 95$) passed through tunnels. Salamanders that encountered fences furthest from the tunnels reached tunnels as successfully as those that encountered the fence closer to the entrances. Once artificial light was provided, salamanders entered and passed through tunnels faster. Tunnels were installed approximately 60 m apart with 30 m long (0.3 m high) drift-fences to direct salamanders to the entrances. Tunnels allowed some rain to enter to maintain moist conditions, but were prevented from flooding. Tunnels and fences were monitored by observations on four nights during spring 1988.

A replicated, site comparison study in 1993–1994 of 56 tunnels under roads in Catalonia, Spain (11) found that amphibians used 23% of circular and 59% of rectangular tunnels. Use was greater for wider tunnels with water within or at entrances. Tunnels with steps or wells at the entrances or within large embankments were used less frequently. A total of 39 circular (1–3 m diameter) and 17 rectangular cross-section (4–12 m diameter) drains/underpasses were surveyed along four 10 km stretches of roads. Tunnels were monitored for four days each season over a year in 1993–1994. Tracks were obtained using marble power across the centre of each structure. Infra-red and photographic cameras were used at entrances.

A replicated study in 1997–1998 of 53 wildlife passages along waterways under roads at over 20 sites in the Netherlands (12) found that 77% of passages were used by amphibians. Amphibian tracks were recorded in 19–22 passages/year. There was no relationship between use and passage width or substrate. Culverts and bridges were adapted for wildlife in the 1990s in the Netherlands. In 1997, 31 passages (0.4–3.5 m wide) were monitored. These included extended banks (unpaved or paved), planks fixed on bridge or culvert walls, planks floating on the water, concrete passageways and plastic gutters covered with sand. In 1998, 22 passages were monitored for the effect of width and substrate. These were wooden passageways fixed on a bridge or culvert wall (0.2–0.6 m wide). Monitoring involved weekly checks of tracks on sandbeds (for 4–7 weeks) and ink pads (12 weeks in 1997, 4 weeks in 1998) across passageways.

A replicated study in 2000 of eight dry and two wet culverts under highways through two wetlands on Vancouver Island, Canada (13) found no amphibian tracks within culverts.

In trials with rough-skinned newts *Taricha granulosa*, a dark culvert was used significantly more than one with daylight (24 vs 6). However, there was no significant difference between use of 0.3 or 0.5 m diameter culverts (11 vs 19 newts), different substrates (bare: 22; cement: 11; soil: 17) or wet or dry culverts (8 vs 7–15 newts). Concentrations of aluminium, zinc, copper and lead within condensation deposits in culverts were 134–124,500 times greater than recommended for protecting freshwater aquatic life. Corrugated steel pipe culverts (29–36 x 0.6–1 m) were constructed in 1995. Aluminium track-plates covered with soot were installed 1–2 m inside each culvert and were monitored 9 times in July–October 2000. There were 3 replicates of each trial (5 for substrate) in which 10 newts had the choice of 3 adjacent culverts (3 x 0.3 m) over 3 days in September–November.

A small, replicated study in 2000–2001 of two amphibian tunnels constructed under a road in a residential development in Santa Cruz County, California, USA (14) found that a small proportion of migrating Santa Cruz long-toed salamanders *Ambystoma macrodactylum croceum* used tunnels. A total of 23 adult salamanders passed through the tunnels. Of the 44 adults marked along the drift-fence, only 4 (9%) were captured on the opposite side of one tunnel and none for the other. The two cement polymer amphibian tunnels were installed in 1999. They were 0.3 x 0.5 m or 0.2 x 0.2 m and 11–12 m long. Entrances were screened with mesh to reduce predator access. Drift-fences (0.4–0.8 m high) were permanently installed at tunnel entrances and along the road to connect tunnels (300 m). Salamanders were monitored by visual survey along the drift-fence on five rainy nights in December–January. Each animal was marked. Pitfall traps captured individuals passing through tunnels.

A replicated study in 1998 of 38 amphibian tunnels at 16 sites, 2 game bridges and 5 game passages in northern Hungary (15) found that 11 amphibian species used the passageways. Some of the passageways were used successfully and others had efficiency below 25%. Problems were considered to include improper design, gaps between the fence and entrance and lack of fencing or maintenance. Population estimates suggested that the mitigation measures helped 1 million to 5 million amphibians across roads annually. Tunnels were circular or square, made of concrete or metal and had diameters of 0.6–1.0 m. Concrete or mesh fences (0.5–0.7 m high) were present in 80% of cases. Day and night road transects were undertaken during spring and summer to count live and dead amphibians. Population sizes in neighbouring habitats were estimated using visual encounter surveys, torching and netting, acoustic surveys and transect counts.

A replicated study in 2000–2001 of nine wildlife culverts with barrier fencing along a highway through coastal lowlands in New South Wales, Australia (16) found that all culverts were used by amphibians. Amphibian tracks made up 14% of those in culverts. Cane toads *Bufo marinus* were observed inside culverts nine times. Twelve additional species were recorded within 2–20 m of entrances. Fifty-five frog (brown-striped frog *Limnodynastes peronii*, dainty green tree frog *Litoria gracilenta*) and 2 cane toad carcasses and 14 live frogs were recorded on the road on one night. The concrete culverts (2.4 m wide, 1.2 m high, 18 m long) lay along a 1.4 km section of highway. A chain-mesh barrier fence (1.8 m high) was installed either side of the bypass. Each culvert was walked through with a spotlight on two wet and two dry nights in January–February 2001. Tracks were recorded on sand across culverts every two days over eight days in spring and autumn. Frog calls were also recorded at entrances.

A replicated study in 2001–2002 of eight culverts underneath a highway through a freshwater marsh in Florida, USA (17) found that 13 frog and 2 salamander species used culverts and road mortality declined. A total of 656 frogs and 6 salamanders were captured using culverts. Following construction of a barrier wall linking culverts, frog species using culverts increased from 5 to 13 and frogs trapped increased from 0.006 to 0.085 trap/night. Ranid frog mortality declined dramatically following installation of the barrier wall-culvert system. However, tree frog mortality appeared to increase (from 149 to 194). Two dry box culverts (1.8 x 1.8 x 44 m) and two partially submerged box culverts (2.4 x 2.4 x 44 m) already existed. In 2001, 4 additional dry/wet cylindrical culverts (0.9 x 44 m) were installed at the same time as a 3 km barrier wall along the highway, parallel to wetland prairie. Culverts were 200–500 m apart along the wall. Monitoring was undertaken on five nights/week from March 2001 to March 2002. Ten wire screen-mesh funnel traps were placed in each box culvert and four crayfish traps in each cylindrical culvert.

A review of studies investigating culverts in Texas and near New York, USA (18) found mixed results. Two tunnels with barrier walls decreased amphibian road deaths by 90%. Eight of the 20 known species were recorded using the tunnels. In contrast, no Houston toads *Bufo houstonensis* used modified drainage culverts and athough diversion fencing reduced road-kills in its vicinity, groups of dead toads were recorded at the ends. Short sections of steel diversion fencing were added to existing drainage culverts to guide toads from known migration routes into the culverts. The culverts were not designed for amphibians and became impassable when flooded. Two concrete tunnels with box openings (1.2 x 1.2 m) and wooden barrier walls were installed along a road adjacent to wetlands in 1999.

A replicated study in 2001, of two experimental tunnels in Pays de la Loire, France (19) found that amphibians used tunnels and preferred the soil-lined to the bare tunnel. Tunnels were preferred to bypassing on the grass by common toads *Bufo bufo* (70%) and edible frogs *Rana esculenta* (68%). However, agile frogs *Rana dalmatina* tended to bypass (70%). The soil-lined tunnel was used by 68% of the animals that used the tunnels. The difference between soil-lined and bare tunnels was significant for both frog species but not common toads. Crossing success was higher for all species in the soil-lined tunnel. Two concrete pipes (2 m long, 0.5 m diameter) were placed side by side within an enclosure (5 x 3 m). One was lined with sand and humus, the other left bare. Two 0.5 m lengths of drift-fence were installed at 45° to the entrances. A single animal was placed 1.2 m in front of the tunnels with male calls playing from the far end. Each trial lasted 10 minutes and was repeated 4 days later. Forty-one common toads, 42 edible frogs and 32 agile frogs captured locally were used.

A replicated, controlled study in 2002–2003 of culverts along small forest streams in the Oregon Coast Range, USA (20, 25) found that culverts were used by a small proportion of larval coastal giant salamander *Dicamptodon tenebrosus*. Complete culvert passage was recorded by 16 larvae at 7 of 9 culvert sites, although only 20% of larvae moved far enough to assess culvert passage. Growth rates and density did not differ significantly, but movements varied in streams with and without culverts. Effects on larval survival were inconclusive. Densities were lower in raised metal pipe culverts than in arch culverts with streambed substrates. Arch culverts and streams had similar densities. Density was associated with the presence of large substrates. In the presence of culverts, the direction and distance moved did not differ significantly (culvert: 3 m; none: 4 m), but larvae moved to the centre of the stream section less frequently. Nine sites with a culvert (four pipe and five arch) and five without were selected. Stream sections (80 m long) and culverts were monitored two to three times in June–August using dip-netting and visual surveys. Culverts were located at the centre of each section. A total of 2,215 larvae were measured and marked.

A study in 2000–2003 of a culvert under a highway by Lake Jackson, Florida, USA (21) found that at least three amphibian species used the culvert. Many leopard frog *Rana sphenocephala*, pig frog *Rana grylio* and American bullfrog *Rana catesbeiana* were observed moving through the culvert. In total, 12 amphibian species were recorded along the fence and road. A temporary fence was installed along the highway to divert animals to an existing metal drainage culvert in April 2000 (700 m; 0.4 m high) and September 2000 (600 m). Monitoring was undertaken 1–4 times daily by walking the fence and checking the road and culvert until November 2003.

A replicated study in 2003–2005 of five amphibian tunnels with guide fencing along a road through Oak Ridges Moraine in Ontario, Canada (22) found that four of the tunnels were used by amphibians but not by the targeted Jefferson salamander *Ambystoma jeffersonianum*. Tunnels were used by a small number of amphibians in 2003, when weather conditions minimized activity. In 2004, 22 amphibians were recorded in or near tunnels. American toad *Bufo americanus*, wood frog *Rana sylvatica*, spring peeper *Pseudacris crucifer* and leopard frog *Rana pipiens*, but not spotted salamander *Ambystoma maculatum* or Jefferson salamander were recorded. Observations were evenly spread across four tunnels; the fifth was waterlogged. Five concrete or steel tunnels, 1.2 m diameter and 25–31 m long, were installed under a new road section in 2001. Each was lined with a sandy substrate and had 30–50 m of guide fencing on each side. Six to eight monitoring visits were undertaken each spring in 2003–2004. Plastic fences directed amphibians to pitfall traps at the tunnel entrances and exits. Fences were also walked by observers at night.

A before-and-after study in 1994–2004 of a brackish and freshwater wetland in southern Tuscany, Italy (23) found that raising a road on a viaduct resulted in a significant decrease in amphibian road deaths. Following construction, no remains of amphibians were found on the road, compared to thousands during some periods pre-construction. For example, after a night rainstorm in July 1997, over 6,500 newly emerged Italian edible frog *Rana hispanica* juveniles were counted on a 100 m stretch of road. Many species used the open space under the viaduct to migrate between wetlands. A viaduct 215 m long was constructed in 2003 to raise a road. The supports of the viaduct (1.6 m high) were built on a bank 1 m higher than potential flood waters to prevent mixing of wetlands. Drift-fencing was installed for 300 m from each end of the viaduct along both sides of the road. Amphibian road kills were monitored before and after construction.

A study in 2004–2008 of a culvert with a barrier wall along a new highway through upland forest in New Hampshire, USA (24) found no evidence that it had been used by amphibians during the first three years. A 'wildlife diversion wall' preventing access to the road and funnelling animals to the culvert did divert amphibians. Small numbers of spotted salamanders *Ambystoma maculatum* and wood frogs *Rana sylvatica* were found moving along the wall. However, small numbers of these species were found crossing the road in areas without a wall or culvert. The culvert was constructed near to the most productive pond for amphibians. It was 17 m long with an opening 1.2 x 1.2 m. Loamy soil material was used and was sloped across the width of the culvert to confine stream flow to one side. The diversion wall (0.3 m high) extended from the culvert to a stone-lined stream channel on one side and a larger pedestrian culvert on the other. Spring amphibian migrations were monitored for three years after construction.

A replicated study in 2005–2006 of tunnels in a Wildlife Management Area in New York, USA (26) found that green frogs *Rana clamitans* and leopard frogs *Rana pipiens* showed some preference for particular tunnel types. Green frogs showed a significant preference for soil (40%) and gravel (38%) linings, compared to concrete (13%) and PVC (9%). Leopard frogs showed no preference (19%; 32%; 29%; 19% respectively). Leopard frogs tended to prefer larger diameters (0.8 m: 35%; 0.6 m: 12%; 0.5 m: 28%; 0.3 m: 25%) and avoid the longest tunnels (9 m: 15%; 6 m: 40%; 3 m: 22–24%). Green frogs showed no preference for diameter (0.8 m: 33%; 0.6 m: 24%; 0.5 m: 27%; 0.3 m: 16%) or length (9 m: 32%; 6 m: 23%; 3 m: 19–26%). Tunnels with the greatest light permeability were preferred (4% light permeability: 39–41%; 1.3% light: 14–17%; 0.6% light: 24–26%; no light: 17–24%).

Choice arenas had 4 different PVC culverts radiating out, which local green frogs (n = 135) and leopard frogs (187) could select to exit through. Frogs were tested in groups of 1–17 individuals, once per arena. Trials lasted 15 minutes, after 5 minutes acclimatization, in June–August 2005–2006. Pitfall traps captured animals at the end of each tunnel.

A replicated study in 2008 of different culvert designs in New York State, USA (27) found that migrating spotted salamanders *Ambystoma maculatum* showed no preference for culverts of particular diameters, length or substrate. However, the concrete-lined culvert was used significantly less than other substrates (concrete: 28%; bare: 35%; sand/gravel: 37%). There was no significant difference in use of culverts of different diameters (0.3 m: 28%; 0.6 m: 33%; 0.9 m: 39%) or lengths (3 m: 30%; 6 m: 32%; 9 m: 39%). Spotted salamanders and American toads *Anaxyrus americanus* did not show a strong preference for crossing near existing culverts under the highway. The 4 test areas were 30–100 m apart, alongside a highway in a forested wetland. Each consisted of two 9 m long fences (1 m high) that funnelled animals towards three choices of PVC culverts. A pitfall trap captured migrating animals at the end of each culvert. A total of 57–139 salamanders were captured per test area. Sampling was undertaken during five nights in March–April 2008.

A review in 2010 of studies monitoring 327 road crossing structures in Australia, Europe and North America (28) found that amphibians used crossing structures in 12 of 14 studies. Amphibians used drainage culverts in four of five studies, adapted culverts in all three studies and pipes in both studies where they were monitored. Wildlife underpasses, bridge underpasses and overpasses were used in the one study that monitored each. Amphibians did not use the one wildlife overpass monitored. Fourteen of the 30 published papers investigated multiple structure types, which resulted in a total of 52 studies of different structure types.

A controlled study in 2009 of wildlife culverts along a new highway through wetlands near Whistler, Canada (29) found that road-kill rates were reduced provided that drift-fencing or barriers were installed to direct animals towards culverts. Road-kill rates were reduced significantly (by over 50%) along road sections with ≥ 50 m of drift-fencing or barriers compared to those with no barriers (2–8 vs 15–17 killed/50 m section). Approximately 400–500 amphibians were still killed annually along the new highway. Amphibians appeared hesitant to use culverts. Eight wildlife culvert underpasses were constructed along the section through the wetland. Drift-fences were installed to funnel animals towards culverts. Barrier walls were also installed to prevent migration along some sections. Amphibians were monitored using roadkill surveys, remote cameras at culvert entrances and a mark-recapture study of red-legged frogs *Rana aurora*.

A small, replicated study in 2009 of four amphibian tunnels in Waterton Lakes National Park, Alberta, Canada (30) found that 8% of the estimated breeding population of long-toed salamanders *Ambystoma macrodactylum* used the tunnels. A total of 104 salamanders were captured in pitfall traps and at least another 26 by cameras in tunnels. Five western toad *Anaxyrus boreas* and seven barred tiger salamander *Ambystoma mavortium* were also recorded in the tunnels. Only one case of snake predation was recorded by cameras. Four concrete tunnels were installed 80–110 m apart under the road (0.6 x 0.5 m, 12 m long). Digital cameras were installed on the ceilings of tunnel entrances to monitor tunnel floors with motion-triggered and timed-interval images. One pitfall trap was installed at each tunnel exit in April–August.

A replicated study in 2011–2012 of 26 wildlife tunnels with guide walls at 3 wetland sites on the Great Hungarian Plain, Hungary (31) found that amphibians used the tunnels in large numbers in the first year. Between 120 and 1,800 amphibians were caught at the end of each tunnel over 2 weeks. European fire-bellied toads *Bombina bombina* and the targeted spadefoot toad *Pelobates fuscus* were recorded in highest numbers. The Danube crested newt *Triturus dobrogicus,* a priority conservation species, also used the tunnels. At one site, ten times more amphibians passed through two new climate tunnels than an existing adjacent concrete culvert. A total of 26 polymer concrete ACO Wildlife Pro climate tunnels, guide walls (300–600 m/tunnel) and stop channels (under side roads that bisected guide walls) were constructed under 3 roads in autumn 2011. Amphibians were monitored using nine pitfall traps/road in April 2012.

A before-and-after study in 2006–2011 of a tunnel with barrier wall along a road in Hungary (32) found that up to 15% of migrating amphibians used the tunnels but road deaths did not decrease significantly. During the two years after construction, there was no significant reduction in road deaths as fewer than 1% of migrating amphibians used the tunnels. Following maintenance, over the next 3 years 9–15% of the amphibians used the tunnels. However, over 10,000 amphibians died on the road section each year in 2009–2011 even though toad rescue was also carried out by volunteers. Seven frog and toad species and two newt species were recorded dead along the road. Almost 90% were common toads *Bufo bufo*. In 2006 a tunnel with barrier system was constructed for amphibians between Hont and Parassapuszta. Maintenance was undertaken in spring 2009 and in 2010 and 2011.

A before-and-after study in 2008–2009 of four amphibian tunnels under a road parallel to a lake in Alberta, Canada (33) found that tunnels were effective at reducing road mortality of long-toed salamanders *Ambystoma macrodactylum*. Road mortality decreased from 10% of the population in 1994 to 2% following installation. In 2009, 104 salamanders were recorded using tunnels; 74% were migrating to the lake. Four gray tiger salamanders *Ambystoma mavortium* and seven western toads *Anaxyrus boreas* were also captured in exit traps. Individual tunnel use differed (7–49%). In May 2008, 4 concrete box culverts (0.6 x 0.5 m) were installed for amphibians 80–110 m apart. They had slots to allow air, moisture and light in. Drift-fences 500 m long were installed either side of the road with pitfall traps checked daily in April–October 2008. In 2009, additional fences (133–274 m) were installed to direct salamanders to tunnels and pitfalls were installed at exits. Road mortality surveys (similar to 1994) and fence surveys were undertaken daily in 2008 and May–June 2009.

(1)   Karthaus G. (1985) Schutzmaßnahmen für wandernde amphibien vor einer gefährdung durch den Staßenverkehr – beobachtungen und erfahrungen. *Natur und Landschaft,* 60, 242–247.
(2)   Brehm K. (1989) *The Acceptance of 0.2-Metre Tunnels by Amphibians During Their Migration to the Breeding Site.* Proceedings of the Amphibians and Roads: Toad Tunnel Conference. Rendsburg, Federal Republic of Germany. pp. 29–42.
(3)   Buck-Dobrick T. & Dobrick R. (1989) *The Behaviour of Migrating Anurans at a Tunnel and Fence System.* Proceedings of the Amphibians and Roads: Toad Tunnel Conference. Rendsburg, Federal Republic of Germany. pp. 137–143.
(4)   Dexel R. (1989) *Investigations into the Protection of Migrant Amphibians from the Threats from Road Traffic in the Federal Republic of Germany – A Summary.* Proceedings of the Amphibians and Roads: Toad Tunnel Conference. Rendsburg, Federal Republic of Germany. pp. 43–49.
(5)   Haslinger H. (1989) *Migration of Toads During the Spawning Season at Stallauer Weiher Lake, Bad Tölz, Bavaria.* Proceedings of the Amphibians and Roads: Toad Tunnel Conference. Rendsburg, Federal Republic of Germany. pp. 181–182.
(6)   Langton T.E.S. (1989) *Tunnels and Temperature: Results from a Study of a Drift Fence and Tunnel System for Amphibians at Henley-on-Thames, Buckinghamshire, England.* Proceedings of the Amphibians and Roads: Toad Tunnel Conference. Rendsburg, Federal Republic of Germany. pp. 145–152.
(7)   Meinig H. (1989) *Experience and Problems with a Toad Tunnel System in the Mittelgebirge Region of West Germany.* Proceedings of the Amphibians and Roads: Toad Tunnel Conference. Rendsburg, Federal Republic of Germany. pp. 59–66.
(8)   Zuiderwijk A. (1989) *Amphibian and Reptile Tunnels in the Netherlands.* Proceedings of the Amphibians and Roads: Toad Tunnel Conference. Rendsburg, Federal Republic of Germany. pp. 67–74.
(9)   Yanes M., Velasco J. M. & Suarez F. (1995) Permeability of roads and railways to vertebrates: the importance of culverts. *Biological Conservation,* 71, 217–222.

(10) Jackson S. D. (1996) *Underpass Systems for Amphibians*. Proceedings of the International Conference on Wildlife Ecology and Transportation. Florida Department of Transportation, Tallahassee. pp. 240–244.

(11) Rosell C., Parpal J., Campeny R., Jove S., Pasquina A. & Velasco J. M. (1997) Mitigation of barrier effect on linear infrastructures on wildlife. In K. Canters (ed) *Habitat Fragmentation & Infrastructure*, Ministry of Transport, Public Works and Water Management., Delft, Netherlands. pp. 367–372.

(12) Veenbaas G. & Brandjes J. (1999) *Use of Fauna Passages Along Waterways Under Highways*. Proceedings of the International Conference on Wildlife Ecology and Transportation. Florida Department of Transportation, Tallahassee. pp. 253–258.

(13) Fitzgibbon K. (2001) An evaluation of corrugated steel culverts as transit corridors for amphibians and small mammals at two Vancouver Island wetlands and comparative culvert trials. MA thesis. Royal Roads University.

(14) Allaback M. L. & Laabs D. M. (2002–2003) Effectiveness of road tunnels for the Santa Cruz long-toed salamander. *Transactions of the Western Section of the Wildlife Society*, 38/39, 5–8.

(15) Puky M. & Vogel Z. (2003) *Amphibian Mitigation Measures on Hungarian Roads: Design, Efficiency, Problems and Possible Improvement, Need for a Co-ordinated European Environmental Education Strategy*. Proceedings of the International Conference on Habitat Fragmentation due to Transportation Infrastructure. Brussels. pp. 1–13.

(16) Taylor B. D. & Goldingay R. L. (2003) Cutting the carnage: wildlife usage of road culverts in north-eastern New South Wales. *Wildlife Research*, 30, 529–537.

(17) Dodd C. K., Barichivich W. J. & Smith L. L. (2004) Effectiveness of a barrier wall and culverts in reducing wildlife mortality on a heavily traveled highway in Florida. *Biological Conservation*, 118, 619–631.

(18) Jochimsen D. M., Peterson C. R., Andrews K. M. & Whitfield Gibbons J. (2004) A literature review of the effects of roads on amphibians and reptiles and the measures used to minimize those effects. Idaho Fish and Game Department and USDA Forest Service Report.

(19) Lesbarrères D., Lodé T. & Merilä J. (2004) What type of tunnel could reduce road kills? *Oryx*, 38, 220–223.

(20) Sagar J. P. (2004) Movement and demography of larval coastal giant salamanders (*Dicamptodon tenebrosus*) in streams with culverts in the Oregon Coast Range. MSc thesis. Oregon State University.

(21) Aresco M. J. (2005) Mitigation measures to reduce highway mortality of turtles and other herpetofauna at a north Florida lake. *Journal of Wildlife Management*, 69, 549–560.

(22) Gartshore R. G., Purchase M., Rook R. I. & Scott L. (2006) *Bayview Avenue Extension, Richmond Hill, Ontario, Canada Habitat Creation and Wildlife Crossings in a Contentious Environmental Setting: A Case Study*. Proceedings of the 2005 International Conference on Ecology and Transportation. pp. 55–76.

(23) Scoccianti C. (2006) Rehabilitation of habitat connectivity between two important marsh areas divided by a major road with heavy traffic. *Acta Herpetologica*, 1, 77–79.

(24) Merrow J. (2007) *Effectiveness of Amphibian Mitigation Measures Along a New Highway*. Proceedings of the 2007 International Conference on Ecology and Transportation. Center for Transportation and the Environment, North Carolina State University. pp. 370–376.

(25) Sagar J. P., Olson D. H. & Schmitz R. A. (2007) Survival and growth of larval coastal giant salamanders (Dicamptodon tenebrosus) in streams in the Oregon coast range. *Copeia*, 1, 123–130.

(26) Woltz H. W., Gibbs J. P. & Ducey P. K. (2008) Road crossing structures for amphibians and reptiles: informing design through behavioral analysis. *Biological Conservation*, 141, 2745–2750.

(27) Patrick D. A., Schalk C. M., Gibbs J. P. & Woltz H. W. (2010) Effective culvert placement and design to facilitate passage of amphibians across roads. *Journal of Herpetology*, 44, 618–626.

(28) Taylor B. D. & Goldingay R. L. (2010) Roads and wildlife: impacts, mitigation and implications for wildlife management in Australia. *Wildlife Research*, 37, 320–331.

(29) Malt J. (2011) *Assessing the Effectiveness of Amphibian Mitigation on the Sea to Sky Highway: Passageway Use, Roadkill Mortality, and Population Level Effects*. Proceedings of the Herpetofauna and Roads Workshop – is there light at the end of the tunnel? Vancouver Island University, Nanaimo, Canada. pp. 17–18.

(30) Pagnucco K. S., Paszkowski C. A. & Scrimgeour G. J. (2011) Using cameras to monitor tunnel use by long-toed salamanders (*Ambystoma macrodactylum*): an informative, cost-efficient technique. *Herpetological Conservation and Biology*, 6, 277–286.

(31) Faggyas S. & Puky M. (2012) Construction and preliminary monitoring results of the first ACO Wildlife Pro amphibian mitigation systems on roads in Hungary. *Állattani Közlemények*, 97, 85–93.

(32) Mechura T., Gémesi D., Szövényi G. & Puky M. (2012) Temporal characteristics of the spring amphibian migration and the use of the tunnel-barrier system along the Hont and Parassapuszta section of the main road No. 2. between 2009 and 2011. *Állattani Közlemények*, 97, 770–84.

(33) Pagnucco K. S., Paszkowski C. A. & Scrimgeour G. J. (2012) Characterizing movement patterns and spatio-temporal use of under-road tunnels by long-toed salamanders in Waterton Lakes National Park, Canada. *Copeia*, 331–340.

4.2   Install barrier fencing along roads

- Seven of eight studies (including one replicated and two controlled studies) in Germany, Canada and the USA found that barrier fencing with culverts decreased amphibian road deaths[2, 5, 7, 10] or decreased deaths provided that the fence length[1, 6] and material[3] were effective. One found that low numbers of amphibians were diverted by barriers during breeding migrations[8].

- One replicated study in the USA[9] found that barriers at least 0.6 m high were required to prevent green frogs and leopard frogs climbing over. Two studies in the Netherlands and USA[4, 5] found that treefrogs and 10% of common toads climbed over barrier fencing during breeding migrations.

Background
Traffic on roads can cause significant mortality of amphibian populations. Barriers can be installed at migration points along roads to try to reduce mortality. These are usually installed in association with underpasses. Studies investigating the use of under road wildlife passages, many of which had barrier fencing, are discussed in 'Install culverts or tunnels as road crossings'.

A study in 1984–1985 of a barrier fence and wildlife tunnel in Lower Saxony, Germany (1) found that many common toads *Bufo bufo* and common frogs *Rana temporaria* went around the end of the barrier fence and were killed on the road during breeding migrations. In 1985, deaths were reduced by lengthening the fence. Initially, fences 350 m long were installed on both sides of the road. A concrete tunnel was located in the centre of the fences. Common toads and common frogs were monitored in March–April. Toads were tagged.

A replicated study in 1986 of 114 sites including at least 60 amphibian barrier fences, 11 road closure sites and 23 hand-collected human assisted crossings in Nordrhein-Westphalia, Germany (2) found that a total of 131,061 amphibians were protected from death on roads. Between 1 and 116,515 individuals of 14 species were recorded at each barrier fence, road crossing or hand-collected crossing. The majority of the 60 barrier fences to protect amphibians were constructed from polythene and averaged 600 m in length (range: 30–3,000 m). Animals were collected by hand alone at 23 sites and at 11 sites roads were closed for migrations. Nine sites had a combination of two of the interventions and for 20 sites it was unknown which of the interventions were used.

A before-and-after study in 1987–1988 of a barrier fence and two amphibian tunnels in the Mittelgebirge region of West Germany (3) found that once an effective fence was installed, numbers of migrating amphibians killed on the road during the breeding migration decreased. Prior to the new fence numbers killed were 109/night, compared to just 20 in 1987 and 30 in 1988. Overall, 85% of amphibians recorded at the fence passed through the tunnels. The total number of individuals captured at the fence and surroundings during the spring migration were 2,432 in 1987 and 2,050 in 1988. Of 211 toads marked at the fence in 1987, 68% were recaptured at tunnel exits within 5 days. Two drain channels with metal grid roofs were installed in the road in 1981. A more effective fence of plastic fabric similar to wire mesh (1 m high) was installed at tunnel entrances and parallel to the road in 1987. Pitfall traps were set at each end of the fence and at tunnel exits.

A study of barrier fencing between 5 amphibian tunnels in Overveen in the Netherlands (4) found that 10% of the population of 2,000–3,000 common toads *Bufo bufo* climbed over the fencing during breeding migrations. The remaining toads walked along the fence, but only 4% used the tunnels. The others were captured in pitfall traps and carried across the road. The cast-iron tunnels had been installed nine years before the study. The road had permanent barrier fencing.

A controlled, before-and-after study in 2001–2002 of a barrier wall linking culverts along a highway in Florida, USA (5) found that the wall significantly decreased amphibian road deaths, apart from treefrogs (Hylidae), which could climb over. A total of 19 amphibian road-kills were found on the 3 km section with the barrier, compared to 326 kills on the 500 m section with no barrier. Treefrogs were excluded from these figures. Treefrog mortality increased after construction of the barrier and culverts (from 149 to 194 over 3 survey sections). In 2001, a 1 m high concrete wall with a 15 cm overhang was erected along the highway, parallel to a wetland prairie. The wall extended 3 km on each side of the road. Concrete culverts under the highway were increased from four to eight. The highway and grass verge were monitored from 200 m before the start of the barrier until 200 m past the end. Monitoring was undertaken on three consecutive days from dawn each week from March 2001 to March 2002.

A study of drainage culverts modified with diversion fencing in Texas, USA (6) found that fencing reduced road-kills in its vicinity, but aggregations of dead toads were recorded at the barrier endpoints. No Houston toads *Bufo houstonensis* used the culverts, which became impassable when flooded. Short sections of steel diversion fencing were added to existing drainage culverts to guide toads from known migration routes into the culverts.

A before-and-after study in 2000–2003 of temporary fencing along a highway to a culvert by Lake Jackson, Florida, USA (7) found that 70% of amphibians and reptiles (not including turtles) were diverted from the highway towards the culvert. Twelve amphibian species were recorded along the barrier. Fences diverted 74% of the 1,088 upland and semi-aquatic amphibians and reptiles from the highway (at fence: 74%; dead on road: 26%). Twenty-two percent of the 299 aquatic animals were also diverted (alive at fence: 22%; dead at fence: 2%; dead on road: 76%). In particular, the fence diverted small frogs and toads. Some species were significantly underestimated. The temporary fence was installed along the highway to divert animals to a culvert in April 2000 (700 m; 0.4 m high) and September 2000 (600 m). Monitoring was undertaken 1–4 times/day by walking the fence and checking the road and culvert until November 2003.

A study in 2004–2008 of a barrier wall leading to a culvert under a new highway through upland forest in New Hampshire, USA (8) found that the wall only diverted small numbers of amphibians towards the culvert. Small numbers of spotted salamanders *Ambystoma maculatum* and wood frogs *Rana sylvatica* were found moving along the wall. However, small numbers were also found crossing the road in areas without a wall or culvert. There was no evidence that amphibians used the tunnel during the first three years. The diversion wall was at least 0.3 m high and extended from the culvert to a stone-lined stream channel on one side and a larger pedestrian culvert on the other. Spring amphibian migrations were monitored for three years after construction.

A replicated study in 2005–2006 of different height barrier fencing in a Wildlife Management Area, New York, USA (9) found that fences of at least 0.6 m excluded most green frogs *Rana clamitans* and leopard frogs *Rana pipiens*. Fences 0.6 m high were more effective at excluding frogs (97–100%) than 0.3 m fences (77–80%). Only one leopard frog climbed over the 0.9 m high fence. Opaque, corrugated plastic fences were used to construct 3 nested, circular enclosures of heights 0.3, 0.6 and 0.9 m. Local green frogs ($n = 135$)

and leopard frogs (n = 187) were placed in the centre of each arena and left for 15 min to attempt to scale the fences.

A controlled study in 2009 of wildlife culverts with barrier fencing along a new highway through wetlands near Whistler, Canada (10) found that drift-fencing or barriers directing amphibians towards culverts significantly reduced road-kills. Road-kill rates were reduced by over 50% along road sections with ≥ 50 m of drift-fencing or barriers compared to those with no barriers (2–8 vs 15–17 killed/50 m section). Additional fencing was therefore installed. Eight wildlife culvert underpasses were constructed along the section through the wetland. Drift-fences were installed to funnel animals towards culverts. Barrier walls were also installed to prevent migration along some sections. Amphibians were monitored using road-kill surveys, remote cameras at culvert entrances and a mark-recapture study of red-legged frogs *Rana aurora*.

(1)   Buck-Dobrick T. & Dobrick R. (1989) *The Behaviour of Migrating Anurans at a Tunnel and Fence System*. Proceedings of the Amphibians and Roads: Toad Tunnel Conference. Rendsburg, Federal Republic of Germany. pp. 137–143.
(2)   Feldmann R. & Geiger A. (1989) *Protection for Amphibians on Roads in Nordrhein-Westphalia*. Proceedings of the Amphibians and Roads: Toad Tunnel Conference. Rendsburg, Federal Republic of Germany. pp. 51–57.
(3)   Meinig H. (1989*) Experience and Problems with a Toad Tunnel System in the Mittelgebirge Region of West Germany*. Proceedings of the Amphibians and Roads: Toad Tunnel Conference. Rendsburg, Federal Republic of Germany. pp. 59–66.
(4)   Zuiderwijk A. (1989) *Amphibian and Reptile Tunnels in the Netherlands*. Proceedings of the Amphibians and Roads: Toad Tunnel Conference. Rendsburg, Federal Republic of Germany. pp. 67–74.
(5)   Dodd C.K., Barichivich W.J. & Smith L.L. (2004) Effectiveness of a barrier wall and culverts in reducing wildlife mortality on a heavily traveled highway in Florida. *Biological Conservation*, 118, 619–631.
(6)   Jochimsen D.M., Peterson C.R., Andrews K.M. & Whitfield Gibbons J. (2004) A literature review of the effects of roads on amphibians and reptiles and the measures used to minimize those effects. Idaho Fish and Game Department and USDA Forest Service Report.
(7)   Aresco M.J. (2005) Mitigation measures to reduce highway mortality of turtles and other herpetofauna at a north Florida lake. *Journal of Wildlife Management*, 69, 549–560.
(8)   Merrow J. (2007) *Effectiveness of Amphibian Mitigation Measures Along a New Highway*. Proceedings of the 2007 International Conference on Ecology and Transportation. Center for Transportation and the Environment, North Carolina State University. pp. 370–376.
(9)   Woltz H.W., Gibbs J.P. & Ducey P.K. (2008) Road crossing structures for amphibians and reptiles: informing design through behavioral analysis. *Biological Conservation*, 141, 2745–2750.
(10) Malt J. (2011) *Assessing the Effectiveness of Amphibian Mitigation on the Sea to Sky Highway: Passageway Use, Roadkill Mortality, and Population Level Effects*. Proceedings of the Herpetofauna and Roads Workshop – is there light at the end of the tunnel? Vancouver Island University, Nanaimo, Canada. pp. 17–18.

4.3   Modify gully pots and kerbs

- One before-and-after study in the UK[1] found that moving gully pots 10 cm away from the kerb decreased the number of great crested newts that fell in by 80%.

Background

Gully pots along roadside kerbs form effective traps for amphibians. Animals crossing roads reach the kerb and often move along its base, until they fall into a gully pot. Once in the gully pot amphibians cannot climb out. A study found that 63% of 636 gully pots in 2 areas in Scotland contained wildlife, of which 91% were amphibians (1,087 animals; Muir 2012).

There are a number of ways in which the impact on amphibians could be reduced, such as moving gully pots, modifying the design of their grills, providing escape ladders or changing the shape of kerb stones (angled or indented).

A before-and-after study in 2005–2006 of gully pots along roads in South Wales, UK (1) found that moving the gully pot 10 cm away from the kerb resulted in 80% fewer great crested newts *Triturus cristatus* falling into the gully pots. Only 65 newts were found in the drains compared to 318 before gully pots were moved. Gully pots were moved in 2005.

(1)   Muir D. (2012) Amphibians in drains project report summary. *Biodiversity News*, 59, 16–18.

4.4   Use signage to warn motorists

* One study in the UK[1] found that despite warning signs and human assistance, over 500 toads were killed on some roads.

Background
The number of amphibians killed by vehicles can be high, particularly where their annual migration routes between overwintering and breeding sites cross roads. Signs to warn motorists of amphibian activity can be installed around the densest migration routes.

A study in 1995 of 76 toad patrol projects, 44 with toad warning road signs in the UK (1) found that despite signs and human assistance in the spring some toads were still killed on the roads. Overall, 65% of patrols reported that up to 100 toads were killed on the road, 28% reported 100–500 were killed and 7% over 500 toads. Only 20% of populations were believed to be stable or increasing. A questionnaire survey of most of the known and established toad patrols was undertaken. Seventy-six replies were obtained.

(1)   Froglife (1996) Toad patrols: a survey of voluntary effort involved in reducing road traffic-related amphibian mortality in amphibians. Froglife Report. Conservation Report No.1.

4.5   Close roads during seasonal amphibian migration

* Two studies (including one replicated study) in Germany found that large numbers of amphibians were protected from death during breeding migrations at road closure sites[1] and at road closure sites with assisted crossings and barrier fences[2].

Background
Road traffic can have significant effects on amphibian populations, particularly where their annual migration routes between overwintering and breeding sites cross roads. In some areas, roads can be closed to protect important migration routes.
   One study showed that reducing traffic on minor roads by creating a highway prevented fragmentation of populations of palmate newts *Lissotriton helveticus* but not midwife toads *Alytes obstetricans* (Garcia-Gonzaleza *et al.* 2012).

Garcia-Gonzaleza C., Campoa D., Polaa I.G. & Garcia-Vazqueza E. (2012) Rural road networks as barriers to gene flow for amphibians: species-dependent mitigation by traffic calming. *Landscape and Urban Planning*, 104, 171–180.

A before-and-after study in 1983 of a road in Oberbergischer Kreis, Germany (1) found that closing the road allowed common toads *Bufo bufo* to successfully cross. While the road was open none of the young amphibians reached the other side. However, one hour after closure about 100,000 toads were found crossing along a 400 m section of the road. The road was closed for eight days until the migration of amphibians was over in spring.

A replicated study in 1986 of 114 sites including at least 11 road closure sites, 60 amphibian barrier fences and 23 hand-collected human-assisted crossings in Nordrhein-Westphalia, Germany (2) found that 131,061 amphibians were protected from death on roads during breeding migrations. Between 1 and 116,515 individuals of 14 species were recorded at the road closure sites, assisted crossings and barrier fences at the 114 sites. Nine sites had a combination of 2 of the interventions and for 20 sites it was unknown which of the interventions were used.

(1)   Karthaus G. (1985) Schutzmaßnahmen für wandernde amphibien vor einer gefährdung durch den Staßenverkehr – beobachtungen und erfahrungen. *Natur und Landschaft*, 60, 242–247.
(2)   Feldmann R. & Geiger A. (1989) *Protection for Amphibians on Roads in Nordrhein-Westphalia*. Proceedings of the Amphibians and Roads: Toad Tunnel Conference. Rendsburg, Federal Republic of Germany. pp. 51–57.

4.6   Use humans to assist migrating amphibians across roads

- Two studies (including one replicated study) in Italy and the UK[4, 6] found that despite assisting toads across roads during breeding migrations, 64–70% of populations declined substantially over 6–10 years.

- One study in the UK[5] found that despite assisting toads across roads during breeding migrations, at 7% of sites over 500 toads were still killed on roads.

- Five studies in Germany[1, 2] the UK[3, 5] and Italy[6] found that large numbers of amphibians were moved across roads by patrols. Numbers ranged from 7,532 toads moved before and after breeding[2] to half a million moved during breeding migrations annually[3]. In the UK, there were over 400 patrols[3] and 71 patrols spent an average of 90 person-hours moving toads and had been active for up to 10 years[5].

Background
Many amphibians are killed by vehicles, particularly where their annual migration routes between breeding and over-wintering habitats cross roads. In some areas local volunteers may try to reduce deaths by collecting animals and releasing them on the other side of the road. Temporary drift-fencing and pitfall traps are often used to capture amphibians so that they can be assisted across the road. Patrols often focus on migrations to breeding sites rather than migrations of adults and juveniles away from those sites.

Ideally evidence of the effectiveness of this intervention would consist of survival rates, counts of animals in the population or numbers killed on the road before and after or at sites with and without human assistance. However, such evidence is rarely available and so here we not only present data on population trends, but also numbers of animals that were moved across roads and numbers and efforts of patrols assisting amphibians.

For other interventions that involve engaging volunteers to help manage amphibian populations or habitats see 'Education and awareness raising – Raise awareness amongst the general public through campaigns and public information' and 'Threat: Agriculture – Engage landowners and volunteers to manage land for amphibians'.

A study in 1986 of 114 sites that included at least 23 human-assisted road crossings, 60 amphibian barrier fences and 11 road closure sites in Nordrhein-Westphalia, Germany (1) found that 131,061 amphibians were protected from death on roads. Between 1 and 116,515 individuals of 14 species were recorded at each hand-assisted, barrier fence or road crossing site. Animals were collected by hand and assisted across roads during breeding migrations at 23 sites. Nine sites had a combination of 2 of the interventions and for 20 sites it was unknown which of the interventions were used.

A study in 1980–1988 of a human-assisted road crossing in Bad Tölz, Bavaria (2) found that thousands of toads were moved across the road each year. In 1980, a total of 15,000 toads were collected and in 1988 the figure was 7,532. Eight tunnels with wire-netting fences on both sides of the road were installed. Animals did not use tunnels and so those at the fence were collected by hand and moved across the road twice a year, before and after breeding.

A review in 1989 of toad patrols in the UK (3) found that by 1988 there were more than 400 human-assisted toad crossings which moved over 500,000 amphibians annually. Most crossings were for populations of common toads *Bufo bufo*, with breeding populations of over 12,000 adults.

A replicated study in 1981–1987 of toad patrols in the Netherlands (4) found that assisting common toads *Bufo bufo* across roads did not prevent the decline of 9 out of 14 (64%) populations over 6 years. About 80% of toad crossings had fences and pitfall traps, from which toads are collected and released on the other side of the road.

A study in 1995 of 76 toad patrol projects in the UK (5) found that 20,000–39,000 toads were moved across roads in the spring by 71 patrols. The most frequent number moved by each patrol was 500–1,000 animals (28% of patrols). Despite human assistance, 65% of patrols reported that up to 100 toads were killed on the road, 28% reported 100–500 were killed and 7% over 500 toads. Many patrols reported that an 'appreciable proportion' of the total number of migrating toads were moved by humans. However, only 20% of populations were believed to be stable or increasing. Patrols involved an average of 90 person-hours, as they tended to have 1–3 volunteers/night (range: 1–14) for 11–20 nights (1–49) each lasting 2 hours (1–7 hours). Most patrols had been active for 3–10 years (49 of 53 patrols). Forty-four sites had toad warning road signs. A questionnaire survey of most of the known and established toad patrols was undertaken. Seventy-six replies were obtained.

A study in 1993–2010 of toad patrols during in Central and Northern Italy (6) found that although 1,042,966 common toads *Bufo bufo* were assisted across roads during breeding migrations, 70% of 30 populations declined substantially from 2000 to 2010. Only 10% of the populations increased over the same period. Data on population trends were gathered mainly from volunteer toad patrol groups, with some from other volunteer groups, herpetologists and the literature. Sampling effort was taken into account when examining population trends.

(1) Feldmann R. & Geiger A. (1989) *Protection for Amphibians on Roads in Nordrhein-Westphalia.* Proceedings of the Amphibians and Roads: Toad Tunnel Conference. Rendsburg, Federal Republic of Germany. pp. 51–57.
(2) Haslinger H. (1989) *Migration of Toads During the Spawning Season at Stallauer Weiher Lake, Bad Tölz, Bavaria.* Proceedings of the Amphibians and Roads: Toad Tunnel Conference. Rendsburg, Federal Republic of Germany. pp. 181–182.
(3) Langton T.E.S. (1989) *Reasons for Preventing Amphibian Mortality on Roads.* Proceedings of the Amphibians and Roads: Toad Tunnel Conference. Rendsburg, Federal Republic of Germany. pp. 75–80.
(4) Zuiderwijk A. (1989) *Amphibian and Reptile Tunnels in the Netherlands.* Proceedings of the Amphibians and Roads: Toad Tunnel Conference. Rendsburg, Federal Republic of Germany. pp. 67–74.
(5) Froglife (1996) Toad patrols: a survey of voluntary effort involved in reducing road traffic-related amphibian mortality in amphibians. Froglife Report. Conservation Report No.1.
(6) Bonardi A., Manenti R., Corbetta A., Ferri V., Fiacchini D., Giovine G., Macchi S., Romanazzi E., Soccini C., Bottoni L., Padoa-Schioppa E. & Ficetola G.F. (2011) Usefulness of volunteer data to measure the large scale decline of 'common' toad populations. *Biological Conservation*, 144, 2328–2334.

5 Threat: Biological resource use

For programmes that may help reduce exploitation of species see 'Education and awareness raising'.

Key messages – hunting and collecting terrestrial animals

Use amphibians sustainably
We captured no evidence for the effects of using amphibians sustainably.

Reduce impact of amphibian trade
One review found that reducing trade through legislation allowed frog populations to recover from over-exploitation.

Use legislative regulation to protect wild populations
One review found that legislation to reduce trade resulted in the recovery of frog populations. One study in South Africa found that the number of permits issued for scientific and educational use of amphibians increased from 1987 to 1990.

Commercially breed amphibians for the pet trade
We captured no evidence for the effects of commercially breeding amphibians for the pet trade on wild amphibian populations.

Key messages – logging and wood harvesting

Thin trees within forests
Six studies, including five replicated and/or controlled studies, in the USA compared amphibians in thinned to unharvested forest. Three found that thinning had mixed effects and one found no effect on abundance. One found that amphibian abundance increased following thinning but the body condition of ensatina salamanders decreased. One found a negative overall response of amphibians. Four studies, including two replicated, controlled studies, in the USA compared amphibians in thinned to clearcut forest. Two found that thinning had mixed effects on abundance and two found higher amphibian abundance or a less negative overall response of amphibians following thinning. One meta-analysis of studies in North America found that partial harvest, which included thinning, decreased salamander populations, but resulted in smaller reductions than clearcutting.

Harvest groups of trees instead of clearcutting
Three studies, including two randomized, replicated, controlled, before-and-after studies, in the USA found that harvesting trees in small groups resulted in similar amphibian abundance to clearcutting. One meta-analysis and one randomized, replicated, controlled, before-and-after study in North America and the USA found that harvesting, which included harvesting groups of trees, resulted in smaller reductions in salamander populations than clearcutting.

Use patch retention harvesting instead of clearcutting

We found no evidence for the effect of retaining patches of trees rather than clearcutting on amphibian populations. One replicated study in Canada found that although released red-legged frogs did not move towards retained tree patches, large patches were selected more and moved out of less than small patches.

Use leave-tree harvesting instead of clearcutting

Two studies, including one randomized, replicated, controlled, before-and-after study, in the USA found that compared to clearcutting, leaving a low density of trees during harvest did not result in higher salamander abundance.

Use shelterwood harvesting instead of clearcutting

Three studies, including two randomized, replicated, controlled, before-and-after studies, in the USA found that compared to clearcutting, shelterwood harvesting resulted in higher or similar salamander abundance. One meta-analysis of studies in North America found that partial harvest, which included shelterwood harvesting, resulted in smaller reductions in salamander populations than clearcutting.

Leave standing deadwood/snags in forests

One randomized, replicated, controlled, before-and-after study in the USA found that compared to total clearcutting, leaving dead and wildlife trees did not result in higher abundances of salamanders. One randomized, replicated, controlled study in the USA found that numbers of amphibians and species were similar with removal or creation of dead trees within forest.

Leave coarse woody debris in forests

Two replicated, controlled studies in the USA found that abundance was similar in clearcuts with woody debris retained or removed for eight of nine amphibian species, but that the overall response of amphibians was more negative where woody debris was retained. Two replicated, controlled studies in the USA and Indonesia found that the removal of coarse woody debris from standing forest did not effect amphibian diversity or overall amphibian abundance, but did reduce species richness. One replicated, controlled study in the USA found that migrating amphibians used clearcuts where woody debris was retained more than where it was removed. One replicated, site comparison study in the USA found that within clearcut forest, survival of juvenile amphibians was significantly higher within piles of woody debris than in open areas.

Retain riparian buffer strips during timber harvest

Six replicated and/or controlled studies in Canada and the USA compared amphibian numbers following clearcutting with or without riparian buffer strips. Five found mixed effects and one found that abundance was higher with riparian buffers. Two of four replicated studies, including one randomized, controlled, before-and-after study, in Canada and the USA found that numbers of species and abundance were greater in wider buffer strips. Two found no effect of buffer width.

Hunting and collecting terrestrial animals

5.1   Use amphibians sustainably

- We found no evidence for the effects of using amphibians sustainably.

Background

Many amphibian species have become popular among collectors in the pet trade. Others are used for food, in traditional medicines or harvested for products such as skin toxins. This means that animals are removed from the wild, which can have significant effects on populations. To avoid overexploitation and the decline of populations, amphibians must be used sustainably.

5.2   Reduce impact of amphibian trade

- One review[1] found that reducing trade in two frog species through legislation allowed populations to recover from over-exploitation.

Background

Amphibians are traded for a number of reasons including consumption, the pet trade, for zoo animals and scientific purposes. For example, it was estimated that 15 million live, wild-caught amphibians entered the USA legally in 1998–2002, millions of which were for the pet trade (Schlaepfer *et al.* 2005). Removal of large numbers of amphibians from the wild can have significant effects on populations.

The movement of animals also increases the risk of spreading infectious diseases. For example, there is increasing evidence that trade is partly responsible for the recent spread of chytridiomycosis *Batrachochytrium dendrobatidis* and amphibian ranaviruses (e.g. Daszak *et al.* 2003; Gratwicke *et al.* 2009; Schloegel *et al.* 2009).

Evidence for interventions designed to reduce the threat from diseases is discussed in 'Threat: Invasive alien and other problematic species – Reduce parasitism and disease'.

Daszak P., Cunningham A.A. & Hyatt A.D. (2003) Infectious disease and amphibian population declines. *Diversity and Distributions*, 9, 141–150.
Gratwicke B., Evans M., Jenkins P., Kusrini M., Moore R., Sevin J. & Wildt D. (2009) Is the international frog legs trade a potential vector for deadly amphibian pathogens? *Frontiers in Ecology and the Environment*, 8, 438–442.
Schlaepfer M.A., Hoover C., Dodd K.D. Jr (2005) Challenges in evaluating the impact of the trade in amphibians and reptiles on wild populations. *Bioscience*, 55, 256–264.
Schloegel L., Picco A., Kilpatrick A., Davies A., Hyatt A. & Daszak P. (2009) Magnitude of the US trade in amphibians and presence of *Batrachochytrium dendrobatidis* and ranavirus infection in imported North American bullfrogs (*Rana catesbeiana*). *Biological Conservation*, 142, 1420–1426.

A review in 2011 (1) found that reducing trade in green pond frog *Euphlyctis Hexadactylus* and the Indian bullfrog *Hoplobatrachus tigerinus* through legislation allowed populations to recover from over-exploitation. Both species were categorized by the International Union for the Conservation of Nature (IUCN) as stable in the 2010 IUCN Red List. Populations of both species had crashed in India and Bangladesh following unsustainable use in the frog leg trade. Over three years of monitoring in India, it was estimated that 9,000 tonnes of frogs were removed from the wild for frogs' legs. In 1985, green pond frogs and Indian bullfrogs were listed in Appendix II of the Convention on International Trade in Endangered Species of Wild Flora and Fauna (CITES). India banned the export of frogs' legs in 1987 and Bangladesh followed in 1989.

(1) Altherr S., Goyenechea A. & Schubert D.J. (2011) Canapés to extinction: the international trade in frogs' legs and its ecological impact. Pro Wildlife Defenders of Wildlife and Animal Welfare Institute Report.

5.3   Use legislative regulation to protect wild populations

• One review[2] found that legislation to reduce trade in two frog species resulted in the recovery of the over-exploited populations.

• One study in South Africa[1] found that the number of permits issued for scientific and educational use of amphibians increased from 1987 to 1990.

Background
Species can be legally protected, either nationally or internationally. Levels of protection vary but can be to prevent capturing, keeping in captivity or trading species. Such activities may be legal for certain species provided that permits are obtained from government licensing authorities.
    Other studies investigating the effect of legally protecting species are discussed in 'Threat: Residential and commercial development – Legal protection for species'.

A study in 1987–1990 of permits issued for amphibians in the Cape Province, South Africa (1) found that the number issued for scientific and educational use increased over the three years. The number issued increased from 100 in 1987 to 380 in 1990. Data were obtained from the governmental licensing authority, Cape Nature Conservation. Permits obtained by scientists from institutions requiring study material and institutions requiring specimens for display or breeding were included. Permits obtained by private individuals to keep species in captivity were not included.

A review in 2011 (2) found that following legislation to reduce trade in green pond frogs *Euphlyctis Hexadactylus* and the Indian bullfrog *Hoplobatrachus tigerinus,* populations recovered from over-exploitation. Both species were categorised by the International Union for the Conservation of Nature (IUCN) as stable in the 2010 IUCN Red List. Populations of both species had crashed in India and Bangladesh following unsustainable use in the frog leg trade. During 3 years of monitoring in India, it was estimated that 9,000 tonnes of frogs were removed from the wild for frogs' legs. In 1985, green pond frogs and Indian bullfrogs were listed in Appendix II of the Convention on International Trade in Endangered Species of Wild Flora and Fauna (CITES). India banned the export of frogs' legs in 1987 and Bangladesh followed in 1989.

(1) Baard E.H.W. (1992) Is legal protection of reptiles and amphibians in the Cape Province contributing to their conservation? *The Journal of the Herpetological Association of Africa,* 41, 92.
(2) Altherr S., Goyenechea A. & Schubert D.J. (2011) Canapés to extinction: the international trade in frogs' legs and its ecological impact. Pro Wildlife Defenders of Wildlife and Animal Welfare Institute Report.

5.4   Commercially breed amphibians for the pet trade

- We found no evidence for the effects of commercially breeding amphibians for the pet trade on wild amphibian populations.

> **Background**
> Global aquaculture of amphibians for the pet and food trade grew from 3,000 tonnes in 1999 to 85,000 tonnes in 2008 (Food and Agriculture Organization 2009). Commercially breeding amphibians for the pet trade can help to reduce the number of animals collected from wild populations.

Food and Agriculture Organization (2009) *Aquaculture Production 2008 – By Species Groups*. In Yearbooks of Fishery. Food and Agriculture Organization, Rome, Italy.

Logging and wood harvesting

5.5   Thin trees within forests

- Five studies (including four replicated and/or controlled studies) in the USA[1-3, 5, 6, 8] compared amphibians in thinned to unharvested forest. Two found mixed effects of thinning on abundance, depending on amphibian species and time since harvest[2, 3, 8]. One found that amphibian abundance increased, except for ensatina salamanders[1]. One found a negative overall response (population, physiological and behavioural) of amphibians[6] and one found that thinning did not affect abundance[5]. A meta-analysis of 24 studies in North America[9] found that partial harvest, which included thinning with 3 other types, decreased salamander populations. One controlled, before-and-after site comparison study in the USA[4] found that high volumes of pre-existing downed wood prevented declines in amphibian populations following thinning.

- Four studies (including two replicated, controlled studies) in the USA[1-3, 6, 8] compared amphibians in thinned to clearcut forest. Two found higher amphibian abundance, apart from ensatina salamanders[1], or a less negative overall response (population, physiological and behavioural) of amphibians[6] in thinned forest. Two found mixed effects on abundance depending on species, life stage and time since harvest[2, 3, 8]. A meta-analysis of 24 studies in North America[9] found that partial harvest, which included thinning with 3 other types, resulted in smaller reductions in salamander populations than clearcutting.

- One randomized, replicated, controlled study in the USA[7] found that migrating amphibians used thinned forest a similar amount, or for one species more than unharvested forest, and that emigrating salamanders, but not frogs, used it significantly more than clearcuts.

- One site comparison study in the USA[1] found that thinning decreased the body condition of ensatina salamanders 10 years after harvest.

Background

Thinning of trees, that is removal of trees to reduce density (by up to 50%), is undertaken in commercial forestry such as in plantations to ensure that stands are made up of healthy, evenly spaced trees. However, it can also be used as a conservation management practice to restore more natural open woodland. It can also increase structural diversity of young even-aged stands and promote development of late-successional characteristics such as larger trees, multi-level canopies and understory vegetation. Such features are often lost due to active fire suppression or loss of populations of large mammal grazers and browsers.

A site comparison study in 2000–2002 of 10 sites within a conifer-hardwood forest in California, USA (1) found that thinning increased amphibian abundance, apart from ensatinas *Ensatina eschscholtzii*, and lowered the body condition of ensatinas, ten years after harvest. Overall, captures were significantly higher in thinned (7/1000 capture nights) compared to unthinned (4) and clearcut forest (4). However, abundance of the dominant species, ensatina, was similar in thinned (148 captures), unthinned (106) and clearcut forests (159). The body condition index of ensatinas was significantly lower in thinned compared to unthinned forests. Five thinned (aged > 10 years) and five unharvested forest stands adjacent to clearcuts (aged 6–25 years) were selected. Forest had been thinned (approximately 50% retained) prior to clearcutting. Amphibians were monitored using seven drift-fences with pitfall traps and artificial coverboards along two 150 m transects/site. Traps were checked weekly in October–December and April–June 2000–2002.

A randomized, replicated, controlled study in 2004–2005 of mixed coniferous and deciduous forest wetlands in Maine, USA (2) found that amphibian abundance in partial (50%) harvest plots tended to be lower than unharvested and higher or similar to clearcuts (see also (8)). The proportion of captures in partial harvest was lower than that in unharvested plots for adults and/or juveniles of eight of nine species including adult wood frogs *Lithobates sylvaticus* (partial: 27%; unharvested: 51%; clearcut: 11%) and juvenile spotted salamanders *Ambystoma maculatum* (partial: 20%; unharvested: 62%; clearcuts: 7–11%). Captures were higher in partial harvests than unharvested plots for adult northern leopard frogs *Lithobates pipiens* (partial: 47%; unharvested: 30%; clearcuts: 7–17%) and red-spotted newts (partial: 44%; unharvested: 25%). Captures in partial harvest were higher than clearcuts for adults of four of nine species, lower for two species and similar for three species. Juvenile captures were higher in partial harvests than clearcuts for seven of nine species. All treatments extended 164 m (2 ha) from each of four created breeding ponds and were cut in 2003–2004. There were two clearcut treatments with and without woody debris retained. Drift-fences with pitfall traps were installed around each pond at 1, 17, 50, 100 and 150 m from the edge. Monitoring was in April–September 2004–2005.

A replicated, site comparison study in 2000–2003 of 12 harvested hardwood forest sites in Maine, USA (3) found that abundance of amphibian species in partially harvested forest was similar or lower than unharvested forest and similar or higher than clearcut forest. Captures in partial harvests were significantly lower than unharvested forest and higher than clearcuts for red-backed salamanders *Plethodon cinereus* (partial: 0.38; clearcut: 0.12; unharvested: 0.61/100 trap nights) and spotted salamanders *Ambystoma maculatum* (partial: 0.03; clearcut: 0.01; unharvested: 0.09). There was no significant difference between treatments for two-lined salamanders *Eurycea bislineata* (partial: 0.12; clearcut: 0.04; unharvested: 0.16), American toads *Bufo americanus* (partial: 1.01; clearcut: 0.49; unharvested: 0.34) or wood frogs

Rana sylvatica (partial: 0.99; clearcut: 0.92; unharvested: 1.54). Twelve headwater streams that had been harvested 4–10 years previously were selected. Treatments were: partial harvest (23–53% removed), clearcut with 23–35 m buffers and unharvested for > 50 years. Monitoring was undertaken in June–September in one year using drift-fences with pitfall traps and visual surveys.

A controlled, before-and-after site comparison study in 1998–2001 at two largely coniferous forest sites in western Oregon, USA (4) found that the amount of pre-existing downed wood affected the response of salamanders to forest thinning. At the site with high volumes of existing downed wood, there was no significant change in capture rates of the dominant species ensatina *Ensatina eschscholtzii* or Oregon slender salamander *Batrachoseps wrighti* following thinning. However, at the site with little downed wood, capture rates declined significantly for the two dominant species, ensatina (40%) and western red-backed salamanders *Plethodon vehiculum* (42%). Captures did not change in unharvested treatments. At the two sites, treatments were unharvested or thinned (80% thinned to 200–240 trees/ha; 10% harvested in groups; 10% patches retained; deadwood was retained) with riparian buffers (6 to ≥70 m). Monitoring was undertaken in May–June before and two years after thinning. Visual count surveys were along 64–142 m transects perpendicular to each stream bank (7–8/treatment).

A replicated, site comparison study in 2005 of three coniferous forest sites in Oregon, USA (5) found that there was no significant difference between amphibian captures in thinned and unharvested sites 5–6 years after harvest. Captures did not differ significantly between treatments for all amphibians, western red-backed salamanders *Plethodon vehiculum* or ensatina *Ensatina eschscholtzii*. Each site (12–24 ha) had two streams within forest that had been thinned (200–600 trees/ha) with riparian buffers (6 m or over 15 m wide) in 2000 and one stream with no harvesting. Amphibians were sampled by visual counts once in April–June within five 5 x 10 m plots at four distances (up to 35 m) from each stream.

A replicated, controlled study in 2003–2009 of 12 ponds in deciduous, pine and mixed-deciduous and coniferous forest in Maine, Missouri and South Carolina, USA (6) found that overall, partially harvesting forest had a negative effect on amphibian population, physiological and behavioural responses, but a smaller negative effect than clearcutting (−7 vs −19 to 32%). Sixteen of 34 response variables were negative, 10 positive and 8 the same as unharvested forest. Four treatments were assigned to quadrats (2–4 ha) around each breeding pond (4/region): partial harvest (opposite control; 50–60% reduction), clearcut with coarse woody debris retained or removed and unharvested. Treatments were applied in 2003–2005. Monitoring was undertaken using drift-fence and pitfall traps, radiotelemetry and aquatic (200–1,000 litres) and terrestrial (3 x 3 m or 0.2 m diameter) enclosures. Different species (*n* = 9) were studied at each of the eight sites. Response variables were abundance, growth, size, survival, breeding success, water loss, emigration and distance moved.

A randomized, replicated, controlled study in 2004–2007 of four seasonal wetlands in pine forest in southeastern USA (7) found that migrating amphibians tended to use thinned forest a similar amount to unharvested forest and that emigrating salamanders, but not frogs, used it more than clearcuts. Proportions of immigrating amphibians and emigrating frogs did not differ between treatments. The proportion of salamanders combined *Ambystoma* spp. and mole salamanders *Ambystoma talpoideum* that emigrated through thinned forest (0.2–0.4) was similar to unharvested forest (0.4–0.5) but significantly higher than clearcuts (0.1–0.2). Significantly higher numbers of ornate chorus frogs *Pseudacris ornata* emigrated through partial harvests than unharvested forest. Significantly more emigrating salamanders, frogs *Rana* spp. and southern toads retreated from clearcuts compared to partial harvests and unharvested sites. There were four wetland sites each surrounded by four randomly assigned treatments extending out 168 m (4 ha): thinning (15% removed), clearcut with or without coarse woody debris retained and unharvested. Harvesting was undertaken in spring 2004. Amphibians were captured using drift-fencing with pitfall traps from February 2004 to July 2007.

In a continuation of a previous study (2), a randomized, replicated, controlled study in 2004–2009 of mixed forest wetlands in Maine, USA (8) found that amphibian abundance in partially (50%) harvested forest was similar to unharvested forest for six of eight amphibian species and significantly lower for two species. Post-breeding, there were significant differences between partial, clearcut and unharvested treatments for wood frog *Lithobates sylvaticus* adults (partial: 0.4; unharvested: 0.5; clearcuts: 0.2) and juveniles (partial: 1.1; unharvested: 1.5; clearcuts: 0.9) and spotted salamander *Ambystoma maculatum* juveniles (partial: 0.4; unharvested: 0.6; clearcuts: 0.2). Abundances during other times of year did not differ significantly for those two species. Post-breeding, partial harvest was used significantly more than clearcuts by the other two forest specialists, eastern red-spotted newts *Notophthalmus viridescens* (partial: 0.10; clearcuts: 0.06–0.08; unharvested: 0.13), red-backed salamanders *Plethodon cinereus* (partial: 0.2; clearcuts: 0.2; unharvested: 0.1). Abundance of four habitat generalist species did not differ between treatments. All treatments extended 164 m (2 ha) from each of four created breeding ponds and were harvested in 2003–2004. Drift-fences with pitfall traps were installed around each pond at 2, 17, 50, 100 and 150 m from the edge. Monitoring was in April–September 2004–2009.

A meta-analysis of the effects of different harvest practices on terrestrial salamanders in North America (9) found that partial harvest, including thinning, cutting individual or groups of trees and shelterwood harvesting, decreased salamander populations, but less so than clearcutting. Reductions in populations were lower following partial harvest (all studies: 31–48%; < 5 years monitoring: 51%; > 10 years monitoring: 29%) compared to clearcutting (all: 54–58%; < 5 years: 62%; > 10 years: 50%). There was no significant effect of the proportion of canopy removed in partial harvests. Sampling methodology influenced perceived effects of harvest. Salamander numbers almost always declined following timber removal, but populations were never lost and tended to increase as forests regenerated. Studies that compared salamander abundance in harvested (partial or clearcut) and unharvested areas were identified. Twenty-four site comparison and before-and-after studies were analysed. Abundance measures included counts, population indices and density estimates.

(1)   Karraker N.E. & Welsh H.H. (2006) Long-term impacts of even-aged timber management on abundance and body condition of terrestrial amphibians in northwestern California. *Biological Conservation*, 131, 132–140.
(2)   Patrick D.A., Hunter M.L. & Calhoun A.J.K. (2006) Effects of experimental forestry treatments on a Maine amphibian community. *Forest Ecology and Management*, 234, 323–332.
(3)   Perkins D.W., Malcolm L. & Hunter J.R. (2006) Effects of riparian timber management on amphibians in Maine. *Journal of Wildlife Management*, 70, 657–670.
(4)   Rundio D.E. & Olson D.H. (2007) Influence of headwater site conditions and riparian buffers on terrestrial salamander response to forest thinning. *Forest Science*, 53, 320–330.
(5)   Kluber M.R., Olson D.H. & Puettmann K.J. (2008) Amphibian distributions in riparian and upslope areas and their habitat associations on managed forest landscapes in the Oregon Coast Range. *Forest Ecology and Management*, 256, 529–535.
(6)   Semlitsch R.D., Todd B.D., Blomquist S.M., Calhoun A.J.K., Whitfield-Gibbons J., Gibbs J.P., Graeter G.J., Harper E.B., Hocking D.J., Hunter M.L., Patrick D.A., Rittenhouse T.A.G. & Rothermel B.B. (2009) Effects of timber harvest on amphibian populations: understanding mechanisms from forest experiments. *BioScience*, 59, 853–862.
(7)   Todd B.D., Luhring T.M., Rothermel B.B. & Gibbons J.W. (2009) Effects of forest removal on amphibian migrations: implications for habitat and landscape connectivity. *Journal of Applied Ecology*, 46, 554–561.

(8)   Popescu V.D., Patrick D.A., Hunter Jr. M.L. & Calhoun A.J.K. (2012) The role of forest harvesting and subsequent vegetative regrowth in determining patterns of amphibian habitat use. *Forest Ecology and Management*, 270, 163–174.
(9)   Tilghman J.M., Ramee S.W. & Marsh D.M. (2012) Meta-analysis of the effects of canopy removal on terrestrial salamander populations in North America. *Biological Conservation*, 152, 1–9.

5.6   Harvest groups of trees instead of clearcutting

- Three studies (including two randomized, replicated, controlled, before-and-after studies) in the USA found that compared to clearcutting, harvesting trees in small groups did not result in higher amphibian[3] or salamander abundance[1, 2, 4]. A meta-analysis of 24 studies in North America[5] found that partial harvest, which included harvesting groups or individual trees, thinning and shelterwood harvesting, resulted in smaller reductions in salamander populations than clearcutting.

- Two studies (including one randomized, replicated, controlled, before-and-after study) in the USA found that compared to no harvesting, harvesting trees in small groups significantly decreased salamander abundance[1, 2, 4] and changed species composition[2].

- One randomized, replicated, controlled, before-and-after study in the USA[4] found that compared to unharvested plots, the proportion of female salamanders carrying eggs were similar and proportion of eggs per female and juveniles similar or lower in harvested plots that included harvest of groups of trees.

Background

Forests naturally undergo disturbances such as storms and lightning that can create open patches. Similarly, harvesting groups of trees rather than clearcutting forest creates a mix of different habitats, allowing a greater range of species to survive in a forest.

A controlled, before-and-after study in 1994–1997 in a hardwood forest in Virginia, USA (1) found that harvesting trees in small groups decreased the relative abundance of salamanders, similar to clearcutting. Captures decreased significantly after group harvesting (before: 14; 1 year after: 11; 3 years: 2/search) and clearcutting (before: 10; 1 year after: 7; 3 years: 1/search). Abundance did not differ significantly within the unharvested plot (before: 10; 1 year after: 10; 3 years: 8). Treatments on 2 ha plots were: group harvesting (3 groups of 0.5 ha), clearcutting (up to 12 wildlife and dead trees retained) and unharvested. Salamanders were monitored along 2 x 15 m transects with artificial cover objects (50/plot).

A randomized, replicated, controlled, before-and-after study in 1993–1999 of five harvested hardwood forests in Virginia, USA (2) found that harvesting trees in groups did not result in higher salamander abundances than clearcutting. Abundance was similar between treatments (groups: 3; clearcut: $1/30$ m^2 respectively; see also (4)). Abundance was significantly lower compared to unharvested plots (6/30 m^2). Species composition differed before and three years after harvest. There were 5 sites with 2 ha plots with each treatment: group harvesting (2–3 small area group harvests with selective harvesting between), clearcutting and an unharvested control. Salamanders were monitored on 9–15 transects (2 x 15 m)/plot at night in April–October. One or 2 years of pre-harvest and 1–4 years of post-harvest data were collected.

A randomized, replicated, controlled, before-and-after study in 1992–2000 of oak-pine and oak-hickory forest in Missouri, USA (3) found that there was no significant difference in amphibian abundance between sites with small group or single tree selection harvesting and those with clearcutting. Abundance of species declined after harvest but also declined on unharvested sites. Nine sites (312–514 ha) were randomly assigned to treatments: small group or single tree selection harvesting (5% area; uneven-aged management), clearcutting in 3–13 ha blocks (10–15% total area) with forest thinning (even-aged), or unharvested controls. Harvesting was in May 1996 and 1997. Twelve drift-fence arrays with pitfall and funnel traps were established/plot. Traps were checked every 3–5 days in spring and autumn 1992–1995 and 1997–2000.

In a continuation of a previous study (2), a randomized, replicated, controlled study in 1994–2007 of 6 hardwood forests in Virginia, USA (4), found that harvesting groups of trees did not result in higher salamander abundance compared to clearcutting up to 13 years after harvest. Abundance was similar between treatments (groups: 4; clearcutting: 2/transect) and significantly lower than unharvested plots (7/transect). Proportions of juveniles and eggs/female were significantly lower in harvested (group harvesting, shelterwoods, leave-tree harvesting and clearcut with wildlife trees or snags left) compared to unharvested treatments for mountain dusky salamander *Desmognathus ochrophaeus* and juveniles for red-backed salamander *Plethodon cinereus*. Proportions of females carrying eggs were similar in harvested and unharvested plots for slimy salamander *Plethodon glutinosus* and southern ravine salamanders *Plethodon richmondii*. There were 6 sites with 2 ha plots randomly assigned to treatments: group harvesting (2–3 small area group harvests with selective harvesting between), clearcutting, other harvested treatments and an unharvested control. Treatments were in 1994–1998 and salamanders were monitored at night along nine 2 x 15 m transects/plot.

A meta-analysis of the effects of different harvest practices on terrestrial salamanders in North America (5) found that partial harvest, which included harvesting groups or individual trees, thinning and shelterwood harvesting, resulted in smaller reductions in salamander populations than clearcutting. Overall, partial harvest produced declines 24% smaller than clearcutting. Average reductions in populations were lower following partial harvest (all studies: 31–48%; < 5 years monitoring: 51%; > 10 years monitoring: 29%) compared to clearcutting (all: 54–58%; < 5 years: 62%; > 10 years: 50%). There was no significant effect of the proportion of canopy removed in partial harvests. Sampling methodology influenced perceived effects of harvest. Salamander numbers almost always declined following timber removal, but populations were never lost and tended to increase as forests regenerated. Twenty-four site comparison and before-and-after studies that compared salamander abundance in harvested (partial or clearcut) and unharvested areas were analysed. Abundance measures included counts, population indices and density estimates.

(1) Harpole D.N. & Haas C.A. (1999) Effects of seven silvicultural treatments on terrestrial salamanders. *Forest Ecology and Management*, 114, 349–356.
(2) Knapp S.M., Haas C.A., Harpole D.N. & Kirkpatrick R.L. (2003) Initial effects of clearcutting and alternative silvicultural practices on terrestrial salamander abundance. *Conservation Biology*, 17, 752–762.
(3) Renken R.B., Gram W.K., Fantz D.K., Richter S.C., Miller T.J., Ricke K.B., Russell B. & Wang X. (2004) Effects of forest management on amphibians and reptiles in Missouri Ozark forests. *Conservation Biology*, 18, 174–188.
(4) Homyack J.A. & Haas C.A. (2009) Long-term effects of experimental forest harvesting on abundance and reproductive demography of terrestrial salamanders. *Biological Conservation*, 142, 110–121.
(5) Tilghman J.M., Ramee S.W. & Marsh D.M. (2012) Meta-analysis of the effects of canopy removal on terrestrial salamander populations in North America. *Biological Conservation*, 152, 1–9.

5.7   Use patch retention harvesting instead of clearcutting

- We found no evidence for the effect of retaining patches of trees rather than clearcutting on amphibian populations.

- One replicated study in Canada[1] found that although released red-legged frogs did not show significant movement towards retained tree patches, large patches were selected more and moved out of less than small patches.

Background

Patch retention harvesting may be used as an alternative to a total clearcutting in commercial forests exploited for timber. Typically, around 10% of trees are retained in patches within a clearcut area. These retained patches can help maintain characteristic forest species and act as reservoirs for recolonization by forest dependent species.

A replicated study in 2000–2001 of red-legged frogs *Rana aurora* in harvested coniferous forest on Vancouver Island, Canada (1) found that although frogs did not show significant movement towards retained patches of trees within the harvested area, large patches of trees were selected more and moved out of less than small patches. Overall, 55% of frogs left patches of trees within 72 hours of being released. However, frogs were less likely to leave with increasing patch size and stream density. Frogs did not tend to move towards patches unless released within 20 m. However, when given a choice, frogs moved towards large patches (0.8 ha) significantly more and small patches (0.3 ha) significantly less than expected. Forest blocks had been harvested two years previously with 5–30% of trees retained. Ten radio-collared frogs were released at the centre of 20 tree patches or at individual trees (canopy areas 1–3 ha) and monitored for 72 hours. Another ten frogs were released at each of four randomly located tree patches and four other random locations and were monitored for six days. Seven frogs were released from each of four points equal distances from three different size patches (0.3–0.8 ha). Ten frogs were released at five distances (5–80 m) from two patches.

(1) Chan-McLeod A.C.A. & Moy A. (2007) Evaluating residual tree patches as stepping stones and short-term refugia for red-legged frogs. *Journal of Wildlife Management*, 71, 1836–1844.

5.8   Use leave-tree harvesting instead of clearcutting

- Two studies, including one randomized, replicated, controlled, before-and-after study, in the USA[1-3] found that compared to clearcutting, leaving a low density of trees during harvest did not result in higher salamander abundance.

- Two studies, including one randomized, replicated, controlled, before-and-after study, in the USA found that compared to no harvesting, leaving a low density of trees during harvest decreased salamander abundance[1-3] and changed species composition[2].

- One randomized, replicated, controlled, before-and-after study in the USA[2, 3] found that compared to unharvested plots, the proportion of female salamanders carrying eggs, eggs per female or proportion of juveniles were similar or lower in harvested plots that included leave-tree harvests, depending on species and time since harvest.

Background
Leave-tree harvest retains a low density of high-quality trees uniformly through the forest stand. Trees can be retained in groups or dispersed and may contain trees with structural characteristics important to wildlife. Compared to clearcutting, this type of management can help maintain forest species.

A controlled, before-and-after study in 1994–1997 in a hardwood forest in Virginia, USA (1) found that leave-tree harvesting decreased relative abundance of salamanders in a similar way to clearcutting. Captures decreased significantly after both leave-tree harvesting (before: 8; 1 year after: 4; 3 years after: 1 amphibian/search) and clearcutting (before: 10; 1 year after: 7; 3 years after: 1/search). Abundance did not differ significantly within the unharvested plot (before: 10; 1 year after: 10; 3 years after: 8). Treatments on 2 ha plots were: leave-tree (up to 16 trees/ha retained), clearcutting (up to 12 wildlife and dead trees retained) and unharvested. Salamanders were monitored along 15 x 2 m transects with artificial cover objects (50/plot).

A randomized, replicated, controlled, before-and-after study in 1993–1999 of five harvested hardwood forests in Virginia, USA (2) found that leave-tree harvesting did not result in higher salamander abundances than clearcutting (see also (3)). Abundance was similar in the leave-tree and clearcut plots (2 vs $1/30$ m^2 respectively). Abundance was significantly lower than unharvested plots (6/30 m^2). Species composition differed before and three years post-harvest. There was no significant difference in the proportion of females carrying eggs or eggs/female for red-backed salamander *Plethodon cinereus* (7 eggs) or mountain dusky salamander *Desmognathus ochrophaeus* (12–13 eggs) in unharvested and harvested treatments (leave-tree, shelterwoods and clearcut with wildlife trees or snags left). The proportion of juveniles was similar except for slimy salamander *Plethodon glutinosus*, which had a significantly lower proportion in harvested plots. There were 5 sites with 2 ha plots with the following treatments: leave-tree harvest (up to 50 trees/ha retained uniformly; average 28%), clearcutting, other harvested treatments and an unharvested control. Salamanders were monitored on 9–15 transects (2 x 15 m)/plot at night in April–October. One or 2 years of pre-harvest and 1–4 years of post-harvest data were collected.

In a continuation of a previous study (2), a randomized, replicated, controlled study in 1994–2007 of 6 hardwood forests in Virginia, USA (3) found that leave-tree harvesting did not result in higher salamander abundance compared to clearcutting up to 13 years after harvest. Abundance was similar between treatments (4 vs 2/transect respectively) and significantly lower than unharvested plots (7/transect). Proportions of juveniles and eggs/female were significantly lower in harvested (leave-tree, shelterwoods, group cutting and clearcut with wildlife trees or snags left) compared to unharvested treatments for mountain dusky salamander *Desmognathus ochrophaeus* and juveniles for red-backed salamander *Plethodon cinereus*. Proportions of females carrying eggs for slimy salamander *Plethodon glutinosus* and southern ravine salamanders *Plethodon richmondii* were similar in harvested and unharvested plots. There were 6 sites with 2 ha plots randomly assigned to treatments: leave-tree harvest (25–45 trees/ha retained), clearcutting, other harvested treatments and an unharvested control. Treatments were in 1994–1998 and salamanders were monitored at night along nine 2 x 15 m transects/site.

(1)   Harpole D.N. & Haas C.A. (1999) Effects of seven silvicultural treatments on terrestrial salamanders. *Forest Ecology and Management*, 114, 349–356.

(2)   Knapp S.M., Haas C.A., Harpole D.N. & Kirkpatrick R.L. (2003) Initial effects of clearcutting and alternative silvicultural practices on terrestrial salamander abundance. *Conservation Biology*, 17, 752–762.
(3)   Homyack J.A. & Haas C.A. (2009) Long-term effects of experimental forest harvesting on abundance and reproductive demography of terrestrial salamanders. *Biological Conservation*, 142, 110–121.

5.9   Use shelterwood harvesting instead of clearcutting

- Three studies (including two randomized, replicated, controlled, before-and-after studies) in the USA found that compared to clearcutting, shelterwood harvesting resulted in higher[1], similar[2] or initially higher and then similar[3, 4] salamander abundance. A meta-analysis of 24 studies in North America[5] found that partial harvest, which included shelterwood harvesting with three other types, resulted in smaller reductions in salamander populations than clearcutting.

- Two of three studies (including two randomized, replicated, controlled, before-and-after studies) in the USA found that compared to no harvesting, shelterwood harvesting decreased salamander abundance[2–4] and changed species composition[3]. One found that shelterwood harvesting did not affect salamander abundance[1].

- One randomized, replicated, controlled, before-and-after study in the USA[3, 4] found that compared to unharvested plots, the proportion of female salamanders carrying eggs, eggs per female or proportion of juveniles were similar or lower in harvested plots that included shelterwood harvested plots, depending on species and time since harvest.

Background

Shelterwood harvesting is a management technique designed to obtain even-aged timber without clearcutting. It involves harvesting trees in a series of partial cuttings, with trees removed uniformly over the plot, which allows new seedlings to grow from the seeds of older trees. This can help maintain characteristic forest species and increase structural diversity of stands.

A randomized, replicated, controlled, before-and-after study in 1993–1995 of forest in Virginia, USA (1) found that shelterwood harvest resulted in higher abundances of otter salamanders *Plethodonh ubrichti* compared to clearcutting. Relative abundance did not differ significantly before and after harvest in the shelterwood (4 vs 4–5) and unharvested sites (7 vs 8). However, numbers declined within clearcuts (5 vs 1). Similarly, population estimates varied over time within the shelterwood (12–50) and unharvested sites (40–103), but declined steadily within clearcuts (from 43 to 8). The proportion of juveniles increased in the unharvested plot (8 to 30%), whereas the proportion remained lower in the shelterwood (4 to 13%) and clearcut sites (3 to 12%). Growth and movement rates were similar between treatments. Treatments were randomly assigned over 12 sites (0.6–1.2 ha): shelterwood harvest (33–64% removed), clearcut and unharvested. Harvest was in May 1994. Salamanders were surveyed up to 8 times a year within one 5 x 5 m plot/site. Mark-recapture was undertaken at one site.

A controlled, before-and-after study in 1994–1997 in a hardwood forest in Virginia, USA (2) found that shelterwood harvesting resulted in a decrease in the relative abundance of salamanders, similar to clearcutting. Captures decreased significantly after shelterwood harvests with 12–15 m^2 basal area retained/ha (before: 9; 1 year after: 6; 3 years: 2/search) or 4–7 m^2 basal area retained/ha (before: 12; 1 year after: 4; 3 years: 1/search) and on clearcut plots (before: 10; 1 year after: 7; 3 years: 1/search). Abundance did not differ significantly within the unharvested plot (before: 10; 1 year after: 10; 3 years: 8). Treatments on 2 ha plots were: two shelterwood harvests, clearcutting (up to 12 wildlife and dead trees retained) and unharvested. Salamanders were monitored along 15 x 2 m transects with artificial cover objects (50/plot).

A randomized, replicated, controlled, before-and-after study in 1993–1999 of five harvested hardwood forests in Virginia, USA (3) found that shelterwood harvesting resulted in significantly higher salamander abundances than clearcutting (3 vs 1/30 m²; see also (4)). However, abundance was significantly lower than unharvested plots (6/30 m²). Species composition differed before and three years after harvest. There was no significant difference in the proportion of females carrying eggs or eggs/female for red-backed salamander *Plethodon cinereus* (7 eggs) or mountain dusky salamander *Desmognathus ochrophaeus* (12–13 eggs) in unharvested and harvested treatments (shelterwoods, leave-tree and clearcut with wildlife trees or snags left). The proportion of juveniles was similar except for slimy salamander *Plethodon glutinosus*, which had a significantly lower proportion in harvested plots. There were 5 sites with 2 ha plots with the following treatments: shelterwoods (41–81% removed), clearcutting, other harvested treatments and an unharvested control. Salamanders were monitored on 9–15 transects (2 x 15 m)/plot at night in April–October. One or 2 years of pre-harvest and 1–4 years of post-harvest data were collected.

In a continuation of a previous study (3), a randomized, replicated, controlled study in 1994–2007 of six hardwood forests in Virginia, USA (4) found that shelterwood harvesting did not increase salamander abundance compared to clearcutting up to 13 years after harvest. Abundance was similar between treatments (4 vs 2/transect respectively) and significantly lower than unharvested plots (7/transect). Proportions of juveniles and eggs/female were significantly lower in harvested (leave-tree and group harvesting and clearcut with wildlife trees or snags left) compared to unharvested treatments for mountain dusky salamander *Desmognathus ochrophaeus* and juveniles for red-backed salamander *Plethodon cinereus*. Proportions of females carrying eggs for slimy salamander *Plethodon glutinosus* and southern ravine salamanders *Plethodon richmondii* were similar in harvested and unharvested plots. There were 6 sites with 2 ha plots randomly assigned to treatments: shelterwood harvest (41% reduction), clearcutting, other harvested treatments and an unharvested control. Treatments were in 1994–1998 and salamanders were monitored at night along nine 15 x 2 m transects/site.

A meta-analysis of the effects of different harvest practices on terrestrial salamanders in North America (5) found that partial harvest, that included shelterwood harvesting, thinning and cutting individual or groups of trees resulted in smaller reductions in salamander populations than clearcutting. Overall, partial harvest produced declines 24% smaller than clearcutting. Average reductions in populations were lower following partial harvest (all studies: 31–48%; < 5 years monitoring: 51%; > 10 years monitoring: 29%) compared to clearcutting (all: 54–58%; < 5 years: 62%; > 10 years: 50%). There was no significant effect of the proportion of canopy removed in partial harvests. Sampling methodology influenced perceived effects of harvest. Salamander numbers almost always declined following timber removal, but populations were never lost and tended to increase as forests regenerated. Twenty-four site comparison and before-and-after studies that compared salamander abundance in harvested (partial or clearcut) and unharvested areas were analysed. Abundance measures included counts, population indices and density estimates.

(1) Sattler P. & Reichenbach N. (1998) The effects of timbering on *Plethodon hubrichti*: short-term effects. *Journal of Herpetology*, 32, 399–404.

(2) Harpole D.N. & Haas C.A. (1999) Effects of seven silvicultural treatments on terrestrial salamanders. *Forest Ecology and Management*, 114, 349–356.
(3) Knapp S.M., Haas C.A., Harpole D.N. & Kirkpatrick R.L. (2003) Initial effects of clearcutting and alternative silvicultural practices on terrestrial salamander abundance. *Conservation Biology*, 17, 752–762.
(4) Homyack J.A. & Haas C.A. (2009) Long-term effects of experimental forest harvesting on abundance and reproductive demography of terrestrial salamanders. *Biological Conservation*, 142, 110–121.
(5) Tilghman J.M., Ramee S.W. & Marsh D.M. (2012) Meta-analysis of the effects of canopy removal on terrestrial salamander populations in North America. *Biological Conservation*, 152, 1–9.

5.10   Leave standing deadwood/snags in forests

*   One randomized, replicated, controlled, before-and-after study in the USA[2, 4] found that compared to total clearcutting, leaving dead or wildlife trees did not result in higher abundances of salamanders.

*   Two studies (including one randomized, replicated, controlled, before-and-after study) in the USA found that compared to no harvesting, leaving dead or wildlife trees during clear-cutting did not prevent a decrease in salamander abundance[1, 2, 4] or a change in species composition[2].

*   One randomized, replicated, controlled study in the USA[3] found that numbers of amphibian species and abundance were similar with removal or creation of dead trees within forest.

*   One randomized, replicated, controlled, before-and-after study in the USA[2, 4] found that compared to unharvested plots, the proportion of female salamanders carrying eggs, eggs per female or proportion of juveniles were similar or lower in harvested plots that included plots where dead and wildlife trees were left during clearcutting, depending on species and time since harvest.

Background

Snags or standing dead trees and other dead wood can provide shelter for amphibians within forest. Retaining these within clearcut forest may help to maintain amphibian populations.

Studies investigating the effect of leaving coarse woody debris during harvest are discussed in 'Leave course woody debris in forests'.

A controlled, before-and-after study in 1994–1997 in a hardwood forest in Virginia, USA (1) found that retaining up to 12 wildlife and dead trees during a clear-cut did not prevent a decrease in the relative abundance of salamanders. Captures decreased significantly after treatment (before: 10; 1 year after: 7; 3 years: 1/search). Abundance did not differ within the unharvested plot (before: 10; 1 year after: 10; 3 years: 8). Treatments were on 2 ha plots. Salamanders were monitored along 2 x 15 m transects with artificial cover objects (50/plot).

A randomized, replicated, controlled, before-and-after study in 1993–1999 of four harvested hardwood forests in Virginia, USA (2) found that leaving up to 12 wildlife or dead trees did not result in higher salamander abundances than clearcutting (see also (4)). Abundance was similar between treatments (2 vs 1/30 m² respectively). Abundance was significantly lower than unharvested plots (6/30 m²). Species composition differed before and three years after harvest. There was no significant difference in the proportion of females carrying eggs or eggs/female for red-backed salamander *Plethodon cinereus*

(7 eggs) or mountain dusky salamander *Desmognathus ochrophaeus* (12–13 eggs) in un-harvested and harvested treatments (clearcut with wildlife trees/snags, shelterwood and leave-tree harvesting). The proportion of juveniles was similar except for slimy salamander *Plethodon glutinosus*, which had a significantly lower proportion in harvested plots. There were 4 sites with 2 ha plots with the following treatments: clearcutting with up to 12 wildlife or dead trees retained (small stems felled and left), clearcutting, other harvested treatments and an unharvested control. Salamanders were monitored on 9–15 transects (2 x 15 m)/plot at night in April–October. One or 2 years of pre-harvest and 1–4 years of post-harvest data were collected.

A randomized, replicated, controlled study in 1998–2005 of pine stands in South Carolina, USA (3) found that amphibian abundance, species richness and diversity did not differ with removal or creation of snags within forest. Abundance, species richness and diversity did not differ significantly between plots with 10-fold increase in snags (1/night; 7; 17 respectively), removal of all snags and downed course woody debris (2; 7; 18) and unmanipulated controls (2; 7; 19). Captures of anurans, salamanders and six individual species did not differ between treatments. Treatments were randomly assigned to 9 ha plots within 3 forest blocks. The first set of treatments was undertaken in 1996–2001 and the second set in 2002–2005. Five drift-fence arrays with pitfall traps/plot were used for sampling in 1998–2005.

In a continuation of a previous study (2), a randomized, replicated, controlled study in 1994–2007 of 6 hardwood forests in Virginia, USA (4) found that leaving scattered wildlife or dead trees did not result in higher salamander abundance compared to clearcutting up to 13-years post-harvest. Abundance was similar between treatments (3 vs 2/transect respectively) and significantly lower than unharvested plots (7/transect). Proportions of juveniles and eggs/female were significantly lower in harvested (clearcut with wildlife trees, shelterwoods, leave-tree and group harvesting) compared to unharvested treatments for mountain dusky salamander *Desmognathus ochrophaeus* and juveniles for red-backed salamander *Plethodon cinereus*. Proportions of females carrying eggs for slimy salamander *Plethodon glutinosus* and southern ravine salamanders *Plethodon richmondii* were similar in harvested and unharvested plots. There were 6 sites with 2 ha plots randomly assigned to treatments: clearcutting with wildlife trees (<10 stems/ha), complete clearcutting, other harvested treatments and an unharvested control. Treatments were in 1994–1998 and salamanders were monitored at night along nine 2 x 15 m transects/site.

(1)   Harpole D.N. & Haas C.A. (1999) Effects of seven silvicultural treatments on terrestrial salamanders. *Forest Ecology and Management*, 114, 349–356.
(2)   Knapp S.M., Haas C.A., Harpole D.N. & Kirkpatrick R.L. (2003) Initial effects of clearcutting and alternative silvicultural practices on terrestrial salamander abundance. *Conservation Biology*, 17, 752–762.
(3)   Owens A.K., Moseley K.R., McCay T.S., Castleberry S.B., Kilgo J.C. & Ford W.M. (2008) Amphibian and reptile community response to coarse woody debris manipulations in upland loblolly pine (*Pinus taeda*) forests. *Forest Ecology and Management*, 256, 2078–2083.
(4)   Homyack J.A. & Haas C.A. (2009) Long-term effects of experimental forest harvesting on abundance and reproductive demography of terrestrial salamanders. *Biological Conservation*, 142, 110–121.

5.11   Leave coarse woody debris in forests

- Two replicated, controlled studies in the USA found that there was no significant difference in abundance in clearcuts with woody debris retained or removed for eight of nine amphibian species[1, 7], but that the overall response (population, physiological and behavioural) of amphibians was more negative where woody debris was retained[4].

- Two replicated, controlled studies in the USA and Indonesia found that the removal of coarse woody debris from standing forest did not decrease amphibian diversity[2] or overall amphibian abundance[2, 6], but did reduce species richness in one study.

- One replicated, controlled study in the USA[5] found that migrating amphibians used clearcuts where woody debris was retained more than where it was removed.

- One replicated, site comparison study in the USA[3] found that within clearcut forest, survival of juvenile amphibians was significantly higher in piles of woody debris than in open areas, and was similar in wood piles to unharvested forest[3].

Background

Coarse woody debris consists of fallen dead trees and cut branches (> 10 cm diameter) that are left during harvesting. Coarse woody debris increases the structural diversity at the forest floor and provides a valuable microhabitat for animals that are moisture and temperature sensitive such as amphibians.

Studies investigating the effect of adding woody debris to forests are discussed in 'Habitat restoration and creation – Create refuges'.

A randomized, replicated, controlled study in 2004–2009 of mixed coniferous and deciduous forest wetlands in Maine, USA (1) found that there was no significant difference in amphibian abundance in clearcuts with woody debris retained or removed for eight of nine amphibian species (see also (7)). Abundance of spotted salamander *Ambystoma maculatum* juveniles was significantly higher in clearcuts with woody debris retained than in those where it was removed (11 vs 7%). Although not significant, captures tended to be higher in clearcuts with woody debris retained for three of nine species and with woody debris removed for five species. Treatments extended 164 m (2 ha) from each of 4 created breeding ponds and were clear-cut in 2003–2004. Drift-fences with pitfall traps were installed around each pond at 1, 17, 50, 100 and 150 m from the edge. Wood frogs were marked. Monitoring was in April–September 2004–2005.

A randomized, replicated, controlled study in 1998–2005 of pine stands in South Carolina, USA (2) found that the removal of coarse woody debris did not effect amphibian abundance, species richness or diversity. Plots with all downed and standing woody debris removed did not differ significantly from controls in terms of abundance (1–2 vs 2), species richness (7 vs 7) or diversity (17–18 vs 19). The southern leopard frog *Rana sphenocephala* had greater capture rates with removal rather than addition of woody debris (0.11 vs 0.02/night). Treatments were randomly assigned to 9 ha plots within 3 forest blocks. The first set of treatments was undertaken in 1996–2001 and a second set in 2002–2005. Control plots had no manipulation of woody debris. Five drift-fence arrays with pitfall traps/plot were used for sampling in 1998–2005.

A replicated, site comparison study in 2005–2006 of microhabitats within clearcut oak–hickory forest in Missouri, USA (3) found that survival rates of juvenile amphibians were significantly higher within piles of woody debris than within open areas in clearcut forest (0.9 vs 0.2). Survival within clearcut brushpile was similar to that within unharvested sites (0.9). The proportion of water loss from animals was lower within woody debris than open areas for American toads *Anaxyrus americanus* (0.2–0.3 vs 0.3–0.6), green frogs *Lithobates clamitans* (0.2–0.4 vs. 0.6–0.7) and wood frogs *Lithobates sylvaticus* (0.1–0.4 vs 0.6–0.7). Water loss in unharvested sites was 0.2–0.4, 0.2–0.3 and 0.1 respectively. Open habitat and piles of coarse woody debris were selected within two clearcuts, where tree crowns had been retained during harvest in 2004. Unharvested forest was used as a reference. Captive-reared American toad and wood frog juveniles and wild-caught green frog metamorphs were placed in individual enclosures within treatments. There were four replicates. Animals were weighed every 6 hours for 24 hours.

A replicated, controlled study in 2003–2009 of 12 ponds in deciduous, pine and mixed-deciduous and coniferous forest in Maine, Missouri and South Carolina, USA (4) found that overall, retaining coarse woody debris during clearcutting had a greater negative effect on amphibian population, physiological and behavioural responses than removing debris, when compared to unharvested forest (−32 vs −19%). However, 14 of 33 response variables were less negative, 4 less positive, 3 more negative and 12 the same when debris was retained compared to removed, when compared to unharvested controls. Four treatments were assigned to quadrats (2–4 ha) around each breeding pond (4/region): partial harvest (opposite control), clearcut with woody debris retained or removed and an unharvested control. Treatments were applied in 2003–2005. Monitoring was undertaken using drift-fence and pitfall traps, radio-telemetry and in aquatic (200–1,000 litres) and terrestrial (3 x 3 m or 0.2 m diameter) enclosures. Different species ($n = 9$) were studied at each of the eight sites. Response variables were abundance, growth, size, survival, breeding success, water loss, emigration and distance moved.

A replicated, controlled study in 2004–2007 of four seasonal wetlands in pine forest in southeastern USA (5) found that migrating amphibians used clearcuts where woody debris had been retained more than where it had been removed. By the final year, the proportion of both salamander species emigrating through clearcut with woody debris retained was significantly higher than in clearcut without woody debris (0.2 vs 0.1). The same was true for immigrating Southern toads *Bufo terrestris* (0.3 vs 0.1) and frogs *Rana* spp. (0 vs 0.5). There were 4 wetland sites, each surrounded by four randomly assigned treatments extending out 168 m (4 ha): partial harvest (15%), clearcut with or without coarse woody debris retained and unharvested. Harvesting was undertaken in spring 2004. Immigrating and emigrating amphibians were captured using drift-fencing with pitfall traps from February 2004 to July 2007.

A replicated, controlled, before-and-after study in 2007–2008 of a cacao plantation in Sulawesi, Indonesia (6) found that removal of woody debris and/or leaf litter did not significantly effect overall amphibian abundance, but did decrease species richness. However, the abundance of *Hylarana celebensis* and Asian toad *Duttaphrynus melanostictus* increased following removal of woody debris and leaf litter. The abundance of Sulawesian toad *Ingerophrynus celebensis* decreased following removal of woody debris. Forty-two plots (40 x 40 m²) were divided into four treatments: removal of woody debris (trunks and branch piles), removal of leaf litter, removal of woody debris plus leaf litter and an unmanipulated control. Monitoring was undertaken twice on two occasions, 26 days before and 26 days after habitat manipulation. Visual surveys were undertaken along both plot diagonals (transects 3 x 113 m).

In a continuation of a previous study (1), a randomized, replicated, controlled study in 2004–2009 of mixed coniferous and deciduous forest wetlands in Maine, USA (7) found that overall there was no significant difference in abundance in clearcuts with woody debris retained or removed for four forest specialist and four generalist amphibian species. This was true for adults and juveniles immigrating and emigrating from breeding ponds. The one exception was that the abundance of spotted salamander *Ambystoma maculatum* metamorphs was significantly higher in clearcuts with woody debris retained than in those where it was removed (2 vs 1). Treatments extended 164 m (2 ha) from each of four created breeding ponds and were cut in 2003–2004. Drift-fences with pitfall traps were installed around each pond at 2, 17, 50, 100 and 150 m from the edge. Monitoring was in April–September 2004–2009.

(1) Patrick D.A., Hunter M.L. & Calhoun A.J.K. (2006) Effects of experimental forestry treatments on a Maine amphibian community. *Forest Ecology and Management*, 234, 323–332.
(2) Owens A.K., Moseley K.R., McCay T.S., Castleberry S.B., Kilgo J.C. & Ford W.M. (2008) Amphibian and reptile community response to coarse woody debris manipulations in upland loblolly pine (*Pinus taeda*) forests. *Forest Ecology and Management*, 256, 2078–2083.
(3) Rittenhouse T.A.G., Harper E.B., Rehard L.E. & Semlitsch R.D. (2008) The role of microhabitats in the desiccation and survival of amphibians in recently harvested oak-hickory forest. *Copeia*, 2008, 807–814.
(4) Semlitsch R.D., Todd B.D., Blomquist S.M., Calhoun A.J.K., Whitfield-Gibbons J., Gibbs J.P., Graeter G.J., Harper E.B., Hocking D.J., Hunter M.L., Patrick D.A., Rittenhouse T.A.G. & Rothermel B.B. (2009) Effects of timber harvest on amphibian populations: understanding mechanisms from forest experiments. *BioScience*, 59, 853–862.
(5) Todd B.D., Luhring T.M., Rothermel B.B. & Gibbons J.W. (2009) Effects of forest removal on amphibian migrations: implications for habitat and landscape connectivity. *Journal of Applied Ecology*, 46, 554–561.
(6) Wanger T.C., Saro A., Iskandar D.T., Brook B.W., Sodhi N.S., Clough Y. & Tscharntke T. (2009) Conservation value of cacao agroforestry for amphibians and reptiles in South-East Asia: combining correlative models with follow-up field experiments. *Journal of Applied Ecology*, 46, 823–832.
(7) Popescu V.D., Patrick D.A., Hunter Jr. M.L. & Calhoun A.J.K. (2012) The role of forest harvesting and subsequent vegetative regrowth in determining patterns of amphibian habitat use. *Forest Ecology and Management*, 270, 163–174.

5.12   Retain riparian buffer strips during timber harvest

• Twelve studies investigated the effectiveness of retaining buffer strips during timber harvest for amphibians.

• Six replicated and/or controlled studies in Canada and the USA compared amphibian numbers following clearcutting with or without riparian buffer strips. Five found mixed effects on abundance depending on species[1, 5, 9, 12, 13] and buffer width[1, 9]. One[2, 4] found that amphibian abundance was significantly higher with buffers.

• Eleven studies, including ten replicated and/or controlled studies in Canada and the USA[1–9, 12, 13] and one meta-analysis[11], compared amphibian numbers in forest with riparian buffers retained during harvest to unharvested forest. Six found mixed effects depending on species[1, 5, 6, 12, 13] or volume of existing downed wood[7]. Four[2–4, 8, 9] found that abundance and species composition were similar to unharvested forest. Two found that numbers of species[2, 4] and abundance[2, 4, 11] were lower than in unharvested forest.

• Two of four replicated studies (including one randomized, controlled, before-and-after study) in Canada and the USA found that numbers of amphibian species[2, 4] and abundance[2, 4, 5] were greater in wider riparian buffer strips. Two[3, 8] found that there was no difference in abundance in buffers of different widths.

Background
Retaining forest strips along water courses or around ponds during timber harvest can help mitigate the effects of habitat loss and disturbance for forest species. They can also help sustain the microclimate and reduce potential problems such as soil erosion. Retained habitat strips also provide corridors for dispersal.

A controlled, before-and-after study in 1988–1991 at three hardwood forest sites in Oregon, USA (1) found that the effects of retaining riparian buffer zones on amphibians were unclear. Three of six species showed no changes in capture rates after total clearcutting and no significant differences in captures in riparian buffers and upslope areas (rough-skin newts *Taricha granulosa*, Dunn's salamanders *Plethodon dunni* and red-legged frogs *Rana aurora*). Capture rates of ensatinas *Ensatinae schscholtzii* and Pacific giant salamanders *Dicamptodon tenebrosus* decreased after clearcutting and tended to be lower in buffers than upslope. Western redback salamanders *Plethodon vehiculum* increased the first year after logging and then decreased. Herbicide treatment had no effect on species. Each site had plots (>8 ha) with each treatment: unharvested control; clearcut and broadcast burned; and clearcut, broadcast burned and sprayed with herbicide (1.3 kg/ha). Clearcuts had 20 m wide untreated riparian buffer strips. Cut sites were planted with fir seedlings. Amphibians were monitored one year before and for two years after treatments using pitfall trapping. Traps were checked daily for eight days in dry and wet seasons.

A replicated, site comparison study in 1994–1995 along streams at 29 forest sites in western Oregon, USA (2, 4) found that clearcut forest with retained riparian buffers had significantly higher amphibian density than total clearcut plots (12 vs 6/1,000 m^2). However, compared to unharvested sites clearcut sites with riparian buffers had significantly lower total salamander abundance (21 vs 30) and species richness (3 vs 5) and abundance of three individual salamander species. Two species did not differ between treatments. Overall and individual species density did not differ significantly within plots with riparian buffers and unharvested sites (amphibians: 12 vs 13/1,000 m^2). Amphibian density was significantly higher within wide (>40 m) compared to narrow (<20 m) buffers (13 vs 5/1,000 m^2). The same was true for species richness (5 vs 2). Seventeen clearcut sites (< 5 years old) with riparian buffers (0–64 m wide) and 12 unharvested sites (> 100 years) were selected. Visual encounter surveys were undertaken in three 20 x 40 m streamside plots/site (within buffers, clearcut, unharvested areas) in April–May and November–December 1994 and March–May 1995.

A replicated, controlled, before-and-after study in 1996–1998 of a mixed wood forest in Alberta, Canada (3) found that forest buffers of 20–200 m around lakes maintained amphibian abundance for three years after harvest. Abundance was not significantly different before and after harvest, within or between buffer widths, or compared to unharvested areas and protected forests. Species composition did not change after harvest. Four lakes were selected in 3 regions and were assigned to buffer strip treatments of 20, 100 or 200 m wide, or were controls within protected forest. Clearcuts were 2–49 ha, with 2 to 4 cuts around each lake in 1996, the remainder was left unharvested. Amphibians were monitored using groups of 3 pitfall traps at 40 m intervals within sampling grids (400 x 100 m) parallel to lakes. Sampling was undertaken in May–June and July–August 1996–1998, for 5–8 days/lake each season.

A randomized, replicated, controlled, before-and-after study in 2000–2003 of forest in Maine, USA (5) found that amphibian abundance tended to be higher when riparian buffers were retained during harvest. Captures were significantly higher with 11 m and 23 m buffers for American toads *Bufo americanus* (clearcut: 0.6; 11 m buffer: 1.0; 23 m buffer: 3.4; unharvested: 0.5/100 trap nights) and wood frogs *Rana sylvatica* (clearcut: 0.8; 11 m: 1.4; 23 m: 2.0; unharvested: 2.2). Red-backed salamanders *Plethodon cinereus* did not differ (0.1–0.3). In forest cut 4–10 years previously, captures of wood frog and American toads were also significantly higher in buffers than clearcuts. Red-backed salamanders showed a similar trend. However, abundance of salamanders and frogs were significantly or tended to be lower in buffers than unharvested forest. Fifteen headwater streams were randomly assigned to 6 ha treatments: clearcut with buffers of 0, 11 or 23 m wide, partial harvest (23–53%) or unharvested. Monitoring was undertaken using drift-fences with pitfall traps and visual surveys in June–September, one year before and two years after harvesting. Twelve sites harvested 4–10 years earlier were also monitored in 1 year. Treatments were: clearcutting with 23–35 m buffers, partial harvest and unharvested (> 50 years).

A replicated, controlled, before-and-after study in 1995–2002 of amphibians in managed forest stands at 11 sites in Oregon, USA (6) found that retaining riparian buffers maintained amphibian abundance in the first two years after tree thinning. There was no significant decrease in four species within buffers following thinning (change: −0.1 to 0.1 animals/m^2). Rough-skinned newts *Taricha granulosa* and coastal giant salamander *Dicamptodon tenebrosus* numbers increased within buffers following thinning (0.007–0.034/m^2) and declined at unthinned control sites (−0.043 to 0.008/m^2). Forty-five streams were assigned riparian buffers of 6, 15, 70 or 145 m on each side within tree thinning areas (from 600 to 200 trees/ha). Thinning took place in 1997–2000. Monitoring was undertaken in spring and summer, before treatment, in 1995–1999 and for two years after treatment, in 1998–2002. Amphibians were sampled in 10 units/stream using hand sampling, electrofishing and visual counts of bank sides (2 m wide). Twenty-three streams within unharvested areas were also monitored.

A controlled, before-and-after, site comparison study in 1998–2001 at two forest sites in western Oregon, USA (7) found that the amount of pre-existing downed wood affected the response of salamanders to forest thinning with riparian buffers. At the site with high volumes of existing downed wood, there was no significant change in amphibian capture rates following thinning with three different buffer widths. However, at the site with little downed wood, capture rates declined following thinning with buffers of ≥6 m or ≥15 m, but not ≥70 m. At the two sites, treatments were unharvested or thinned (to 200 trees/ha; 10% cut in groups; 10% patches retained; deadwood retained) with riparian buffer widths of ≥6 m (streamside-retention), ≥15 m (variable-width) or ≥70 m. Monitoring was undertaken in May–June before and two years after thinning. Visual count surveys were along 102 m transects perpendicular to each stream bank (7–8/treatment).

A replicated, site comparison study in 2005 of 3 forest sites in Oregon, USA (8) found that there was no significant difference between amphibian captures in riparian buffers and unharvested forest 5–6 years after harvest. Captures did not differ significantly between thinned and unharvested, or between two buffer widths (6 and >15 m) for all amphibians, western red-backed salamanders *Plethodon vehiculum* or ensatina *Ensatina eschscholtzii*. However, captures did decrease significantly with distance from stream for all amphibians and red-backed salamanders. Captures varied with distance for ensatina. Overall, 60% of captures occurred within 15 m of the stream. Each 12–24 ha site had 2 streams within forest that had been thinned (600 to 200 trees/ha) with riparian buffers (6 m or >15 m wide) retained in 2000 and one stream with no harvesting. Amphibians were sampled by visual counts once in April–June within five 5 x 10 m plots at 4 distances from each stream (up to 35 m).

A replicated, controlled study in 2005–2007 of salamanders in five headwater streams in North Carolina, USA (9) found that retaining 30 m riparian buffers during timber harvest maintained salamander populations. Two-lined salamander *Eurycea wilderae* larvae were significantly more abundant within 30 m buffers (413 larvae) and unharvested streams (171–533) than in streams with 9 m or no buffers (72–73). However, black-bellied salamanders *Desmognathus quadramaculatus* showed no difference in abundance between treatments (25–34 larvae). Treatments were timber harvest with riparian buffers of 0, 9 or 30 m retained on both sides of the stream. The two controls were no harvest. Timber was harvested in 2005–2006. Salamanders were monitored within three 40 m sampling blocks along streams in

May–August 2006 (9 m buffer and controls) and 2007 (all sites). Animals were captured using 48 leaf litter bags/site each 1–2 weeks.

A randomized, replicated, controlled study in 2003–2005 of 11 forest ponds in east-central Maine, USA (10) found that the impact of buffer zones on spotted salamander *Ambystoma maculatum* migration behaviour depended on weather conditions. Migration rate and distance of salamanders from ponds did not differ significantly between treatments. However, the probability of migration differed significantly between the 100 m buffer and unharvested, but not 30 m buffer treatments. If rainfall was low, salamanders were more likely to move in the 100 m compared to unharvested treatment, above 390 mm of cumulative rainfall the opposite was true. Ponds were randomly assigned to treatments: clearcut with 30 m or 100 m buffers or unharvested. Concentric 100 m wide clearcuts were created around buffers surrounding ponds in 2003–2004. Salamanders were captured in pitfall traps along drift-fences as they left breeding ponds in spring. Forty salamanders were radio-tracked (6–21/treatment) in April–November 2004–2005.

A meta-analysis of global studies of amphibians in harvested forests (11) found that riparian buffers were not effective at maintaining amphibian abundance. Amphibian abundance was significantly lower in buffers compared to unharvested areas. Frogs and toads (15 studies) showed greater differences between buffers and unharvested sites (both positive and negative) compared to salamanders (16 studies). There was no significant effect of buffer width or time since buffer establishment on the size of the difference in abundance between buffers and unharvested sites (amphibians, birds, small mammals and arthropods combined). Wider buffers did not result in greater similarity between buffer and unharvested sites. A meta-analysis was undertaken using published data from 31 studies comparing abundance of species in riparian buffers and unharvested riparian sites.

A replicated, controlled, site comparison study in 2001 of amphibians in 41 forest streams in Washington, USA (12) found that where buffers were retained during clearcutting, densities of two of three species were significantly higher. Densities were significantly higher with buffers than without for tailed frogs *Ascaphus truei* (0.4 vs 0/m^2) and cascade torrent salamander *Rhyacotriton cascadae* (0.5 vs 0.2). For both species, densities were significantly higher in unharvested forests (0.7 and 1.5/m^2 respectively) but not secondary forests (0.2 and 0.6). In contrast, giant salamander *Dicamptodon* spp. densities were significantly lower in buffered (0.2/m^2) than unbuffered streams and secondary forests (0.3/m^2). Densities in unharvested forests (0.2) were significantly lower than the average for managed forests. Nine to 12 streams in each of 4 management types were sampled: clearcuts (≤10 years old) with 5–23 m wide buffers or without buffers, second-growth forest (≥35 years old) and unharvested forest. Amphibians were monitored within six 2 m long plots within 45–55 m sub-sections of streams in June–August 2001.

A replicated, controlled, before-and-after study in 1992–2004 of conifer plantations in Washington, USA (13) found that retaining riparian buffers during harvest had mixed effects on amphibians. Western red-backed salamanders *Plethodon vehiculum* and ensatinas *Ensatina eschscholtzii* appeared to benefit from riparian buffers. However, coastal tailed frogs *Ascaphus truei* declined significantly immediately after harvest at sites with wide buffers and 10 years after treatment the species was almost locally extinct at narrow and wide buffered sites. For other species there was suggestion of treatment effects, but analyses were confounded by patterns of natural population changes. In 1992, 18 sites (33–50 ha) were selected and assigned to 3 treatments: forest harvested with a riparian buffer of approximately 8 m or a wider buffer (plus wildlife reserve trees/logs) and control sites of previously logged second-growth forests. Streams were 2–6 m wide and had clear-cutting of 15 ha either side. Amphibians were monitored in October–November before harvest (1992–1993), 2-years after (1995–1996) and 10-years after harvest (2003–2004). Eighteen pairs of pitfall traps were placed in buffers and adjacent habitat.

(1)   Cole E.C., McComb W.C., Newton M., Chambers C.L. & Leeming J.P. (1997) Response of amphibians to clearcutting, burning, and glyphosate application in the Oregon Coast Range. *Journal of Wildlife Management*, 61, 656–664.
(2)   Vesely D.G. (1997) Terrestrial amphibian abundance and species richness in headwater riparian buffer strips, Oregon Coast Range. MSc thesis. Oregon State University.
(3)   Hannon S.J., Paszkowski C.A., Boutin S., DeGroot J., Macdonald S.E., Wheatley M. & Eaton B.R. (2002) Abundance and species composition of amphibians, small mammals, and songbirds in riparian forest buffer strips of varying widths in the boreal mixedwood of Alberta. *Canadian Journal of Forest Research-Revue Canadienne de Recherche Forestiere*, 32, 1784–1800.
(4)   Vesely D.G. & McComb W.C. (2002) Salamander abundance and amphibian species richness in riparian buffer strips in the Oregon Coast Range. *Forest Science*, 48, 291–297.
(5)   Perkins D.W., Malcolm L. & Hunter J.R. (2006) Effects of riparian timber management on amphibians in Maine. *Journal of Wildlife Management*, 70, 657–670.
(6)   Olson D.H. & Rugger C. (2007) Preliminary study of the effects of headwater riparian reserves with upslope thinning on stream habitats and amphibians in western Oregon. *Forest Science*, 53, 331–342.
(7)   Rundio D.E. & Olson D.H. (2007) Influence of headwater site conditions and riparian buffers on terrestrial salamander response to forest thinning. *Forest Science*, 53, 320–330.
(8)   Kluber M.R., Olson D.H. & Puettmann K.J. (2008) Amphibian distributions in riparian and upslope areas and their habitat associations on managed forest landscapes in the Oregon Coast Range. *Forest Ecology and Management*, 256, 529–535.
(9)   Peterman W.E. & Semlitsch R.D. (2009) Efficacy of riparian buffers in mitigating local population declines and the effects of even-aged timber harvest on larval salamanders. *Forest Ecology and Management*, 257, 8–14.
(10)   Veysey J.S., Babbitt K.J. & Cooper A. (2009) An experimental assessment of buffer width: implications for salamander migratory behavior. *Biological Conservation*, 142, 2227–2239.
(11)   Marczak L.B., Sakamaki T., Turvey S.L., Deguise I., Wood S.L.R. & Richardson J.S. (2010) Are forested buffers an effective conservation strategy for riparian fauna? An assessment using meta-analysis. *Ecological Applications*, 20, 126–134.
(12)   Pollett K.L., MacCracken J.G. & MacMahon J.A. (2010) Stream buffers ameliorate the effects of timber harvest on amphibians in the Cascade Range of southern Washington, USA. *Forest Ecology and Management*, 260, 1083–1087.
(13)   Hawkes V.C. & Gregory P.T. (2012) Temporal changes in the relative abundance of amphibians relative to riparian buffer width in western Washington, USA. *Forest Ecology and Management*, 274, 67–80.

6 Threat: Human intrusions and disturbance

In addition to large-scale disturbances from activities such as agriculture, building developments, energy production and biological resource use, disturbance of amphibian populations can come from smaller scale human intrusions.

Key messages

Use signs and access restrictions to reduce disturbance
We captured no evidence for the effects of using signs and access restrictions to reduce disturbance on amphibian populations.

6.1 Use signs and access restrictions to reduce disturbance

- We found no evidence for the effects of using signs and access restrictions to reduce disturbance on amphibian populations.

Background
Amphibian species are able to tolerate different levels of disturbance. For particularly sensitive species or populations, or in areas subject to high levels of disturbance, it may be possible to reduce human disturbance with signs or access restrictions. Reducing access helps to reduce the risk of human introduction of non-native plants, animals or disease.

7 Threat: Natural system modifications

Key messages

Use prescribed fire or modifications to burning regime
Eight of 15 studies, including three randomized, replicated, controlled studies, in Australia, North America and the USA found no effect of prescribed forest fires on amphibian abundance or numbers of species. Four found that fires had mixed effects on abundance. Four found that abundance, numbers of species or hatching success increased and one that abundance decreased. Two of three studies, including one replicated, before-and-after study, in the USA and Argentina found that prescribed fires in grassland decreased amphibian abundance or numbers of species. One found that spring, but not autumn or winter burns in grassland, decreased abundance.

Use herbicides to control mid-storey or ground vegetation
Three studies, including two randomized, replicated, controlled studies, in the USA found that understory removal using herbicide had no effect or negative effects on amphibian abundance. One replicated, site comparison study in Canada found that following logging, abundance was similar or lower in stands with herbicide treatment and planting compared to those left to regenerate naturally.

Mechanically remove mid-storey or ground vegetation
One randomized, replicated, controlled study in the USA found that mechanical understory reduction increased numbers of amphibian species, but not amphibian abundance.

Regulate water levels
Three studies, including one replicated, site comparison study, in the UK and USA found that maintaining pond water levels, in two cases with other habitat management, increased or maintained amphibian populations or increased breeding success. One replicated, controlled study in Brazil found that keeping rice fields flooded after harvest did not change amphibian abundance or numbers of species, but changed species composition. One replicated, controlled study in the USA found that draining ponds increased abundance and numbers of amphibian species.

7.1 Use prescribed fire or modifications to burning regime

Background
Prescribed fires are undertaken to reduce the amount of combustible fuel in an attempt to reduce the risk of more extensive, potentially more damaging 'wildfires'. They may also be used in the maintenance or restoration of habitats historically subject to occasional 'wildfires' that have been suppressed through management.

In forests, fires may remove large amounts of woody material from the understorey and result in increased grasses and herbaceous vegetation. Such changes can affect forest amphibians. For example, one study found that frog and toad species richness was not affected by the interval between fires, but six species showed some response to the number of fires at the site (Westgate *et al.* 2012).

Westgate M.J., Driscoll D.A. & Lindenmayer D.B. (2012) Can the intermediate disturbance hypothesis and information on species traits predict anuran responses to fire? *Oikos*, 121, 1516–1524.

7.1.1 Forests

- Eight of 14 studies (including 3 randomized, replicated, controlled studies) in Australia, North America and the USA found no effect of prescribed forest fires on amphibian abundance[7, 10, 12, 13, 15, 17] or numbers of species[2, 7, 10–12]. Four found that forest fires had mixed effects on amphibian abundance depending on species[8], species and year[4, 5] or season of burn[16]. Three found that fires increased amphibian abundance[1, 2, 11] or numbers of species[1]. One found that abundance decreased with fires[3].

- Two studies (including one randomized, replicated, controlled study) in the USA found that numbers of amphibian species and abundance increased[9] or abundance decreased[18] with time since prescribed forest fires.

- One before-and-after study in the USA found that spotted salamander hatching success increased following a prescribed forest fire[14].

A controlled, site comparison study in 1982–1983 of sandhill-scrub habitat in west central Florida, USA (1) found that controlled burns resulted in higher species diversity and abundance of amphibians. The 7-year burn cycle plot had the greatest number of species in both years (7-year cycle: 16–20; 2-year: 10–15; 1-year: 14–16; unburned: 10–15). Although burn plots had greater fluctuations in species diversity over the two years than the unburned plot, numbers of captures were higher. Captures tended to be highest in 7- and 1-year burn plots (7 years: 115–307; 2 years: 102–187; 1 year: 126–203; unburned: 71–125). The 1-year cycle was most consistent for supporting high numbers of individuals and species. A 1 ha plot was established for each burn cycle in adjacent strips. These were compared to a plot unburned for 20 years. Burns were in May–June. Five drift-fence arrays with pitfall traps and an artificial cover board were established/plot. Traps were checked 5–6 times/week in April–October 1983–1984.

A replicated, controlled, site comparison study in 1994 of native forest managed near Brisbane, Australia (2) found that prescribed fires in native forest resulted in increased amphibian abundance but not species richness. In native forest there was a significantly higher number captured in 5-year burn cycles than unburned sites (5-year cycle: 127; 3-year: 85; unburned: 51). In plantations, numbers were similar (burned 7 years ago: 37; burned 2 years ago: 48; unburned: 39). There was no significant difference in species richness between treatments (native: 3–4; plantation: 6). Treatments in native forest (1.5 ha; 2 replicates) were: burned in autumn–winter on a 3-year cycle (burned 1991), in winter–spring on a 5-year cycle (burned 1993) or unburned (since 1973). In the plantation (25 ha) treatments were: burned two or seven years ago or unburned. Drift-fencing with pitfall traps and active searching were used for monitoring in January or March 1994 (75–180 trap nights/treatment).

A controlled study in 1992–1993 of pine stands in Maryland, USA (3) found that annual prescribed burns resulted in significantly lower amphibian abundance. Captures were significantly lower in the burned compared to unburned stand for total amphibians (74 vs 391), salamanders (8 vs 105), ranid species (6 vs 20) and frogs and toads (66 vs 214). The same was true for two of ten frog and toad species, adults of two of four salamander species and young of the year for three frog species. The other species showed no significant difference between treatments. Study sites were an unburned mixed pine-hardwood stand (5 ha) and a pine stand (4 ha) that had been burned annually since 1981, with alternating thirds being burned from 1988. Monitoring was undertaken using three drift-fences with pitfall and funnel traps per site in March–July 1992–1993.

A replicated, controlled study in 1995–1996 in a national forest in Carolina, USA (4) found that prescribed fires did not tend to affect the abundance of salamanders. There were no significant difference in numbers of blue ridge two-line salamanders *Eurycea wilderae*, Jordan's salamanders *Plethodon jordani* or mountain dusky salamanders *Desmognathus ochrophaeus* captured in burned and unburned areas. Seepage salamander *Desmognathus aeneus* captures were significantly lower in the riparian zone of the burned compared to unburned areas in 1996 (0.2 vs 1.3). Monitoring was undertaken for two weeks immediately before an April burn and after the burn in June 1995 and August 1996 at two sites. Drift-fencing with pitfalls and snap-traps were installed at three locations in the upper slope, mid-slope and riparian zone at each site. Visual searches were also undertaken. An unburned area at one of the sites was monitored in the same way.

A randomized, replicated, controlled study in 1997–1998 of pine sandhills in Florida, USA (5) found that prescribed burning resulted in similar or lower abundance of amphibians compared to unburned sites. In 1997 there was no significant difference between treatments for any species. In 1998, capture rates were significantly lower in prescribed burn plots and herbicide understory removal plots than fire suppressed (control) plots for southern toad *Bufo terrestris* (burn: 0; understory: 0.002; no burn: 0.008; reference: 0.003 captures/trap days). Capture rates did not differ between burned, understory removal or fire suppressed treatments for oak toad *Bufo quercicus* or eastern narrowmouthed toad *Gastrophryne carolinensis*. In 1997 (not 1998), similarity indices indicated that burned plots were significantly more similar to reference (frequently burned) sites than understory removal or fire suppressed plots (burn: 0.76; understory: 0.49; no burn: 0.49). Treatments were in randomly assigned 81 ha plots within 4 replicate blocks in spring 1997. Data were also collected from four frequently burned reference sites. Monitoring was with drift-fencing and pitfall traps in April–August 1997–1998.

A replicated, site comparison study in 1994–1996 of mature pine forest in Georgia, USA (6) found that there was no apparent difference between amphibian abundance or numbers of species in forest burned in the growing or dormant season. Total amphibian captures and numbers of species were similar between plots burned in the growing season (abundance: 32; species: 7) and dormant season (abundance: 19; species: 4). Captures were higher in unburned hardwood forest (abundance: 101; species: 14). Sample sizes were considered too small for statistical analysis. Three plots burned in the 1994 growing season (April–August; 3-year cycle) and three burned in the dormant season (January–March) were selected. Three adjacent hardwood plots were also surveyed. Three drift-fences with 12 pitfall traps and 4 artificial cover boards were installed within each plot. Monitoring was undertaken over four weeks, four times in 1995–1996.

A replicated, controlled study in 2001 of bottomland hardwood forest in Georgia, USA (7) found that amphibian abundance, diversity and richness were similar in burned and unburned stands. Abundance did not differ significantly at burned and unburned sites for all amphibians (43 vs 62), salamanders (2 vs 6) or frogs and toads (39 vs 50). The same was true for species richness overall (8 vs 8 species), for salamanders (2 vs 2) or frogs and toads (6 vs 6). The volume of coarse woody debris was similar in burned and unburned stands (60 vs 128 m³/ha). Amphibians were monitored in three winter-burned and unburned stands from July to October 2001. Drift-fencing with pitfall traps, artificial cover boards and PVC pipe refugia were randomly placed within each site.

A review in 2003 of the effects of prescribed fire on amphibians in North America (8) found that results were mixed. Four studies found that amphibian abundance or abundances of some species were lower in burned compared to unburned stands. One study found that abundance of certain species was higher following burning, two found mixed results depending on species and two found no significant differences between treatments. One of two studies found that species richness was greatest in 5–7 year burn cycles and the other found no difference between burned and unburned stands. The majority of studies focused on short-term responses (1–3 years post-burn), with only 1 of 10 investigating longer-term effects (5 years post-burn).

A site comparison study of 15 ponds in a pine forest in South Carolina, USA (9) found that amphibian abundance and species richness increased with time since prescribed burns. Abundance of all amphibians and frogs and toads increased significantly with time since burning. This was not the case for salamanders. Amphibian species richness also increased significantly over time following burns. This was likely to be because salamanders were rarely encountered at sites burned within two years, but became more abundant with time. Amphibians were monitored at 15 ponds with 5 different prescribed burn (in winter/spring) histories: 0, 1, 3, 5 and 12 years after burns. Drift-fences, tree-frog shelters, calling censuses, minnow trapping and visual surveys were used.

A randomized, replicated, controlled study in 1995–1996 of shelterwood-harvested oak stands in Virginia, USA (10) found that prescribed burns did not affect amphibian abundance or species richness. There were no significant differences in relative abundances between burned and unburned sites for all amphibians (burned: 10–15; unburned: 6), eastern red-backed salamanders *Plethodon cinereus* (7–11 vs 3) or American toads *Bufo americanus* (3 vs 2). Amphibian species richness did not differ significantly between burned and unburned sites (2–3 vs 5). Three replicates (2–5 ha) of 4 randomly assigned treatments were applied in 1995: burning in February, April or August, or unburned. Three uncut reference sites were also monitored. Amphibians were monitored using pitfall traps (20/site) for 53 nights in June, July and October 1996.

A replicated, controlled study in 2003–2004 of pine savanna in Mississippi, USA (11) found that prescribed burning resulted in a greater abundance but similar diversity of amphibians compared to unburned sites. Greater numbers of amphibians were found at burned than unburned sites (275 vs 90). However, species diversity was similar (burned: 13; unburned: 10). Some species were significantly more abundant in burned compared to unburned areas including oak toads *Bufo quercicus* (125 vs 9) and southern leopard frogs *Rana utricularia* (51 vs 2). In comparison, a small number of species were more common in unburned sites including the pig frog *Rana grylio* (13 vs 2). A low intensity burn was undertaken over a large proportion of a National Wildlife Refuge in 2003. From January to June 2004, amphibians were monitored at three burned and three unburned sites. Visual encounter surveys (200 m transects), minnow traps (6/site) and PVC tubes (5/site) were used.

A randomized, replicated, controlled study in 2001–2004 in hardwood forest in Carolina, USA (12) found that prescribed burns did not increase overall amphibian abundance or species richness, but did increase abundance of frogs and toads. The relative abundance of total amphibians, salamanders and green frog *Rana clamitans* did not differ significantly between treatments. However, abundances of anurans (frogs and toads) and American toads *Bufo americanus* were significantly higher in burn treatments compared to controls (anurans:

52–54 vs 8; American toads: 50 vs 10 captured/100 nights). Species richness did not differ significantly (burned: 5; burned with understory reduction: 5; control: 3). There were three 14 ha replicates of each randomly assigned treatment: prescribed burn, burn and mechanical understory reduction and controls. Understory reduction was undertaken in winter 2001–2002 and burns in March 2003. Drift-fences with pitfall and funnel traps were used for monitoring in August–October 2001 and May–September 2002–2004.

A replicated, site comparison study in 1999–2001 of pine woodland in western Arkansas, USA (13) found that controlled burning did not affect amphibian species abundance. There was no significant difference between numbers of captures in burned and unburned plots for all amphibians (73 vs 59), all frogs and toads (71 vs 55), individual species or salamanders (2 vs 4). The most abundantly caught species, the western slimy salamander *Plethodon albagula*, was captured almost exclusively in unmanaged woodland (28 of 29 captures). Nine plots (11–42 ha) that had been thinned (1980–1990) and then burned at least 3 times at 3–5-year intervals were sampled. These were compared to three unmanaged, unburned plots. Controlled fires were in March–April. Three drift-fence arrays with pitfall and box traps were established/plot. Traps were checked weekly in April–September 1999–2001.

A before-and-after study in 2005–2007 of a pond in restored mixed forest in Illinois, USA (14) found that prescribed burning resulted in increased hatching success for spotted salamander *Ambystoma maculatum*. Eggs failed to hatch in 2005, but following burning, hatching success of egg masses was 29% in 2006 and 53% in 2007. Restoration started in 2000 and included destruction of drainage tiles, clearing of invasive plants, prescribed burning and removal of leaf litter. The burn was in autumn 2005. An egg mass was placed in 2 mesh enclosures (56 x 36 x 36 cm) in the pond. Eggs were monitored every five days until hatching was complete.

A controlled, before-and-after study in 2001–2006 of ponderosa pine forest in Idaho, USA (15) found that a prescribed fire had no significant effect on the density of rocky tailed frog tadpoles *Ascaphus montanus*. During the study, the density of tadpoles decreased by 50% in both burned (pre-burn: 2.3; post-burn: $1.1/m^2$) and unburned catchments (pre: 2.7; post: 1.6). A prescribed burn was undertaken in May 2004 and burned 12% of 1 catchment. Four nearby unburned catchments were monitored for comparison. Tadpoles were monitored using kick-sampling in 30 transects (1 m wide) per stream in 2001–2006.

A replicated, before-and-after study in 1988–2008 of 25 wetlands in forest and grassland reserves in Indiana, USA (16) found that the relative abundance of salamanders declined following prescribed spring, but not autumn or winter burns. The six forest species declined significantly (82–100%) following spring burns and took an average of five years to recover to pre-burn levels. Declines were not associated with autumn or winter burns and tiger salamander *Ambystoma tigrinum* and eastern newt *Notophthalmus viridescens* increased at two sites after an autumn burn. Monitoring was undertaken the year before and after burns. Each site was visited monthly for three months in spring and one in summer or autumn. Visual searches, minnow traps, dipnets and seines were used to survey entire small ponds (< 0.25 ha) and 50 m of adjacent upland habitat, or along transects for larger ponds.

A replicated, controlled before-and-after study in 2001–2007 of hardwood forest in West Virginia, USA (17) found that although population responses were difficult to interpret following two prescribed fires, results suggested that there was no significant affect on the salamander assemblage. Mountain dusky salamanders *Desmognathus ochrophaeus* and red-backed salamander *Plethodon cinereus* counts were greater following burns compared to before burns or unburned controls. However, the authors considered that this was due to increased use of artificial cover boards in response to reduced leaf litter following fires. Treatments were burn plots on upper slopes or lower slopes ($n = 20$), half of which were fenced and control plots that were unburned and unfenced ($n = 4$). Burns were in 2002–2003 and 2005. Cover board arrays were used to monitor salamanders before and after two fires in April–October in 2001–2007.

A randomized, replicated study in 1999–2001 of nine restored pine woodlands in western Arkansas, USA (18) found that overall numbers of amphibians were highest in the first year after burns compared to the following two years. This was true for total amphibians (1st year: 114; 2nd year: 53; 3rd year: 51/stand) and anurans (1st: 112; 2nd: 51; 3rd: 49). However, this trend was largely due to high numbers of dwarf American toads *Bufo americanus charlessmithi* in the first year (83 vs 27–31). Fowler's toads *Bufo fowleri* were also captured most often in year one stands (2.0 vs 0.1–0.2). Salamander captures did not differ between years after burn. In 1999–2000, stands (11–42 ha) were burned on a 3-year cycle, so 3 were burned each year in March–April. Stands had been thinned at least 9 years previously and had undergone 3–7 prescribed burns at 2–5 year intervals. Monitoring was undertaken using three drift-fence arrays per stand (15 m) connected to central funnel traps in April–September in 1999–2001.

(1)   Mushinsky H.R. (1985) Fire and the Florida sandhill herpetofaunal community: with special attention to responses of *Cnemidophorus sexlineatus*. *Herpetologica*, 41, 333–342.
(2)   Hannah D.S. & Smith G.C. (1995) Effects of prescribed burning on herptiles in southeastern Queensland. *Memoirs of the Queensland Museum*, 38, 529–531.
(3)   McLeod R.F. & Gates J.E. (1998) Response of herpetofaunal communities to forest cutting and burning at Chesapeake Farms Maryland. *American Midland Naturalist*, 139, 164–177.
(4)   Ford W.M., Menzel M.A., McGill D.W., Laerm J. & McCay T.S. (1999) Effects of a community restoration fire on small mammals and herpetofauna in the southern Appalachians. *Forest Ecology and Management*, 114, 233–243.
(5)   Litt A.R., Provencher L., Tanner G.W. & Franz R. (2001) Herpetofaunal responses to restoration treatments of longleaf pine sandhills in Florida. *Restoration Ecology*, 9, 462–474.
(6)   Miller K.V., Chapman B.R. & Ellington K.K. (2001) Amphibians in pine stands managed with growing-season and dormant-season prescribed fire. *Journal of the Elisha Mitchell Scientific Society*, 117, 75–78.
(7)   Moseley K.R., Castleberry S.B. & Schweitzer S.H. (2003) Effects of prescribed fire on herpetofauna in bottomland hardwood forests. *Southeastern Naturalist*, 2, 475–486.
(8)   Pilliod D.S., Bury R.B., Hyde E.J., Pearl C.A. & Corn P.S. (2003) Fire and amphibians in North America. *Forest Ecology and Management*, 178, 163–181.
(9)   Schurbon J.M. & Fauth J.E. (2003) Effects of prescribed burning on amphibian diversity in a southeastern U.S. National Forest. *Conservation Biology*, 17, 1338–1349.
(10)   Keyser P.D., Sausville D.J., Ford W.M., Schwab D.J. & Brose P.H. (2004) Prescribed fire impacts to amphibians and reptiles in shelterwood-harvested oak-dominated forests. *Virginia Journal of Science*, 55, 159–168.
(11)   Langford G.J., Borden J.A., Major C.S. & Nelson D.H. (2007) Effects of prescribed fire on the herpetofauna of a southern Mississippi pine savanna. *Herpetological Conservation and Biology*, 2, 135–143.
(12)   Greenberg C.H. & Waldrop T.A. (2008) Short-term response of reptiles and amphibians to prescribed fire and mechanical fuel reduction in a southern Appalachian upland hardwood forest. *Forest Ecology and Management*, 255, 2883–2893.
(13)   Perry R.W., Rudolph D.C. & Thill R.E. (2009) Reptile and amphibian responses to restoration of fire-maintained pine woodlands. *Restoration Ecology*, 17, 917–927.
(14)   Sacerdote A.B. & King R.B. (2009) Dissolved oxygen requirements for hatching success of two Ambystomatid salamanders in restored ephemeral ponds. *Wetlands*, 29, 1202–1213.
(15)   Arkle R.S. & Pilliod D.S. (2010) Prescribed fires as ecological surrogates for wildfires: a stream and riparian perspective. *Forest Ecology and Management*, 259, 893–903.
(16)   Brodman R. (2010) The importance of natural history, landscape factors, and management practices in conserving pond-breeding salamander diversity. *Herpetological Conservation and Biology*, 5, 501–514.
(17)   Ford W.M., Rodrigue J.L., Rowan E.L., Castleberry S.B. & Schuler T.M. (2010) Woodland salamander response to two prescribed fires in the central Appalachians. *Forest Ecology and Management*, 260, 1003–1009.
(18)   Perry R.W., Rudolph D.C. & Thill R.E. (2012) Effects of short-rotation controlled burning on amphibians and reptiles in pine woodlands. *Forest Ecology and Management*, 271, 124–131.

7.1.2 Grassland

- Two studies, including one before-and-after, site comparison study, in the USA and Argentina found that annual prescribed fires in grassland decreased numbers of amphibian species and abundance[3] or, along with changes in grazing regime, increased rates of species loss[1].

- One replicated, before-and-after study in the USA[2] found that spring, but not autumn or winter burns, decreased salamander abundance.

A before-and-after study in 1989–2003 of tallgrass prairie in Kansas, USA (1) found that rates of species loss were significantly higher during burn years compared to non-burn years (0.04 vs 0.00). However, the authors considered that strong conclusions could not be reached because of confounding effects of changes in both burning and grazing. From 1989 to 1998, management was traditional season-long stocking (0.6 cattle/ha) with burning in alternate years. From 1999, management changed to intensive-early cattle stocking (1.0 cattle/ha) for three months from late spring combined with annual burning. Amphibians were surveyed in April annually along a 4 km transect.

A replicated, before-and-after study in 1988–2008 of 25 wetlands in grassland and forest reserves in Indiana, USA (2) found that the relative abundance of salamanders declined following prescribed spring, but not autumn or winter burns. There was a significant decline (33–63%) in the abundance of three of four species following spring burns. Open habitat (grassland and savanna) salamanders took two years to recover and abundance often exceeded that before the burn. Declines were not associated with autumn or winter burns and tiger salamander *Ambystoma tigrinum* and eastern newt *Notophthalmus viridescens* increased at two sites after an autumn burn. Monitoring was undertaken the year before and after burns. Each site was visited monthly for three months in spring and one in summer or autumn. Visual searches, minnow traps, dipnets and seines were used to survey entire small ponds (< 0.25 ha) and 50 m of adjacent upland habitat, or along transects for larger ponds.

A site comparison study in 2006 of cattle pasture in Corrientes, Argentina (3) found that amphibian diversity, species richness and abundance was significantly lower following annual prescribed fires. Species richness and abundance was significantly lower with annual prescribed fire with or without grazing (richness: 7–9; abundance: 17–23) compared to sites that had not been burned for 3 or 12 years (richness: 10; abundance: 46–49). Diversity was significantly lower at the site with annual prescribed fire and grazing (1.3 vs 1.9–2.1). Species composition differed most between the unburned site and that with annual prescribed fire and grazing (Sorensen's similarity index = 0.58). Only 2 of 12 species showed significant differences between treatments. The 4 historic treatments (≥ 400 ha) were: annual prescribed fire (August–September) without or with grazing (3 ha/cattle unit), 3 years since a prescribed fire, and no fire or grazing for 12 years. Monitoring was undertaken using drift-fencing with pitfall traps in January–April 2006.

(1) Wilgers D.J., Horne E.A., Sandercock B.K. & Volkmann A.W. (2006) Effects of rangeland management on community dynamics of the herpetofauna of the tall grass prairie. *Herpetologica*, 62, 378–388.
(2) Brodman R. (2010) The importance of natural history, landscape factors, and management practices in conserving pond-breeding salamander diversity. *Herpetological Conservation and Biology*, 5, 501–514.
(3) Cano P.D. & Leynaud G.C. (2010) Effects of fire and cattle grazing on amphibians and lizards in northeastern Argentina (Humid Chaco). *European Journal of Wildlife Research*, 56, 411–420.

7.2   Use herbicides to control mid-storey or ground vegetation

- Three studies (including two randomized, replicated, controlled studies) in the USA found that understory removal using herbicide had no effect[1, 3, 5] or some negative effects[2] on amphibian abundance.

- One replicated, site comparison study in Canada[4] found that following logging American toad abundance was similar and wood frogs lower in stands with herbicide treatment and planting compared to stands left to regenerate naturally.

Background

Herbicides can be used as a substitute for prescribed fire to eliminate competing mid-storey or ground vegetation. Although herbicides do not have the multiple ecosystem functions provided by fire, they have some advantages such as increased selectivity and decreased risk of offsite fire damage.

Other studies that control mid-storey or ground vegetation are discussed in 'Mechanically remove mid-storey or ground vegetation'.

A controlled, before-and-after study in 1994–1997 in a hardwood forest in Virginia, USA (1) found that understory removal using herbicide did not affect the relative abundance of salamanders. Captures did not differ significantly before and after understory removal (9 vs 11/search). Abundance did not differ significantly within the untreated plot over time (1994: 10; 1995–1997: 8–10). Treatment was within a 2 ha plot. Salamanders were monitored along 15 x 2 m transects using artificial cover objects (50/plot).

A randomized, replicated, controlled study in 1997–1998 of pine sandhills in Florida, USA (2) found that understory removal using herbicide did not result in increased abundance of amphibians. In 1998, capture rates were significantly lower in understory removal plots and prescribed burning plots than fire suppressed (control) plots for southern toad *Bufo terrestris* (herbicide: 0.002; burn: 0; no burn: 0.008; reference: 0.003 captures/trap days). However, capture rates did not differ between understory removal, burned or fire suppressed treatments for oak toad *Bufo quercicus* or eastern narrowmouthed toad *Gastrophryne carolinensis* in 1998, or any species in 1997. In 1997 (not 1998), herpetofauna similarity indices indicated that burned plots were significantly more similar to reference (frequently burned) sites than understory removal or fire-suppressed plots (burn: 0.76; herbicide: 0.49; no burn: 0.49). Treatments were in randomly assigned 81 ha plots within 4 replicate blocks in spring 1997. Data were also collected from four frequently burned reference sites. Monitoring was undertaken using drift-fencing and pitfall traps in April–August 1997–1998.

A randomized, replicated, controlled study in 1993–1999 of four harvested forests in Virginia, USA (3) found that salamander abundance was similar in plots with and without herbicide treatment (7 vs 6/30 m²; see also (5)). Four sites had 2 ha plots with herbicide application (Garlon4) to reduce woody shrubs and a control with no management. Salamanders were monitored on 9–15 transects (2 x 15 m)/plot at night in April–October. Monitoring was undertaken 1–2 years before and 1–4 years after treatment.

A replicated, site comparison study in 2001–2002 of boreal forest stands in Ontario, Canada (4) found that herbicide treatment and planting after logging did not result in higher amphibian abundance compared to stands left to regenerate naturally. Wood frogs *Rana sylvatica* were significantly less abundant in 20–30-year-old stands that had been managed by planting and herbicide treatment with or without tree scarring (0.06 captures/trap/

night) compared to those that had been left to regenerate naturally (0.09). Capture rates in 32–50-year-old managed stands (0.07) did not differ significantly from naturally regenerated (0.12) and uncut stands (0.06). For American toads *Bufo americanus*, there was no significant difference in capture rates between treatments or ages of stands (managed: 0.02–0.04; natural regeneration: 0.02–0.03; uncut: 0.03). Nineteen stands that had received each treatment and five uncut stands were surveyed. Drift-fencing with pitfall traps were used for monitoring in August–September 2001–2002.

In a continuation of a previous study (3), a randomized, replicated, controlled study in 1994–2007 of 6 hardwood forests in Virginia, USA (5) found that salamander abundance was similar in plots with mid-storey herbicide treatment and without up to 13-years post-harvest (8 vs 7/transect). There were 6 sites with 2 ha plots randomly assigned to treatments: herbicide application (triclopyr and imazapyr) to reduce woody shrubs and a control with no management. Treatments were in 1994–1998 and salamanders were monitored at night along nine 15 x 2 m transects/site.

(1) Harpole D.N. & Haas C.A. (1999) Effects of seven silvicultural treatments on terrestrial salamanders. *Forest Ecology and Management*, 114, 349–356.
(2) Litt A.R., Provencher L., Tanner G.W. & Franz R. (2001) Herpetofaunal responses to restoration treatments of longleaf pine sandhills in Florida. *Restoration Ecology*, 9, 462–474.
(3) Knapp S.M., Haas C.A., Harpole D.N. & Kirkpatrick R.L. (2003) Initial effects of clearcutting and alternative silvicultural practices on terrestrial salamander abundance. *Conservation Biology*, 17, 752–762.
(4) Thompson I.D., Baker J.A., Jastrebski C., Dacosta J., Fryxell J. & Corbett D. (2008) Effects of post-harvest silviculture on use of boreal forest stands by amphibians and marten in Ontario. *Forestry Chronicle*, 84, 741–747.
(5) Homyack J.A. & Haas C.A. (2009) Long-term effects of experimental forest harvesting on abundance and reproductive demography of terrestrial salamanders. *Biological Conservation*, 142, 110–121.

7.3   Mechanically remove mid-storey or ground vegetation

• One randomized, replicated, controlled study in the USA[1] found that numbers of amphibian species, but not abundance, were significantly higher in plots with mechanical understory reduction compared to those without.

Background

Removing vegetation can be used as a substitute for prescribed fire to eliminate competing mid-storey or ground vegetation. Although this technique does not have the multiple ecosystem functions provided by fire, it has advantages, such as increased selectivity and decreased risk of offsite fire damage.

Other studies that control mid-storey or ground vegetation are discussed in 'Use herbicides to control mid-storey or ground vegetation'.

A randomized, replicated, controlled study in 2001–2004 of upland hardwood forest in North Carolina, USA (1) found that mechanical understory reduction significantly increased amphibian species richness, but not abundance. Species richness was significantly higher in understory reduction plots compared to controls (6 vs 3). However, there was no significant difference in the relative abundance of total amphibians compared to controls (18 vs 17 captured/100 nights), total anurans (frogs and toads; 11 vs 10), salamanders (8 vs 4), American toads *Bufo americanus* (10 vs 10) or green frog *Rana clamitans* (2 vs 1). There were three randomly assigned replicates of treatment and control plots. Mechanical removal of shrubs was undertaken in winter 2001–2002 using chainsaws. Drift-fences with pitfall and funnel traps were used for monitoring in August–October 2001 and May–September 2002–2004.

(1) Greenberg C.H. & Waldrop T.A. (2008) Short-term response of reptiles and amphibians to prescribed fire and mechanical fuel reduction in a southern Appalachian upland hardwood forest. *Forest Ecology and Management*, 255, 2883–2893.

7.4   Regulate water levels

• Two studies (including one replicated, site comparison study) in the UK[1, 5] found that habitat management that included maintaining pond water levels increased natterjack toad populations[5] or maintained newt populations[1]. One replicated, controlled study in Brazil[4] found that keeping rice fields flooded after harvest changed amphibian species composition, but not numbers of species or abundance.

• One replicated, controlled study in the USA[2] found that draining ponds, particularly in the summer, significantly increased abundance and numbers of amphibian species.

• One before-and-after study in the USA[3] found that maintaining pond water levels enabled successful breeding by dusky gopher frogs.

Background
Drying of amphibian breeding sites before terrestrial life stages have developed can have significant detrimental effects on populations. In some cases it may be possible to maintain water levels until after metamorphosis by using a local water source or by bringing in water from an outside source.
   Occasional drying of breeding sites can increase diversity, as it can help control predators, non-native species or more dominant species.
   Studies that manipulated water levels to restore wetlands are discussed in 'Habitat restoration and creation – Restore wetlands'.

A before-and-after study in 1986–1995 of a pond within a housing development near Peterborough, England, UK (1) found that deepening the pond and regulating water levels maintained great crested newt *Triturus cristatus* and smooth newt *Triturus vulgaris* populations. Before the development, numbers varied for great crested newts (1–9) and smooth newts (1–2). Adults of both species returned to breed in 1989–1995 following the development (crested: 10–20; smooth: 9–57). However, production of metamorphs failed in 1990 due to pond drying. Larval catches increased in 1991 following maintenance of water levels (crested: 62; smooth: 22), but then decreased (crested: 15 to 0; smooth: 27 to 2). Development was undertaken in 1987–1989. The pond (800 m²) was deepened in 1988 and water pumped to the pond from 1991. A 1 ha area was retained around the pond. Newts were counted by torch and larvae netted once or twice in 1986–1987 and 3–4 times in March–May 1988–1995.
   A replicated, controlled study of 12 created ponds in forest in South Carolina, USA (2) found that draining ponds resulted in a significant increase in amphibian abundance and species richness. Species richness increased 50% in created wildlife ponds and 100% in construction ponds, compared to those left undrained. Draining in summer resulted in larger increases than draining in winter. Amphibian abundance was also significantly higher in drained ponds compared to those undrained. Created wildlife ponds and ponds created

following removal of construction material were drained in summer, winter, both or never. Each treatment was replicated three times.

A before-and-after study in 2001 of a pond in southern Mississippi, USA (3) found that maintaining the water level to stop the pond drying resulted in the first successful breeding by dusky gopher frog *Rana sevosa* for three years. Complete death of the larvae from the 36 egg masses laid in March was avoided as rather than drying by mid-May, water levels were successfully maintained until heavy rainfall in June. Metamorphs were produced for the first time since 1998, although at 130, numbers were lower than in 1997 (221) and 1998 (2,248). Over seven weeks from mid-April 2001, 366,000 litres of water was pumped from three nearby wells to stop the 440 m circumference pond drying. One of the wells was dug specifically, 50 m from the pond. Irrigation hoses and tanker trucks were used to bring in water.

A replicated, controlled study in 2005–2006 of rice fields in southern Brazil (4) found that keeping fields flooded after harvest did not result in increased amphibian species richness or abundance, but did change species composition. Mean species richness and abundance did not differ between flooded and drained fields (species: 2–8; abundance: 3–66). However, species composition did differ between flooded and dry fields, and a natural wetland. Mean species richness and abundance was lower in flooded and drained fields than the natural wetland (species: 5–8; abundance: 54–139). Abundance at all sites was higher in the growing seasons. Amphibians were monitored in six randomly selected rice fields (1 ha), three that were kept flooded after harvest and three that were drained dry. Three surveys were undertaken in a natural wetland (10 km^2). Each field was surveyed 6 times at night using 6 random 15 minute visual transects in June 2005 to June 2006.

A replicated, site comparison study in 1985–2006 of 20 sites in the UK (5) found that natterjack toad *Bufo calamita* populations increased with species specific habitat management including maintenance of water levels. In contrast, long-term trends showed population declines at unmanaged sites. Individual types of habitat management (aquatic, terrestrial or common toad *Bufo bufo* management) did not significantly affect trends, but length of management did. Overall, 5 of the 20 sites showed positive population trends, five showed negative trends and 10 trends did not differ significantly from zero. Data on populations (egg string counts) and management activities over 11–21 years were obtained from the Natterjack Toad Site Register. Habitat management for toads was undertaken at seven sites. Management varied between sites, but included maintaining water levels, pond creation, adding lime to acidic ponds, vegetation clearance and implementing grazing schemes. Translocations were also undertaken at 7 of the 20 sites using wild-sourced (including head-starting) or captive-bred toads.

(1) Cooke A.S. (1997) Monitoring a breeding population of crested newts (*Triturus cristatus*) in a housing development. *Herpetological Journal*, 7, 37–41.
(2) Fauth J.E. (2002) Restoring amphibian diversity in manufactured ponds: if you drain it, they will come. *Ecological Society of America Annual Meeting Abstracts*, 87, 125–126.
(3) Seigel R. A., Dinsmore A. & Richter S. C. (2006) Using well water to increase hydroperiod as a management option for pond-breeding amphibians. *Wildlife Society Bulletin*, 34, 1022–1027.
(4) Machado I.F. & Maltchik L. (2010) Can management practices in rice fields contribute to amphibian conservation in southern Brazilian wetlands? *Aquatic Conservation*, 20, 39–46.
(5) McGrath A. L. & Lorenzen K. (2010) Management history and climate as key factors driving natterjack toad population trends in Britain. *Animal Conservation*, 13, 483–494.

8 Threat: Invasive alien and other problematic species

Invasive and other problematic species of animals, plants and diseases have caused significant declines in many amphibian species worldwide. Invasive species may prey on amphibians, provide competition for resources, alter habitats or infect them with new diseases. For example, the fungal disease chytridiomycosis, caused by *Batrachochytrium dendrobatidis*, is considered to have been responsible for the decline or extinction of up to 200 species of frogs (Forzan *et al.* 2008). This chapter describes the evidence from interventions designed to reduce the threat from invasive and other problematic species.

Forzan M.J., Gunn H. & Scott P. (2008). Chytridiomycosis in an aquarium collection of frogs, diagnosis, treatment, and control. *Journal of Zoo and Wildlife Medicine*, 39, 406–411.

Key messages – reduce predation by other species

Remove or control mammals
One controlled study in New Zealand found that controlling rats had no significant effect on numbers of Hochstetter's frog. Two studies, one of which was controlled, in New Zealand found that predator-proof enclosures enabled or increased survival of frog species.

Remove or control fish population by catching
Four of six studies, including two replicated, controlled studies, in Sweden, the USA and the UK found that removing fish by catching them increased amphibian abundance, survival and recruitment. Two found no significant effect on newt populations or toad breeding success.

Remove or control fish using rotenone
Three studies, including one replicated study, in Sweden, the UK and the USA found that eliminating fish using rotenone increased numbers of amphibians, amphibian species and recruitment. One review in Australia, the UK and the USA found that fish control that included using rotenone increased breeding success. Two replicated studies in Pakistan and the UK found that rotenone use resulted in frog deaths and negative effects on newts.

Remove or control fish by drying out ponds
One before-and-after study in the USA found that draining ponds to eliminate fish increased numbers of amphibian species. Four studies, including one review, in Estonia, the UK and the USA found that pond drying to eliminate fish, along with other management activities, increased amphibian abundance, numbers of species and breeding success.

Exclude fish with barriers
One controlled study in Mexico found that excluding fish using a barrier increased weight gain of axolotls.

Encourage aquatic plant growth as refuge against fish predation
We captured no evidence for the effects of encouraging aquatic plant growth as refuge against fish predation on amphibian populations.

Remove or control invasive bullfrogs
Two studies, including one replicated, before-and-after study, in the USA and Mexico found that removing American bullfrogs increased the size and range of frog populations. One replicated, before-and-after study in the USA found that following bullfrog removal, frogs were found out in the open more.

Remove or control invasive viperine snake
One before-and-after study in Mallorca found that numbers of Mallorcan midwife toad larvae increased after intensive, but not less intensive, removal of viperine snakes.

Remove or control non-native crayfish
We captured no evidence for the effects of removing or controlling non-native crayfish on amphibian populations.

Key messages – reduce competition with other species

Reduce competition from native amphibians
One replicated, site comparison study in the UK found that common toad control did not increase natterjack toad populations.

Remove or control invasive cane toads
We captured no evidence for the effects of removing or controlling invasive cane toads on amphibian populations.

Remove or control invasive Cuban tree frogs
One before-and-after study in the USA found that removal of invasive Cuban tree frogs increased numbers of native frogs.

Key messages – reduce adverse habitat alteration by other species

Prevent heavy usage or exclude wildfowl from aquatic habitat
We captured no evidence for the effects of preventing heavy usage or excluding wildfowl from aquatic habitat on amphibian populations.

Control invasive plants
One before-and-after study in the UK found that habitat and species management that included controlling swamp stonecrop increased a population of natterjack toads. One replicated, controlled study in the USA found that more Oregon spotted frogs laid eggs in areas where invasive reed canarygrass was mown.

Key messages – reduce parasitism and disease – chytridiomycosis

Sterilize equipment when moving between amphibian sites
We found no evidence for the effects of sterilizing equipment when moving between amphibian sites on the spread of disease between amphibian populations or individuals. Two randomized, replicated, controlled studies in Switzerland and Sweden found that Virkon S disinfectant did not affect survival, mass or behaviour of eggs, tadpoles or hatchlings. However, one of the studies found that bleach significantly reduced tadpole survival.

Use gloves to handle amphibians
We found no evidence for the effects of using gloves on the spread of disease between amphibian populations or individuals. A review for Canada and the USA found that there were no adverse effects of handling 22 amphibian species using disposable gloves. However, three replicated studies in Australia and Austria found that deaths of tadpoles were caused by latex, vinyl and nitrile gloves for 60–100% of species tested.

Remove the chytrid fungus from ponds
One before-and-after study in Mallorca found that drying out a pond and treating resident midwife toads with fungicide reduced levels of infection but did not eradicate chytridiomycosis.

Use zooplankton to remove zoospores
We captured no evidence for the effects of using zooplankton to remove chytrid zoospores on amphibian populations.

Add salt to ponds
One study in Australia found that following addition of salt to a pond containing the chytrid fungus, a population of green and golden bell frogs remained free of chytridiomycosis for over six months.

Use antifungal skin bacteria or peptides to reduce infection
Three of four randomized, replicated, controlled studies in the USA found that introducing antifungal bacteria to the skin of chytrid infected amphibians did not reduce infection rate or deaths. One found that it prevented infection and death. One randomized, replicated, controlled study in the USA found that adding antifungal skin bacteria to soil significantly reduced chytridiomycosis infection rate in salamanders. One randomized, replicated, controlled study in Switzerland found that treatment with antimicrobial skin peptides before or after infection with chytridiomycosis did not increase toad survival.

Use antifungal treatment to reduce infection
Twelve of 16 studies, including four randomized, replicated, controlled studies, in Europe, Australia, Tasmania, Japan and the USA found that antifungal treatment cured or increased survival of amphibians with chytridiomycosis. Four studies found that treatments did not cure chytridiomycosis, but did reduce infection levels or had mixed results. Six of the eight studies testing treatment with itraconazole found that it was effective at curing chytridiomycosis. One found that it reduced infection levels and one found mixed effects. Six studies found that specific fungicides caused death or other negative side effects in amphibians.

Use antibacterial treatment to reduce infection

Two studies, including one randomized, replicated, controlled study, in New Zealand and Australia found that treatment with chloramphenicol antibiotic, with other interventions in some cases, cured frogs of chytridiomycosis. One replicated, controlled study found that treatment with trimethoprim-sulfadiazine increased survival time but did not cure infected frogs.

Use temperature treatment to reduce infection

Four of 5 studies, including 4 replicated, controlled studies, in Australia, Switzerland and the USA found that increasing enclosure or water temperature to 30–37°C for over 16 hours cured amphibians of chytridiomycosis. One found that treatment did not cure frogs.

Treating amphibians in the wild or pre-release

One before-and-after study in Mallorca found that treating wild toads with fungicide and drying out the pond reduced infection levels but did not eradicate chytridiomycosis.

Immunize amphibians against infection

One randomized, replicated, controlled study in the USA found that vaccinating mountain yellow-legged frogs with formalin-killed chytrid fungus did not significantly reduce chytridiomycosis infection rate or mortality.

Key messages – reduce parasitism and disease – ranaviruses

Sterilize equipment to prevent ranaviruses

We captured no evidence for the effects of sterilizing equipment to prevent ranavirus on the spread of disease between amphibian individuals or populations.

Reduce predation by other species

8.1   Remove or control mammals

- One controlled study in New Zealand[2] found that controlling rats had no significant effect on numbers of Hochstetter's frog.

- One controlled study in New Zealand[3] found that survival of Maud Island frogs was significantly higher in a predator-proof enclosure than in the wild. One study in New Zealand[1] found that at 58% of translocated Hamilton's frogs survived the first year within a predator-proof enclosure.

Background

Predation of amphibians by mammal species can have a significant effect on populations, particularly if the mammal species is not native or the amphibian population is small.

There is a large amount of literature that is not included here examining the success of controlling non-native mammal predators, which may be undertaken for the conservation of a range of taxa including amphibians (e.g. Genovesi 2005; Morley 2006).

Genovesi, P. (2005) Eradications of invasive alien species in Europe: a review. *Biological Invasions*, 7, 127–133.

Morley C.G. (2006) Removal of feral dogs Canis familiaris by befriending them, Viwa Island, Fiji. *Conservation Evidence*, 3, 3–4.

A study in 1990–1993 of endangered Hamilton's frog *Leiopelma hamiltoni* on Stephens Island, New Zealand (1) found that at least 7 of 12 translocated frogs survived the first year within a predator-proof exclosure. The 7 frogs were recaptured 27 times by June 1993. There was no control and so the frogs may have survived without the exclosure. In May 1992, frogs were translocated 40 m to a new habitat (a rock-filled pit 72 m²) created in May–October 1991 in a nearby forest remnant. A predator-proof fence was built around the new habitat to exclude tuatara *Sphenodon punctatus* and the area was 'seeded' with invertebrate prey. Frogs were surveyed regularly from November 1990 to May 1992 (90 visits).

A controlled study in 2002–2009 at two stream catchments within secondary forest in the Waitakere Ranges, New Zealand (2) found that control of invasive rats had no significant effect on the abundance of Hochstetter's frog *Leiopelma hochstetteri*. In 2008–2009, abundance was 5–7/20 m in the treatment area compared to 4–6/20 m in the non-treatment area. Snout–vent lengths were also similar (treatment: 9–45 mm; non-treatment: 11–45 mm). The rat abundance index decreased from eight in 2002 to three in 2009. Abundance in the non-treatment area was 73. Poison bait was placed at 50 m intervals along lines spaced 100 m apart over the entire 200 ha treatment area. These were restocked with 125 g of brodifacoum in spring and autumn. Rats were monitored at 7 locations using 60 tracking tunnels in the treatment area and three locations using 20 tunnels in the non-treatment area. Frogs were sampled on two 20 m transects along five small streams/site in summer 2008–2009.

A controlled study in 2006–2009 of translocated Maud Island frogs *Leiopelma pakeka* in Zealandia, New Zealand (3) found that survival was significantly higher in a predator-proof enclosure than in the wild. Survival in the enclosure was 93%. In the wild, numbers observed declined significantly, where house mice *Mus musculus* and little spotted kiwis *Apteryx owenii* were known predators. In the enclosure, two males bred successfully in 2008. Sixty frogs were translocated from Maud Island and placed in a 2 x 4 m predator-proof mesh enclosure in 2006. In April 2007, 29 were retained in the enclosure and 28 released into the adjacent forest.

(1) Brown D. (1994) Transfer of Hamilton's frog, *Leiopelma hamiltoni*, to a newly created habitat on Stephens Island, New Zealand. *New Zealand Journal of Zoology*, 21, 425–430.
(2) Nájero-Hilman E., King P., Alfaro A. C. & Breen B.B. (2009) Effect of pest-management operations on the abundance and size-frequency distribution of the New Zealand endemic frog *Leiopelma hochstetteri*. *New Zealand Journal of Zoology*, 36, 389–400.
(3) Bell B.D., Bishop P.J. & Germano J.M. (2010) Lessons learned from a series of translocations of the archaic Hamilton's frog and Maud Island frog in central New Zealand. In P.S. Soorae (ed) *Global Reintroduction Perspectives: 2010. Additional case studies from around the globe*, IUCN/SSC Re-introduction Specialist Group, Gland, Switzerland. pp. 81–87.

8.2   Remove or control fish population by catching

• Four studies (including two replicated, controlled studies) in the USA[3–6] found that removing fish by catching them significantly increased abundance of salamanders[3] and frogs[4–6] and increased recruitment, survival and population growth rate of cascades frog[6]. One before-and-after study in the UK[2] found that fish control had no significant effect on great crested newt populations and fish remained or returned within a few years.

• One replicated, before-and-after study in Sweden[1] found that fish control did not increase green toad breeding success and fish were soon reintroduced.

Background

Predatory fish can have negative impacts on amphibian populations, often through direct predation on embryos and larvae. This is particularly the case if the fish are invasive species, often introduced for fishing. For example, a systematic review found that evidence indicates that newts, salamanders and some frog species are less likely to be found in water bodies stocked with salmonids, such as salmon and trout than those with no stocking (Stewart *et al.* 2007).

There is a large amount of literature that is not included here examining the success of controlling fish by catching, which may be undertaken specifically for the conservation of amphibian species (e.g. Knapp *et al.* 2004).

Knapp R.A. & Matthews K.R. (2004) Eradication of nonnative fish by gill netting from a small mountain lake in California. *Restoration Ecology*, 6, 207–213.

Stewart G.B., Bayliss H.R., Showler D.A., Sutherland W.J. & Pullin A.S. (2007) What are the effects of salmonid stocking in lakes on native fish populations and other fauna and flora? Part A: Effects on native biota. Systematic Review No.13. Report.

A replicated, before-and-after study in 1986–1993 of ponds on the island of Samsø, Sweden (1) found that fish and eel *Anguilla anguilla* control was short-term and did not tend to increase breeding success by green toads *Bufo viridis*. Breeding was successful in two and failed in two of six ponds with just fish removal. One of the ponds was colonized by adults two years after fish and eels were removed, but breeding was not recorded. Only one male was seen in one of the ponds that was enlarged and had fish removed. Fish or eels were reintroduced to ponds within 1–2 years. In winter (1986–1993), fish were removed from six ponds (three twice). Seven ponds had fish removed and were enlarged. Ponds were monitored by call and torch surveys and by counting tadpoles and metamorphs during 4–6 visits in April–September.

A before-and-after study in 1992–2000 at two sites in England, UK (2) found that fish control by catching and treatment with rotenone had no significant effect on great crested newt *Triturus cristatus* populations. At one site, there was no significant increase in great crested newt numbers in the three years following fish removal, which the authors considered to have been only partially effective. At the second site, although great crested newt adults and eggs were recorded following fish control, no larvae were seen. Over 2,000 sticklebacks were removed from the pond, but they were observed again a few years after treatment. Electrofishing and treatment with rotenone were undertaken at a forest pond in 1996. At the other site, a pond (600 m²) was netted twice to remove trout in autumn 1997. Great crested newts were surveyed at that site in 1992–2000.

A before-and-after, site comparison study in 1993–2003 of two lakes in a National Park in Washington, USA (3) found that northwestern salamanders *Ambystoma gracile* increased significantly following elimination of non-native brook trout *Salvelinus fontinalis*. Day surveys showed that numbers of egg masses increased from 11 to 25–107/150 m and larvae from 5 to 18–90/150 m. Numbers increased to similar to those in the existing fishless lake (egg masses 65–165/150 m; larvae: 57–114/150 m). Night surveys showed a similar pattern with larvae increasing from 72 to 172/150 m and becoming similar to the fishless lake (50–145/150 m). Trout were removed from June to September 1993–2002 using gill nets (42 m long, 2 m tall). One to four nets were set once to several times during a field season. Salamanders were monitored using snorkel surveys along 25 m transects (four nearshore and two offshore) once or twice annually from July to September. Five night and 17–18 day larvae/neotene surveys and 10 egg mass surveys were completed per lake.

A replicated, controlled, before-and-after study in 1996–2005 of 21 lakes in California, USA (4) found that mountain yellow-legged frogs *Rana muscosa* increased following fish removal. One year after removal, numbers had increased for frogs (0.1 to 1.0/10 m) and tadpoles (0.1 to 8.1). Following removal, numbers were significantly greater than in lakes with fish (frogs: 0.1; tadpoles: 0.1/10 m). Within 3 years there was no significant difference between numbers within removal lakes and fishless control lakes (frogs: 7 vs 5; tadpoles: 10 vs 30/10 m). Trout *Oncorhynchus mykiss*, *Salvelinus fontinalis* were eliminated from three and greatly reduced in two removal lakes. Fish were removed by gill-netting starting in 1997–2001. Frog visual encounter surveys along shorelines and snorkelling surveys were undertaken in trout removal lakes ($n = 5$), fish-containing lakes ($n = 8$) and fishless lakes ($n = 8$) every two weeks in 1997–2001 and 2–3 times in 2002–2003.

A replicated, before-and-after study in 1996–2005 in six lakes in California, USA (5) found that mountain yellow-legged frog *Rana muscosa* densities increased significantly following predatory fish removal. In 3 lakes, densities increased significantly from the first 5 (1996–2002) to last 5 surveys (2004–2005) for tadpoles (0–12 to 4–91/10 m) and frogs (1–2 to 24–29). Increases were significantly greater than in fishless control lakes for tadpoles (+35 vs +2) and frogs (+25 vs +1). Within 1–3 years of starting fish control, frogs were detected in three lakes where they were previously absent (frogs: 3–67; tadpoles: 0). Complete eradication of fish was achieved from three lakes within 3–4 years, in the other three small numbers remained because of connecting streams. Non-native trout (*Oncorhynchus* sp., *Salmo* sp., *Salvelinus* sp.) were removed using 3–13 sinking gill nets (36 m long x 1.8 m high) set continuously in each lake. Netting was continued until catch rates fell to zero for an entire summer. Fish were eliminated from connecting streams when they dried out, using gill nets and electro-fishing. Frogs and tadpoles were recorded using visual surveys of lake perimeters before and 1–6 times after fish eradication started, up until 2005.

A replicated, controlled study in 2003–2006 of 16 lakes in northern California, USA (6) found that cascades frog *Rana cascadae* density, survival, recruitment and population growth rate increased following elimination of fish. Initially, frog densities were similar in the 12 treatment lakes (2 frogs/100 m). However, following fish elimination, densities were significantly higher in removal lakes (frogs: 5–20/100 m; larvae: 12–40/100 m) than in fish stocked and stocking-suspended lakes (frogs: 2; larvae: 1–2). By 2006, there was no significant difference in frog densities in removal lakes and four existing fishless lakes. By 2006, survival estimates of frogs at removal lakes (94%) were higher than those in fishless (64%) and fish-containing lakes (75%). The same was true for population growth rates (removal: 1.7–3.0; fishless: 1.2–1.4; with fish: 0.9–1.2) and recruitment rates (removal: 0.8–1.8; fishless: 0.4–0.6; fish: 0.2–0.5). Twelve lakes were randomly assigned as fish-removal, stocking-suspended or continually stocked lakes. An additional four lakes were fishless. Trout were removed from autumn 2003 to spring 2004 with multiple, repeated sets of sinking gill nets. Frogs were surveyed in 2003 and every two weeks from June to September in 2004–2006. Visual encounter surveys of the shoreline and capture-mark-recapture surveys were undertaken.

(1) Amtkjær J. (1995) Increasing populations of the green toad *Bufo viridis* due to a pond project on the island of Samsø. *Memoranda Societatis pro Fauna et Flora Fennica*, 71, 77–81.
(2) Watson W.R.C. (2002) Review of fish control methods for the great crested newt species action plan. Countryside Council for Wales Report. Contract Science Report No 476
(3) Hoffman R.L., Larson G.L. & Samora B. (2004) Responses of *Ambystoma gracile* to the removal of introduced non-native fish from a mountain lake. *Journal of Herpetology*, 38, 578–585.
(4) Vredenburg V.T. (2004) *Reversing introduced species effects: experimental removal of introduced fish leads to rapid recovery of a declining frog*. Proceedings of the National Academy of Sciences of the USA, 101, 7646–7650.
(5) Knapp R.A., Boiano D.M. & Vredenburg V.T. (2007) Removal of non-native fish results in population expansion of a declining amphibian (mountain yellow-legged frog, *Rana muscosa*). *Biological Conservation*, 135, 11–20.

(6)   Pope K.L. (2008) Assessing changes in amphibian population dynamics following experimental manipulations of introduced fish. *Conservation Biology*, 22, 1572–1581.

8.3   Remove or control fish using rotenone

- Three studies (including one replicated study) in Sweden, the UK and the USA found that eliminating fish using rotenone increased numbers of amphibian species, abundance and recruitment[5, 7, 8] or newt populations[2, 3].

- One review in Australia, the UK and the USA[4] found that fish control, which included using rotenone, increased breeding success for four amphibian species.

- Two replicated studies in Pakistan[1] and the UK[6] found that when rotenone was applied, many frogs died and a small number of newts showed symptoms of negative effects.

Background

Rotenone is used as a broad-spectrum pesticide to control fish and insects. It is derived from the roots of plants in the bean family and is rapidly broken down in soil and water.

There is a large amount of literature that is not included here examining the success of controlling fish using rotenone, which may be undertaken specifically for the conservation of amphibian species (e.g. Piec 2006; Willis & Ling 2000).

Piec D. (2006) Rotenone as a conservation tool in amphibian conservation. A case study of fish control operation undertaken at Orton Pit SSSI, Peterborough, UK. Froglife report.
Willis K. & Ling N. (2000) Sensitivities of mosquitofish and black mudfish to a piscicide: could rotenone be used to control mosquitofish in New Zealand wetlands? *New Zealand Journal of Zoology*, 27, 85–91.

A replicated, before-and-after study in 1970 of three ponds in Mymensingh, Pakistan (1) found that rotenone treatment to eradicate fish resulted in death of frogs. It was reported that many frogs died following application of rotenone, but that a similar number escaped death by moving to the shore. Fish were affected within 5 minutes of application. Approximately 110 kg of fish were removed from the ponds. There was no significant difference between the effects of the three treatment concentrations. Rotenone was added to 3 ponds in concentrations of 1.0, 1.5 and 2.0 parts per million in May 1970. Fish were collected and ponds monitored for two days following treatment.

A before-and-after study in 1992 of an artificial pond in woodland in England, UK (2) found that great crested newts *Triturus cristatus* and smooth newts *Triturus vulgaris* colonized following the removal of sticklebacks *Gasterosteus aculeatus* using rotenone. Larvae of both species were observed in the pond two months after treatment. Released toad tadpoles survived and metamorphosed in the pond. The concrete tank had sloping walls and a water depth of 90 cm. It contained approximately 2,000–3,000 sticklebacks. Rotenone was applied (5%; 0.2 mg/L) in May 1992 and 7 days later the pond was dredged to remove dead fish. Over 100 toad tadpoles were then released into the pond. Aquatic plants were also introduced.

A controlled study in 1977–1984 in two lakes in south western Sweden (3) found that fish elimination using rotenone resulted in a rapid increase in the smooth newt *Triturus vulgaris* population. Newts colonized within two years of fish removal. Between 1977 and 1980 the breeding population increased from 2,000 to almost 10,000 individuals. Following fish stocking in 1979 with 2,000 roach *Rutilus rutilus*, newt numbers declined to below 900 by 1984. No newts were found in an adjacent (50 m) lake without fish removal. Rotenone was applied in 1973. Newts were sampled using a capture-recapture survey from May to June in 1977–1984.

Forty-two cage traps were uniformly distributed around the removal lake. Traps were set in the untreated lake from 1978–1983.

A review of fish control programmes from 1992 to 1998 of two ponds in England, UK and one in Australia and Alabama, USA (4) found that breeding success increased for dusky gopher frogs *Rana sevosa*, green and golden bell frogs *Litoria aurea*, great crested newts *Triturus cristatus* and smooth newts *Lissotriton vulgaris*. Egg masses of the gopher frogs increased from 10 to 150. At one site in England both newt species re-colonized and reproduced in a treated pond in the first year following stickleback (Gasterosteidae) elimination (2,000–3,000 fish). At the second site in England, although great crested newt adults and eggs were recorded following stickleback removal, no larvae were seen. Fish were recorded at two of the sites within a few years of treatment. At the first English site, rotenone (5%) was applied, dredge netting undertaken and aquatic plants introduced to an isolated concrete pond (104 m^2) in May 1992. At the other site, rotenone and electrofishing were undertaken in 1996. In Alabama a pond was drained, fish removed and rotenone added in 1992. On Kooragang Island, Australia, rotenone was added to a pond to remove non-native plague minnows *Gambusia holbrooki* in 1998.

A replicated, before-and-after site comparison study in 2000–2002 of four ponds in a Nature Preserve in Illinois, USA (5) found that amphibian abundance and recruitment increased after fish control using rotenone (see also (7, 8)). Overall, numbers of amphibians increased by 411% in the 2 treated ponds compared to 165% in 2 existing fishless ponds. Recruitment increased by 873% in treated and 219% in historically fishless ponds. Abundance increases were greater in treated compared to fishless ponds for smallmouth salamanders *Ambystoma texanum* (610 vs 82%), American toad *Bufo americanus* (206 vs 190%), bullfrog *Rana catesbeiana* (101 vs 40%) and southern leopard frog *Lithobates sphenocephalus* (950 vs 325%). Wood frog *Rana sylvatica* increased by the same amount in treatment and controls (188 vs 188). Rotenone was applied to the 2 ponds (3–7 parts per million) with introduced native fish in December 2001. Amphibians were monitored in these two ponds and two without fish by using drift-fencing and pitfall traps from May 2000 to December 2002. Call surveys were also undertaken.

A replicated study in 2005–2006 of 39 ponds in a nature reserve in England, UK (6) found that rotenone application to eliminate sticklebacks *Pungitius pungitius* had a direct negative effect on a small number of newts at the time of application. Nine great crested newts *Triturus cristatus* (1 adult; 8 larvae) and 12 smooth newts *Triturus vulgaris* (7 adult; 5 larvae) were negatively affected, 19 from one pond. Additional newts were potentially affected but not found. Eight of the affected newts (38%; 5 crested newts) survived a 48-hour observation period in clean water and were released into nearby untreated ponds. Populations in the nature reserve were estimated at 30,000 adult great crested newts and several thousand smooth newts. Rotenone was applied (2.5%; 3 parts per million) in December 2005 using sprayers. Seventeen ponds received a second application (2 parts per million) in January 2006. Most ponds were hand netted prior to treatment in an attempt to remove newts; 14 newts were found in 5 ponds.

A continuation of a replicated, before-and-after, site comparison study (5) in 2000–2002 (7) found that recruitment of three amphibian species increased after fish elimination using rotenone (see also (8)). Recruitment (emerging metamorphs per breeding adult) increased significantly for smallmouth salamanders *Ambystoma texanum* (from 0 to 1–11), wood frog *Rana sylvatica* (0 to 1–2) and in one of two ponds American toad *Bufo americanus* (0 to 15). Recruitment tended to become higher than in two historically fishless ponds (salamanders: 0–1; wood frog: 0–0.5; American toad: 1–10). Numbers of emerging metamorphs increased significantly at experimental ponds for salamanders (0 to 20–205), wood frog (0–2 to 2–15) and American bullfrog *Rana catesbeiana* (35–42 to 47–50), but not American toad (0–2500 to 100–1700). Numbers of adults captured did not differ with treatment in experimental (before: 2–24; after: 5–44) and fishless ponds (before: 4–68; after: 16–84), apart from American toad which decreased in treatment ponds (before: 20–130; after: 2–80). Amphibians were monitored before (2001) and after (2002) treatment using drift-fencing with pitfall traps (7.5 m apart). Fish were eliminated, apart from bullhead catfish *Ameiurus melas* in one pond.

A continuation of a replicated, before-and-after, site comparison study (5, 7) in 2001–2004 (8) found that amphibian diversity and smallmouth salamander recruitment increased significantly after fish elimination using rotenone. Species relative abundance increased from 0.2 to 0.7 and became similar to that in historically fishless ponds (0.5–0.6). Small-mouth salamanders became the most abundant species in both treatment (41%) and fishless ponds (54%). American toad had been most abundant before fish removal (treatment: 91%; fishless: 67%). Although fish elimination did not result in increased salamander size at metamorphosis (42 vs 37 mm), it resulted in a significantly longer larval period (12% increase) and increased reproductive success (proportion of juveniles to breeding females: 0.3 vs 16.0). In fishless ponds larval period decreased 7% and recruitment was similar (0.2 vs 2.5). Numbers of juveniles increased significantly in treated (12 to 861) and fishless ponds (29 to 400). Amphibians were monitored before (2001) and after (2002–2004) treatment. One pond received a second application of rotenone to eliminate black bullhead catfish *Ameiurus melas* in January 2003.

(1)   Haque K.A. (1971) Rotenone and its use in eradication of undesirable fish from ponds. *Pakistan Journal of Scientific and Industrial Research*, 14, 385–387.
(2)   McLee A.G. & Scaife R.W. (1992/1993) The colonisation by great crested newts (*Triturus cristatus*) of a water body following treatment with a piscicide to remove a large population of sticklebacks (*Gasterosteus aculeatus*). *British Herpetological Society Bulletin*, 42, 6–9.
(3)   Aronsson S. & Stenson J.A.E. (1995) Newt-fish interactions in a small forest lake. *Amphibia-Reptilia*, 16, 177–184.
(4)   Watson W.R.C. (2002) Review of fish control methods for the great crested newt species action plan. Countryside Council for Wales Report. Contract Science Report No 476
(5)   Mullin S.J., Towey J.B. & Szafoni R.E. (2004) Using Rotenone to enhance native amphibian breeding habitat in ponds. *Ecological Restoration*, 22, 305–306.
(6)   Piec D. (2006) Rotenone as a conservation tool in amphibian conservation. A case study of fish control operation undertaken at Orton Pit SSSI, Peterborough, UK. Froglife Report.
(7)   Towey J.B. (2007) Influence of fish presence and removal on woodland pond breeding amphibians. MSc thesis. Eastern Illinois University.
(8)   Walston L.J. & Mullin S.J. (2007) Responses of a pond-breeding amphibian community to the experimental removal of predatory fish. *American Midland Naturalist*, 157, 63–73.

8.4   Remove or control fish by drying out ponds

- One before-and-after study in the USA[4] found that draining ponds to eliminate fish increased numbers of amphibian species. One replicated, before-and-after study in Estonia[5] found that pond restoration, which sometimes included drying to eliminate fish, and pond creation increased numbers of species and breeding populations of common spadefoot toads and great crested newts compared to no management.

- Three studies (including one review) in the UK and the USA found that pond drying to eliminate fish, along with other management activities in some cases, increased breeding success of frog[2, 3] and newt[1] species.

Background

Occasional drying of ponds can help control predators including native or non-native fish species.

A before-and-after study in 1986–1995 of a pond within a housing development near Peterborough, England, UK (1) found that fish removal by pond drying, along with pond deepening, maintained populations of great crested newts *Triturus cristatus* and smooth newts *Triturus vulgaris* seven years after the development. Larval catches increased the year after fish removal (crested: 37; smooth: 13) and then varied (crested: 1–14; smooth: 1–22). Although adults of both species reproduced after the development (crested: 41–102; smooth: 7–68), production of metamorphs failed in 1990 due to introduction of three-spined stickle-backs *Gasterosteus aculeatus*. Development was undertaken in 1987–1989. The pond (800 m²) was deepened in 1988 and fish were removed by pond drying in 1990. A 1 ha area was retained around the pond. Newts were counted by torch and larvae netted once or twice in 1986–1987 and 3–4 times in March–May 1988–1995.

A review of fish control programmes from 1992 to 2001 at a pond in England, Australia and Alabama, USA (2) found that breeding success increased for two frog species following pond draining. At the Australian site, green and golden bell frogs *Litoria aurea* bred success-fully the year after a reduction of non-native plague minnows *Gambusia holbrooki*. In Alabama, breeding success of dusky gopher frogs *Rana capito sevosa* increased following draining and rotenone treatment (egg masses: 10 to 150). In England, one great crested newt *Triturus cristatus* colonized a pond in the first year following elimination of sticklebacks (Gastero-steidae). A pond (690 m²) in England was drained down to 20 cm and bottom sediments agitated to release gases in 2001. A pond on Kooragang Island, Australia was drained in 1997. A pond in Alabama was drained, fish removed and then rotenone added in 1992.

A replicated, before-and-after study in 1998–2003 of seven ponds in California, USA (3) found that the reproductive success of California red-legged frogs *Rana draytonii* increased significantly following elimination of non-native fish by pond drying. Adult numbers were similar after fish elimination (0–40 to 1–41/pond), but juveniles increased significantly (0–15 to 1–650). Fish were eliminated during the first draining, or for two ponds with mosquitofish *Gambusia affinis* on the second draining. Seven ponds were drained in autumn in 1998–2001. Pumps were used to drain the water to a depth of 50 cm and then below 3 cm. Seines, throw nets and dip nets were used to remove all fish. Mud was smoothed and a small amount of household bleach applied to eliminate mosquitofish. Ponds were filled from ground water springs. Red-legged frogs and fish were surveyed six times per year in 1998–2001.

A before-and-after study in 1999–2001 of a seasonal wetland bay in South Carolina, USA (4) found that removing fish by drying the bay increased amphibian species richness. Before removal the bay supported only cricket frogs *Acris gryllus*. After fish removal the bay suppor-ted nine amphibian species including the Carolina gopher frog *Rana capito*. Amphibians were sampled in 1999 before fish removal and in the spring of 2001.

A replicated, before-and-after site comparison study of 450 existing ponds, 22 of which were restored, and 208 created ponds in 6 protected areas in Estonia (5) found that within three years amphibian species richness was higher in both restored ponds, some of which had been drained to eliminate fish, and created ponds than unmanaged ponds (3 vs 2 species/ pond). The proportion of ponds occupied also increased for targeted common spadefoot toad *Pelobates fuscus* (2 to 15%) and great crested newt *Triturus cristatus* (24 to 71%), as well as the other five species present (15–58% to 41–82%). Breeding occurred at increasing numbers of pond clusters from one to three years after restoration/creation for crested newt (39% to 92%) and spadefoot toad (30% to 81%). Prior to management, only 22% of ponds were considered high quality for breeding. In 2005, 405 existing ponds were sampled by dip-netting. In au-tumn 2005–2007, ponds were restored and created for great crested newts and spadefoot toads in 27 clusters. Restoration included clearing vegetation, extracting mud, levelled banks and for fish elimination pond drying and ditch blocking. Post-restoration monitoring in 2006–2008 comprised an annual visual count and dip-netting survey.

(1)   Cooke A.S. (1997) Monitoring a breeding population of crested newts (*Triturus cristatus*) in a housing development. *Herpetological Journal*, 7, 37–41.
(2)   Watson W.R.C. (2002) Review of fish control methods for the great crested newt species action plan. Countryside Council for Wales Report. Contract Science Report No 476.
(3)   Alvarez J.A., Dunn C. & Zuur A.F. (2002/2003) Response of California red-legged frogs to removal of non-native fish. *Transactions of the Western Section of the Wildlife Society*, 38/39, 9–12.
(4)   Scott D.E., Metts B.S. & Whitfield Gibbons J. (2008) Enhancing amphibian biodiversity on golf courses with seasonal wetlands. In J.C. Mitchell, R.E. Jung Brown & B. Bartholomew (eds) *Urban Herpetology*, SSAR, Salt Lake City. pp. 285–292.
(5)   Rannap R., Lõhmus A. & Briggs L. (2009) Restoring ponds for amphibians: a success story. *Hydrobiologia*, 634, 87–95.

8.5   Exclude fish with barriers

• One controlled study in Mexico[1] found that excluding fish using a barrier increased weight gain of axolotls.

Background
Fish can have negative impacts on amphibian populations, either through predation of eggs and larvae or through competition for food. In some cases barriers can be constructed within water bodies to create refuge areas for amphibians.

A controlled study in 2009 of a canal within agricultural land in Xochimilco, Mexico (1) found that filters to exclude competitive fish and improve water quality resulted in increased weight gain in axolotls *Ambystoma mexicanum*. Only 4 of 12 previously marked axolotls were recaptured; however, their weight had increased by 16%. Weight gain was greater than that of axolotls in control colonies over the same period. Farmers traditionally created canals linking lakes and wetlands. Working with farmers in 2009, one canal used as a refuge by axolotls was isolated from the main system using filters made of wood to exclude fish and improve water quality.

(1)   Valiente E., Tovar A., Gonzalez H., Eslava-Sandoval D. & Zambrano L. (2010) Creating refuges for the axolotl (*Ambystoma mexicanum*). *Ecological Restoration*, 28, 257–259.

8.6   Encourage aquatic plant growth as refuge against fish predation

• We found no evidence for the effects of encouraging aquatic plant growth as refuge against fish predation on amphibian populations.

Background
Vegetation can be planted or managed to provide refuge for amphibians against predatory fish. However, vegetation can also provide habitat for predators.

8.7   Remove or control invasive bullfrogs

• One replicated, before-and-after study in the USA[1] found that removing American bullfrogs significantly increased a population of California red-legged frogs.

- One before-and-after study in the USA and Mexico[2] found that eradicating bullfrogs from the area increased the range of leopard frogs. One replicated, before-and-after study in the USA[1] found that once bullfrogs had been removed, California red-legged frogs were found out in the open twice as frequently.

Background

The American bullfrog *Rana catesbeiana* has been introduced to many parts of the world. The species is relatively large and adaptable and has significant effects on some native species through competition for resources and predation.

There is additional literature that is not included here examining the success of controlling bullfrogs, which may be undertaken for the conservation of a range of taxa including amphibians (e.g. Banks *et al.* 2000; Orchard 2011; Louette 2012). For example, one modelling study found that culling bullfrog metamorphs in autumn was the most effective method of decreasing population growth rate (Govindarajulu *et al.* 2005). A review suggested that an indirect approach, by managing habitat rather than directly controlling bullfrogs, may be a more effective way to reduce the effects of bullfrogs on native amphibians (Adams & Pearl 2007).

Adams M.J. & Pearl C.A. (2007) Problems and opportunities managing invasive bullfrogs: is there any hope? In F. Gherardi (ed) *Biological Invaders in Inland Waters: Profiles, Distribution and Threats*, Springer, Dordrecht, The Netherlands. pp. 679–693.
Banks B., Foster J., Langton T. & Morgan K. (2000) British bullfrogs? *British Wildlife*, 11, 327–330.
Govindarajulu P., Altwegg R. & Anholt B.R. (2005) Matrix model investigation of invasive species control: bullfrogs on Vancouver Island. *Ecological Applications*, 15, 2161–2170.
Louette G. (2012) Use of a native predator for the control of an invasive amphibian. *Wildlife Research*, 39, 271–278.
Orchard S.A. (2011) Removal of the American bullfrog *Rana (Lithobates) catesbeiana* from a pond and a lake on Vancouver Island, British Columbia, Canada. In C.R. Veitch, M.N. Clout & D.R. Towns (eds) *Island Invasives: Eradication and Management*, IUCN, Gland, Switzerland. pp. 217–221.

A replicated, before-and-after study in 2004–2007 of 12 ponds in California, USA (1) found that there was a significant increase in adult California red-legged frogs *Rana draytonii* in ponds in the two years after American bullfrog *Rana catesbeiana* removal. Counts increased from 8 to 11 frogs in removal ponds. Numbers did not change in control ponds. Adult frogs were less visible when bullfrogs were present. Frogs used willows significantly less as cover, and were found on bare shores twice as much when adult bullfrogs were absent. Invasive American bullfrogs were removed from 12 ponds in 2004–2007. They were captured by hand, Hawaiian slings (spears) and seine netting (for tadpoles). Six ponds without bullfrogs in an adjacent field were monitored for comparison. Amphibians were monitored three times each week until October 2007.

A before-and-after study in 2008–2011 of leopard frogs in Arizona, USA and Mexico (2) found that eradication of bullfrogs *Rana catesbeiana* resulted in an increase in range of chiricahua leopard frogs *Lithobates chiricahuensis* and lowland leopard frogs *Lithobates yavapaiensis*. Surveys in 2010–2011 showed that chiricahua leopard frogs had dispersed into eight and lowland leopard frogs into three sites that had previously been unsuitable due to presence of bullfrogs. Chiricahua leopard frogs dispersed over 8 km to a site further north than it had recently been documented in the region. Bullfrogs were eradicated between 2008 and 2010.

(1) D'Amore A., Kirby E. & McNicholas M. (2009) Invasive species shifts ontogenetic resource partitioning and microhabitat use of a threatened native amphibian. *Aquatic Conservation: Marine and Freshwater Ecosystems*, 19, 534–541.
(2) Sredl M.J., Akins C.M., King A.D., Sprankle T., Jones T.R., Rorabaugh J.C., Jennings R.D., Painter C.W., Christman M.R., Christman B.L., Crawford C., Servoss J.M., Kruse C.G., Barnitz J. & Telles A. (2011) Re-introductions of Chiricahua leopard frogs in southwestern USA show promise, but highlight problematic threats and knowledge gaps. In P.S. Soorae (eds) *Global Re-introduction Perspectives: 2011. More case studies from around the globe*, IUCN/SSC Re-introduction Specialist Group & Abu Dhabi Environment Agency, Gland, Switzerland. pp. 85–90.

8.8   Remove or control invasive viperine snake

- One before-and-after study in Mallorca[1] found that numbers of Mallorcan midwife toad larvae increased after intensive, but not less intensive, removal of viperine snakes.

Background
Introduced species can have significant effects on native species, particularly on oceanic islands. For example, the viperine snake *Natrix maura* is an invasive species on Mallorca and as one of the main predators of the threatened midwife toad *Alytes muletensis*, contributed towards its decline (Guicking *et al.* 2006).

Guicking D., Griffiths R.A., Moore R.D., Joger U. & Wink M. (2006) Introduced alien or persecuted native? Resolving the origin of the viperine snake (*Natrix maura*) on Mallorca. *Biodiversity and Conservation*, 15, 3045–3054.

A before-and-after study in 1991–2002 of Mallorcan midwife toads *Alytes muletensis* in Mallorca (1) found that abundance increased at one of two sites after removal of viperine snakes *Natrix maura*. At the site with intensive control over 3 years, no snakes were seen from 1997 and larval toad counts increased from 1,300 in 1991 to 2,200 in 1999. Control was not considered successful by the authors at the second site due to the large snake population and more open habitat. Viperine snakes were controlled by capturing intensive at one site in 1991–1993 and by capturing less intensively at the second site in 2002.

(1) Román A. (2003) El ferreret, la gestión de una especie en estado crítico. *Munibe*, 16, 90–99.

8.9   Remove or control non-native crayfish

- We found no evidence for the effects of removing or controlling non-native crayfish on amphibian populations.

Background
Signal crayfish *Pacifastacus leniusculus* and red swamp crayfish *Procambarus clarkia* have been introduced to many parts of the world. Signal crayfish reproduce and grow fast and so can reach high densities. Non-native crayfish have direct effects on amphibians through predation of eggs but also affect aquatic communities by consuming aquatic plants and competing with and introducing disease to native crayfish.
   There is additional literature that is not included here examining the success of controlling crayfish, which may be undertaken for the conservation of a range of taxa including amphibians (e.g. Aquiloni *et al.* 2009; Aquiloni & Gherardi 2010).

Aquiloni L., Becciolini A., Berti R., Porciani S., Trunfio C. & Gherardi F. (2009) Managing invasive cray-fish: use of X-ray sterilisation of males. *Freshwater Biology*, 54, 1510–1519.
Aquiloni L. & Gherardi F. (2010) The use of sex pheromones for the control of invasive populations of the crayfish *Procambarus clarkii*: a field study. *Hydrobiologia*, 649, 249–254.

Reduce competition with other species

8.10 Reduce competition from native amphibians

- One replicated, site comparison study in the UK[1] found that natterjack toad populations did not increase following common toad control.

Background
Management for threatened amphibian species can sometimes include reducing numbers of a common amphibian species that compete for resources.

A replicated, site comparison study in 1985–2006 of 20 sites in the UK (1) found that natterjack toad *Bufo calamita* populations did not increase following control of common toads *Bufo bufo*. However overall, natterjack population trends were positive at sites that had received species-specific management that included aquatic and terrestrial habitat management and common toad control. Trends were negative at unmanaged sites. Five of the 20 sites showed positive population trends, 5 showed negative trends and 10 trends were not significantly different from zero. Data on populations (egg string counts) and management activities over 11–21 years were obtained from the Natterjack Toad Site Register. Habitat management for toads was undertaken at seven sites. Management varied between sites, but included pond creation, adding lime to acidic ponds, maintaining water levels, vegetation clearance and implementation of grazing schemes. Translocations were also undertaken at 7 of the 20 sites.

(1) McGrath A.L. & Lorenzen K. (2010) Management history and climate as key factors driving natterjack toad population trends in Britain. *Animal Conservation*, 13, 483–494.

8.11 Remove or control invasive cane toads

- We found no evidence for the effects of removing or controlling invasive cane toads on amphibian populations.

Background
Cane toads *Bufo marinus* have been introduced to many places including Australia and Pacific and Caribbean islands. The species can have significant effects on native species, particularly those that prey on the cane toads as they contain a lethal toxin. They may also affect native amphibians through competition at the tadpole stage and through predation of eggs or tadpoles.
   There is additional literature that is not included here examining the success of controlling cane toads, which may be undertaken for the conservation of a range of taxa including amphibians (e.g. Nakajima *et al*. 2005; Shanmuganathan *et al*. 2010; Ward-Fear *et al*. 2010; Wingate 2011).

Nakajima T., Toda M., Aoki M. & Tatara M. (2005) The project for control of the cane toad *Bufo marinus* on Iriomote Island, Okinawa prefecture. *Bulletin of the Herpetological Society*, 2005, 179–186.

Shanmuganathan T., Pallister J., Doody S., McCallum H., Robinson T., Sheppard A., Hardy C., Halliday D., Venables D., Voysey R., Strive T., Hinds L. & Hyatt A. (2010) Biological control of the cane toad in Australia: a review. *Animal Conservation Biology*, 13, 16–23.

Ward-Fear G., Brown G.P. & Shine R. (2010) Using a native predator (the meat ant, *Iridomyrmex reburrus*) to reduce the abundance of an invasive species (the cane toad, *Bufo marinus*) in tropical Australia. *Journal of Applied Ecology*, 47, 273–280.

Wingate D.B. (2011) The successful elimination of cane toads, *Bufo marinus*, from an island with breeding habitat off Bermuda. *Biological Invasions*, 13, 1487–1492.

8.12   Remove or control invasive Cuban tree frogs

- One before-and-after study in the USA[1] found that the abundance of squirrel tree frogs and green tree frogs increased after removal of invasive Cuban tree frogs.

Background

Invasive amphibians such as Cuban tree frogs *Osteopilus septentrionalis* can have significant impacts on native amphibian species if they compete for resources. For example, a study found that survival and growth rates of tadpoles of the dominant native species, southern toad *Bufo terrestris*, decreased significantly in the presence of Cuban tree frog tadpoles and that the invasive tadpoles became dominant (Smith 2006). The same study found that the effects of Cuban tree frogs on southern toads were reduced if predatory eastern newts were also present.

Smith K.G. (2006) Keystone predators (eastern newts, *Notopthalmus viridescens*) reduce the impacts of an aquatic invasive species. *Oecologia*, 148, 342–349.

A before-and-after study in 2001–2003 in Florida, USA (1) found that the abundance of squirrel tree frogs *Hyla squirella* and green tree frogs *Hyla cinerea* increased after removal of Cuban tree frogs *Osteopilus septentrionalis*. Squirrel tree frog abundance in the wet season doubled following Cuban tree frog removal at 1 site (20 removed; abundance: 109 vs 200). However, survival rates did not differ (0.9). Green tree frogs also increased at 1 site where 589 Cuban tree frogs were removed (7 vs 24). Other species and sites were not compared due to small sample sizes. A total of 693 Cuban tree frogs were removed (10–589/site). Tree frogs were captured in 84–99 refuges/site, which were checked each week or month. Refuges were 1 m long, 5 cm diameter polyvinyl chloride pipes hung 1 m from the ground and with a cap at the bottom to retain water. Tree frogs were marked and from 2002 all Cuban tree frogs captured were removed.

(1) Rice K.G., Waddle J.H., Miller M.W., Crockett M.E., Mazzotti F.J. & Percival H.F. (2011) Recovery of native treefrogs after removal of non-indigenous Cuban treefrogs *Osteopilus septentrionalis*. *Herpetologica*, 67, 105–117.

Reduce adverse habitat alteration by other species

8.13   Prevent heavy usage or exclude wildfowl from aquatic habitat

- We found no evidence for the effects of preventing heavy usage or excluding wildfowl from aquatic habitat on amphibian populations.

Background
High densities of wildfowl can strip aquatic vegetation from ponds and their banks, reducing shelter habitat and egg-laying sites for amphibians. Water quality may also be reduced through defecation and continual stirring up of sediments. Wildfowl can also prey on adult amphibians and their eggs. They are also potential environmental reservoirs for *Batrachochytrium dendrobatidis*, the cause of chytridiomycosis. A study in Belgium found that 15% of wild geese tested were positive for the fungus (Garmyn *et al.* 2012).

Garmyn A., Van Rooij P., Pasmans F., Hellebuyck T., Van Den Broeck W., Haesebrouck F. & Martel A. (2012) Waterfowl: potential environmental reservoirs of the chytrid fungus *Batrachochytrium dendrobatidis*. *PLoS ONE*, 7, e35038.

8.14   Control invasive plants

- One before-and-after study in the UK[1] found that aquatic and terrestrial habitat management that included controlling swamp stonecrop, along with release of captive-reared toadlets, tripled a population of natterjack toads.

- One replicated, controlled study in the USA[2] found that Oregon spotted frogs laid eggs in areas where invasive reed canarygrass had been mown more frequently than where it was not mown.

Background
Non-native plant species can be introduced into or naturally invade terrestrial habitat or water bodies, where they can out-compete native species altering the habitats. For example, in the UK swamp stonecrop *Crassula helmsii* can out-compete native plant species and form thick mats covering whole ponds. In parts of the USA, invasive reed canarygrass *Phalaris arundinacea* is widespread and develops dense, tall stands in shallow wetland habitats. Invasive water fern *Azolla filiculoides* has been found to cause declines in amphibian populations (Gratwicke & Marshall 2001) and Japanease knotweed *Fallopia japonica* to reduce foraging success of green frogs *Rana clamitans* (Maerz *et al.* 2005).

Gratwicke B. & Marshall B.E. (2001) The impact of *Azolla filiculoides* Lam. on animal biodiversity in streams in Zimbabwe. *African Journal of Ecology*, 39, 216–218.
Maerz, J.C., Blossey, B. & Nuzzo, V. (2005) Green frogs show reduced foraging success in habitats invaded by Japanese knotweed. *Biodiversity & Conservation*, 14, 2901–2911.

A before-and-after study in 1972–1991 of ponds on heathland in Hampshire, England, UK (1) found that pond restoration and creation with swamp stonecrop *Crassula helmsii* control, vegetation clearance, liming and captive-rearing and releasing toadlets resulted in a three-fold increase in natterjack toad *Bufo calamita* populations. Spawn string counts, which relate to the female breeding population, increased from 15 to 43. Swamp stonecrop was eliminated from two of six new ponds it invaded and controlled in two others. Nine small ponds (< 1,000 m^2) were created and four restored by excavation. Swamp stonecrop was pulled up and treated with herbicide. In addition, one pond was treated with limestone (1983–1989), scrub was cleared by cutting and uprooting (40 ha) and bracken was treated with herbicide

(12 ha). Captive-reared toadlets were released in 1975 (8,800), 1979, 1980 and 1981 (1,000 each). Each year, toads were monitored every 10 days in March and August.

A replicated, controlled study in 2000–2001 of a wetland in Washington, USA (2) found that Oregon spotted frogs *Rana pretiosa* laid eggs in more plots than expected by chance following mowing of invasive reed canarygrass *Phalaris arundinacea*. No eggs were laid in unmown plots. Egg mass clusters (1–18 egg masses) were recorded in 2 of 32 mown plots. Three egg mass clusters (5–20 masses) were also recorded outside study plots in habitat that appeared structurally similar to mown plots. Breeding sites were located using systematic searches within the reed canarygrass dominated wetland. Four of seven sites found were selected and used as the centre of a 30 m diameter circle. Within each circle, eight pairs of randomly located 3 m diameter plots were created. One of each pair was mown close to the ground in August 2000. Breeding was monitored in February–March 2001 using visual encounter surveys.

(1)   Banks B., Beebee T.J.C. & Denton J.S. (1993) Long-term management of a natterjack toad (*Bufo calamita*) population in southern Britain. *Amphibia-Reptilia*, 14, 155–168.
(2)   Kapust H.Q.W., Mcallister K.R. & Hayes M.P. (2012) Oregon spotted frog (*Rana pretiosa*) response to enhancement of oviposition habitat degraded by invasive reed canary grass (*Phalaris arundinacea*). *Herpetological Conservation and Biology*, 7, 358–366.

Reduce parasitism and disease – chytridiomycosis

Chytridiomycosis is caused by the fungus *Batrachochytrium dendrobatidis*, which colonizes amphibian skin. The disease is highly infectious and has an almost global distribution. It has significant, long-term effects on some amphibian populations and is thought to be responsible for the decline or extinction of up to 200 species of frogs (Forzan *et al.* 2008). Interventions to prevent the spread or to treat the disease in the wild and captivity are therefore the focus of many amphibian conservation efforts.

Captive assurance populations have been established for some species that are at serious risk of extinction in the wild because of chytridiomycosis (e.g. Zippel 2002; Gratwicke 2012; McFadden 2012). The aim is to maintain disease-free breeding populations in captivity to provide animals for release at disease-free sites or release once the threat has been removed. Studies investigating the success of captive breeding are discussed in 'Species management – Captive breeding, rearing and releases (*ex-situ* conservation)'.

There is a large amount of research currently being undertaken on chytridiomycosis and so the amount of evidence for the effectiveness of interventions should increase over the next few years.

Forzan M.J., Gunn H. & Scott P. (2008). Chytridiomycosis in an aquarium collection of frogs, diagnosis, treatment, and control. *Journal of Zoo and Wildlife Medicine*, 39, 406–411.
Gratwicke B. (2012) Amphibian rescue and conservation project – Panama. *Froglog*, 102, 17–20.
McFadden M. (2012) Captive-bred southern corroboree frog eggs released. *Amphibian Ark Newsletter*, 19, 10.
Zippel K.C. (2002) Conserving the Panamanian golden frog: Proyecto Rana Dorada. *Herpetological Review*, 33, 11–12.

8.15   Sterilize equipment when moving between amphibian sites

• We found no evidence for the effects of sterilizing equipment when moving between amphibian sites on the spread of disease between amphibian populations or individuals.

- Two randomized, replicated, controlled studies in Switzerland and Sweden found that Virkon S disinfectant did not affect survival, mass or behaviour of common frog or common toad tadpoles[1] or moor frog embryos or hatchlings[2]. One of the studies found that bleach significantly reduced survival of common frog and common toad tadpoles[1].

Background

The movement of field biologists increases the risk of spreading wildlife diseases such as the chytrid fungus *Batrachochytrium dendrobatidis*. For example, the chytrid fungus has been found to survive in lake water for seven weeks and tap water for three weeks after introduction (Johnson & Speare 2003). Precautions therefore need to be taken to reduce the risk of spreading diseases between sites and populations. This is also the case within and between captive populations.

We found no evidence for the effects of sterilizing equipment when moving between amphibian sites on the spread of disease between amphibian populations. The studies captured here examine the effect of different types of disinfectants on amphibians.

There is additional literature examining the effectiveness of using a range of disinfectants to kill the chytrid fungus *Batrachochytrium dendrobatidis*. Most chemicals killed 100% of chytrid zoospores when used at certain concentrations (e.g. sodium chloride, household bleach, potassium permanganate, formaldehyde solution, Path-XTM agricultural disinfectant, quaternary ammonium compound 128, Dithane, Virkon, ethanol and benzalkonium chloride; Johnson *et al.* 2003; Webb *et al.* 2007). Complete drying of the fungus or heating above 37°C for at least four hours also resulted in 100% mortality (Johnson *et al.* 2003).

Johnson M.L., Berger L., Philips L. & Speare R. (2003) Fungicidal effects of chemical disinfectants, UV light, desiccation and heat on the amphibian chytrid *Batrachochytrium dendrobatidis*. *Diseases of Aquatic Organisms*, 57, 255–260.
Johnson M. & Speare R. (2003) Survival of *Batrachochytrium dendrobatidis* in water: quarantine and disease control implications. *Emerging Infectious Diseases*, 9, 922–925.
Webb R., Mendez D., Berger L. & Speare R. (2007) Additional disinfectants effective against the amphibian chytrid fungus Batrachochytrium dendrobatidis. *Diseases of Aquatic Organisms*, 74, 13–16.

A randomized, replicated, controlled study in 30 artificial pools in Switzerland (1) found that Virkon S disinfectant did not affect survival, mass or behaviour of common frog *Rana temporaria* and common toad *Bufo bufo* tadpoles, but bleach did. Survival did not differ between Virkon treatments for frogs (untreated: 70–100%; low dose: 90–100%; high dose: 40–100%) or toads (untreated: 90–100%; low dose: 100%; high dose: 70–100%). All tadpoles died within 1–2 days in high dose bleach. Survival was significantly lower in low dose bleach than untreated water for frogs (20–100 vs 70–100%) and toads (40–100 vs 90–100%). Frog tadpole mass was significantly higher in low dose bleach (0.5–0.6 g) than other treatments (0.3–0.5 g). Toad tadpole mass did not differ (0.2–0.4 g). The proportion of tadpoles feeding did not differ significantly for frogs (0.4–0.9) or toads (0.6–0.9). Local leaves, phytoplankton, zooplankton and a snail were added to artificial pools (80 L). Disinfectants (bleach 2%; Virkon 10 g/L) that would be used for boots and field equipment were applied to pools once a week at high (0.04 L) or low doses (0.004 L), with 0.060 L or 0.096 L of water respectively. Water was added as the control. Treatments were replicated five times and assigned randomly to tubs. Ten frog and toad tadpoles were added to each treatment.

A randomized, replicated, controlled study in 2011 of captive moor frogs *Rana arvalis* at Uppsala University, Sweden (2) found that Virkon S disinfectant had no significant effects on moor frog embryos and hatchlings, but did reduce hatching success. Embryonic survival was significantly lower in the low (92%), but not high concentration of Virkon S (94%) compared to the control (99%). Abnormalities were infrequent in all treatments (low: 3%; high: 4%; control: 1%). Hatchling body length did not differ between treatments (5 mm). However, hatching success was lower with Virkon S compared to without, suggesting that it may have weak negative effects on amphibian embryos. Embryos and hatchlings were reared at 19°C in high (5 mg/L) and low (0.5 mg/L) Virkon S concentrations and in a control of water. One embryo and six hatchlings from each of six clutches were used per treatment. Survival was recorded daily until the free swimming stage and hatchling length for seven days.

(1) Schmidt B.R., Geiser C., Peyer N., Keller N. & von Rütte M. (2009) Assessing whether disinfectants against the fungus *Batrachochytrium dendrobatidis* have negative effects on tadpoles and zooplankton. *Amphibia-Reptilia*, 30, 313–319.
(2) Hangartner S. & Laurila A. (2012) Effects of the disinfectant Virkin S on early life-stages of the moor frog (*Rana arvalis*). *Amphibia-Reptilia*, 33, 349–353.

8.16   Use gloves to handle amphibians

- We found no evidence for the effects of using gloves on the spread of disease between amphibian populations or individuals.

- A review for Canada and the USA[4] found that there were no adverse effects of handling 22 amphibian species using disposable gloves. However, three replicated studies (including one controlled study) in Australia and Austria[1–3] found that deaths of tadpoles were caused by latex gloves for all four species tested, by vinyl gloves for three of five species[1–3] and by nitrile gloves for the one species tested[3].

Background
Precautions need to be taken to reduce the risk of spreading diseases such as chytridiomycosis between amphibian individuals, populations and habitats. One way to minimize the risk is to wear disposable gloves when handling individual amphibians.

We found no evidence for the effects of using gloves on the spread of disease between amphibian populations. The studies captured here investigate the effect of different types of gloves on amphibians.

There is additional literature examining the effectiveness of disposable gloves acting as a fungicide on the chytrid fungus *Batrachochytrium dendrobatidis*. For example one study found that nitrile gloves (and bare hands), but not latex, polyethylene or vinyl gloves were effective in killing the chytrid fungus (Mendez *et al.* 2008).

Mendez D., Webb R., Berger L. & Speare R. (2008) Survival of the amphibian chytrid fungus *Batrachochytrium dendrobatidis* on bare hands and gloves: hygiene implications for amphibian handling. *Diseases of Aquatic Organisms*, 82, 97–104.

A small, replicated study in a laboratory (1) found that latex, but not vinyl gloves caused death in African clawed frog *Xenopus laevis* tadpoles. All tadpoles exposed to unrinsed and rinsed latex gloves died within 24 hours, most within 2 hours. None of the tadpoles exposed to vinyl gloves showed adverse effects. Four of 12 tadpoles in tanks cleaned with latex gloves died within 4 hours of exposure. Between 10 and 20 tadpoles were placed in each of three 700 ml beakers containing water at 20°C. One of the following gloves was partially immersed for 24 hours in each beaker: unrinsed latex (powder-free); rinsed latex; or rinsed vinyl gloves. Rinsing was done in deionized distilled water to remove any powder.

A small, replicated study in a laboratory in Austria (2) found that mortality of African clawed frog *Xenopus laevis* and common frog *Rana temporaria* tadpoles increased with increasing concentrations of latex and vinyl glove contaminated water. All African clawed frog tadpoles died within 12 hours when exposed to dilutions of 1:350 or less and 50% died in dilutions of 1:425 (i.e. one glove in 128 litres). Surviving tadpoles showed no symptoms. All common frog tadpoles died in dilutions of 1:600 or less (i.e. 1 glove in 195 litres). African clawed frog tadpoles survived in vinyl glove dilutions lower than 1:4, but showed 100% mortality in dilutions of 1:3 or less. The latex gloves used in the experiment were the most toxic of the materials (latex, vinyl, nitril) and brands tested. Ten latex and vinyl gloves were soaked in water for 24 hours at 20°C. Solutions were further diluted to a maximum of 1:900 using tap water. Ten African clawed frog and 10 common frog tadpoles were placed in each solution (water volume 700 ml). Mortality was scored after 12 hours of exposure.

A replicated, controlled study in the laboratory and in the field in Australia (3) found that unrinsed latex or nitrile gloves caused death of green-eyed tree frog *Litoria genimaculata* and cane toad *Bufo marinus* tadpoles and unrinsed vinyl gloves caused death of waterfall frogs *Litoria nannotis*. Direct or indirect contact with unrinsed latex gloves caused 72% mortality of green-eyed tree frog tadpoles (*n* = 36). Unrinsed latex or nitrile gloves caused 10–100% mortality of non-native cane toad tadpoles (*n* = 10). Rapid, localized tissue damage was observed at the point of contact. In the laboratory, no adverse effects were seen 24 hours after handling with unrinsed vinyl gloves in green-eyed tree frogs (*n* = 23), cane toads (*n* = 20) or waterfall frogs *Litoria nannotis* (*n* = 32). However, in the field 40% of waterfall frogs handled with unrinsed gloves died within one hour. The remainder and those handled with rinsed vinyl gloves showed no effects. Cane toad tadpoles handled with unrinsed vinyl gloves or bare hands (*n* = 10–20) showed no adverse effects. In the laboratory, tadpoles were handled for 30–90 seconds with unrinsed latex or vinyl gloves, and nitrile or no gloves for cane toads. In the field, 30 waterfall frog tadpoles were handled with unrinsed or rinsed vinyl gloves or bare hands.

A review of 22 amphibian species in laboratory experiments, in the field and in zoo settings in Canada and the USA (4) found that there were no adverse effects of handling amphibians using disposable gloves. No effects were noticed in wood frogs *Rana sylvatica* (*n* = 240), Arizona tiger salamanders *Ambystoma tigrinum nebulosum* (*n* = 1372) or gray tiger salamanders *Ambystoma tigrinum diaboli* (*n* = 397) handled for up to three minutes, weekly for 4–20 weeks in laboratories. The same was true for wood frogs (*n* = 32), western toads *Bufo boreas* (*n* = 98), boreal choral frogs *Pseudacris maculata* (*n* = 4) and Arizona tiger salamanders *Ambytoma tirgrinum nebulosum* (*n* = 2309) handled for up to two minutes in the field. In addition, no symptoms or deaths were ever detected in the larvae of 17 amphibian species that had been repeatedly handled with gloves at Detroit Zoo.

(1) Sobotka J.M. & Rahwan R.G. (1994) Lethal effect of latex gloves on *Xenopus laevis* tadpoles. *Journal of Pharmacological and Toxicological Methods*, 32, 59.
(2) Gutleb A.C., Bronkhorst M., Van denberg J.H.J. & Murk A.J. (2001) Latex laboratory-gloves: an unexpected pitfall in amphibians toxicity assays with tadpoles. *Environmental Toxicology and Pharmacology*, 10, 119–121.
(3) Cashins S.D., Alford R.A. & Skerrati L.F. (2008) Lethal effects of latex, nitrile, and vinyl gloves on tadpoles. *Herpetological Review*, 39, 298–301.
(4) Greer A.L., Schock D.M., Brunner J.L., Johnson R.A., Picco A.M., Cashins S.D., Alford R.A., Skerratt L.F. & Collins J.P. (2009) Guidelines for the safe use of disposable gloves with amphibian larvae in light of pathogens and possible toxic effects. *Herpetological Review*, 40, 145–147.

8.17   Remove the chytrid fungus from ponds

• One before-and-after study in Mallorca[1] found that pond drying and fungicidal treatment of resident midwife toads reduced levels of infection but did not eradicate chytridiomycosis.

> **Background**
>
> The chytrid fungus *Batrachochytrium dendrobatidis* has been found to survive in lake water for seven weeks after introduction (Johnson & Speare 2003). Treatment of the aquatic environment may help to reduce the effect of the disease on amphibians. One potential method is completely drying ponds, as a study found that complete drying of the chytrid fungus resulted in 100% mortality (Johnson *et al.* 2003).

Johnson M.L., Berger L., Philips L. & Speare R. (2003) Fungicidal effects of chemical disinfectants, UV light, desiccation and heat on the amphibian chytrid *Batrachochytrium dendrobatidis*. *Diseases of Aquatic Organisms*, 57, 255–260.

Johnson M. & Speare R. (2003) Survival of *Batrachochytrium dendrobatidis* in water: quarantine and disease control implications. *Emerging Infectious Diseases*, 9, 922–925.

A before-and-after study in 2009–2010 of a pond in Mallorca (1) found that drying out the pond and treating resident Mallorcan midwife toads *Alytes muletensis* with a fungicide reduced the prevalence but did not eradicate chytridiomycosis. All samples from tadpoles came back positive for the chytrid fungus the spring after pond drying and treatment. However, the number of spores detected on each swab was lower than the previous year, suggesting a lower level of infection. Healthy-looking toads were seen breeding in the pond following pond drying and treatment. Over 2,000 toad tadpoles were removed from the pond in March–August 2009. The pond was emptied and left to dry over the summer. Tadpoles were taken to a laboratory and given daily five minute baths in the fungicide itraconazole for one week. They were held in captivity for up to seven months. Once the pond refilled in autumn, tadpoles were released. The following spring tadpoles were swabbed to test for chytridiomycosis.

(1)   Lubick N. (2010) Emergency medicine for frogs. *Nature*, 465, 680–681.

8.18   Use zooplankton to remove zoospores

• We found no evidence for the effects of using zooplankton to remove chytrid zoospores on amphibian populations.

> **Background**
>
> Zooplankton such as water fleas (Cladocera), copepods (Copepoda) and seed shrimps (Ostracoda) consume the aquatic zoospores of the chytrid fungus *Batrachochytrium dendrobatidis* (e.g. Buck *et al.* 2011). They may therefore play a role in regulating the fungus and so could help to reduce the risk of amphibian infection in aquatic environments. Copepods have successfully been used as biological control agents in other disease systems (Marten 2000).

Buck J.C., Truong L. & Blaustein A.R. (2011) Predation by zooplankton on *Batrachochytrium dendrobatidis*: biological control of the deadly amphibian chytrid fungus? *Biodiversity and Conservation*, 20, 3549–3553.
Marten, G.G. (2000) Dengue hemorrhagic fever, mosquitoes, and copepods. *Journal of Policy Studies (Japan)*, 9, 131–141.

8.19   Add salt to ponds

- One study in Australia[1] found that following addition of salt to a pond containing the chytrid fungus, a population of green and golden bell frogs remained free of chytridiomycosis for at least six months.

Background

The chytrid fungus *Batrachochytrium dendrobatidis* has been found to survive in lake water for seven weeks after introduction (Johnson & Speare 2003). Treating the aquatic environment may help to reduce the effect of the disease on amphibians. Salt is often used for fungal diseases in aquaculture and for veterinary treatments of fish and amphibians (Wright & Whitaker 2001; Mifsud & Rowland 2008) and has been found to kill the chytrid fungus (Johnson *et al.*2003).

Johnson M.L., Berger L., Philips L. & Speare R. (2003) Fungicidal effects of chemical disinfectants, UV light, desiccation and heat on the amphibian chytrid *Batrachochytrium dendrobatidis*. *Diseases of Aquatic Organisms*, 57, 255–260.
Johnson M. & Speare R. (2003) Survival of Batrachochytrium dendrobatidis in water: quarantine and disease control implications. *Emerging Infectious Diseases*, 9, 922–925.
Mifsud C. & Rowland S.J. (2008) Use of salt to control ichthyophthiriosis and prevent saprolegniosis in silver perch, *Bidyanus bidyanus*. *Aquaculture Research*, 39, 1175–1180.
Wright K.M. & Whitaker B.R. (2001) Pharmacotherapeutics. Pages 309–330 in: K. M. Wright, B. R. Whitaker & F. L. Malabar (eds) *Amphibian Medicine and Captive Husbandry*, Krieger Publishing Company.

A study in 2000–2001 of captive green and golden bell frogs *Litoria aurea* in Sydney, Australia (1) found that following addition of salt to a constructed pond the population remained free of chytridiomycosis for at least six months. Thirty-three of 40 green and golden bell frog tadpoles released survived to juvenile frogs in the salted pond. However, growth appeared slower in salt water than fresh water (first metamorph: 49 vs 43 days; last metamorph: 123 vs 76–80 days). Following addition of salt, the two striped marsh frogs *Limnodynastes peroni* tested were negative for chytridiomycosis. Striped marsh frogs had introduced chytridiomycosis to the pond and it had killed all but one of the previous green and golden bell frog population. Following the initial outbreak of chytridiomycosis, uniodized table salt was added to the pond to achieve 1 parts per trillion (ppt) sodium chloride (3% sea water) in December 2000. Forty tadpoles were then released into the pond and were monitored weekly.

(1)   White A.W. (2006) A trial using salt to protect green and golden bell frogs from chytrid infection. *Herpetofauna*, 36, 93–96.

8.20   Use antifungal skin bacteria or peptides to reduce infection

- Three of four randomized, replicated, controlled studies in the USA found that adding antifungal bacteria to the skin of salamanders or frogs exposed to the chytrid fungus did not reduce chytridiomycosis infection rate[2] or death[3, 5]. One found that adding antifungal bacteria to frogs prevented infection and death[1]. One randomized, replicated, controlled study in the USA[4] found that adding antifungal skin bacteria to soil significantly reduced chytridiomycosis infection rate of red-backed salamanders.

- One randomized, replicated, controlled study in Switzerland[5] found that treatment with antimicrobial skin peptides before or after infection with chytridiomycosis did not significantly increase survival of common toads.

- Three randomized, replicated, controlled studies in the USA[1, 2, 5] found that adding antifungal skin bacteria to chytrid infected amphibians reduced weight loss.

Background

The chytrid fungus *Batrachochytrium dendrobatidis* infects the outer layer of amphibian skin. A number of bacterial species of amphibian skin have been found to inhibit the chytrid fungus in experiments (e.g. Harris *et al.* 2006; Becker & Harris 2010; Lam *et al.* 2011). There is also some evidence that anti-microbial peptides, which are secreted into mucus and thought to help protect against colonization by skin pathogens, may provide some resistance to chytrid infections (e.g. Pask *et al.* 2012; 2013). It is therefore possible that adding such anti-fungal species or peptides to amphibian skin or to their environment may reduce the effects of the disease.

Becker, M.H. & Harris, R.N. 2010. Cutaneous bacteria of the redback salamander prevent morbidity associated with a lethal disease. *PLoS One*, 5, e10957.

Harris R.N., James T.Y., Lauer A., Simon M.A. & Patel A. (2006) The amphibian pathogen *Batrachochytrium dendrobatidis* is inhibited by the cutaneous bacteria of amphibian species. *EcoHealth*, 3, 53–56.

Lam B.A., Walton D.B. & Harris R.N. (2011) Motile zoospores of *Batrachochytrium dendrobatidis* move away from antifungal metabolites produced by amphibian skin bacteria. *EcoHealth*, 8, 36–45.

Pask J.D., Cary T.L. & Rollins-Smith L. A. (2013) Skin peptides protect juvenile leopard frogs (*Rana pipiens*) against chytridiomycosis. *Journal of Experimental Biology*, 216, 2908–2916.

Pask J.D., Woodhams D.C. & Rollins-Smith L.A. (2012) The ebb and flow of antimicrobial skin peptides defends northern leopard frogs (*Rana pipiens*) against chytridiomycosis. *Global Change Biology*, 18, 1231–1238.

A randomized, replicated, controlled study in a laboratory in California, USA (1) found that adding antifungal bacteria (*Janthinobacterium lividum*) to the skins of mountain yellow-legged frog *Rana muscosa* prevented death from chytridiomycosis. Infected frogs treated with the antifungal skin bacteria all survived, gained 33% body mass and had no chytrid zoospores on their skin. In contrast, five of six exposed to chytrid zoospores alone lost weight and died; the sixth had severe chytridiomycosis. Treatment with *Janthinobacterium lividum* increased colonization by the skin bacteria and did not result in reduced growth or death. There were three treatments each with 6 frogs: exposure to chytrid zoospores (300 zoospores/15 ml for 24 h); exposure to antifungal skin bacteria (26 x 106 cells/ml for 30 min) and exposure to skin bacteria and 48 hours later chytrid zoospores. There were also 10 untreated control frogs. Before treatments, animals were rinsed in 3% hydrogen peroxide and sterile Provosoli medium to reduce natural skin bacteria. Frogs were weighed and tested for antifungal skin bacteria and chytrid before and every 2 weeks after treatment until day 139.

A randomized, replicated, controlled study in a laboratory in Virginia, USA (2) found that the severity, but not the infection rate, of chytridiomycosis was reduced by adding chytrid-inhibiting skin bacteria to the skin of red-backed salamanders *Plethodon cinereus*. Infection rate did not differ significantly between those with added bacteria (*Pseudomonas reactans*; 80%) and those with chytrid alone (60%). Numbers of zoospore equivalents on infected individuals were also similar (with bacteria: 6; chytrid alone: 10). However, by day 46, salamanders with the bacteria had lost significantly less body mass (15%) than those with chytrid alone (30%) and a similar amount to controls (bacteria or medium alone: 8%). Following inoculation with skin bacteria, 89% of 18 individuals tested positive for the bacteria. Individuals were randomly assigned to one of four exposure treatments: anti-chytrid skin bacteria, chytrid zoospores, bacteria followed by chytrid zoospores three days later or solution alone. Sample sizes were 5, 20, 20 and 5 respectively. Individuals were tested for chytrid on day 1 and 14 and for skin bacteria on day 1 and 10. Salamanders were bathed with 5 ml of solution containing bacteria (3×10^9 cells/ml) for 2 hours and/or a solution with chytrid (3×10^6 zoospores/5 ml) for 24 hours.

A randomized, replicated, controlled study in a laboratory in the USA (3) found that although the chytrid-inhibiting skin bacteria *Janthinobacterium lividum* colonized skin temporarily, it did not reduce or delay death of chytrid infected Panamanian golden frogs *Atelopus zeteki*. All infected frogs died within four months, whereas all control frogs survived. Although mortality and overall chytrid load did not differ between frogs exposed and not exposed to the bacteria, at death those exposed had significantly lower numbers of chytrid zoospores (1.5×10^5 vs 1.3×10^6). Colonization by the bacteria was successful on 95% of frogs. However, by day 39 bacterial cell counts had declined ($<2.8 \times 10^5$ cells/frog), infection with chytrid had increased (>13,000 zoospore equivalents/frog) and frogs began to die. Frogs were randomly assigned to one of four exposure treatments: anti-chytrid skin bacteria, chytrid zoospores, bacteria followed by chytrid or water alone. Sample sizes were 7, 20, 20 and 7 respectively. Bacteria were isolated from four-toed salamanders *Hemidactylium scutatum*. Frogs were swabbed every 2 weeks for 120 days to test for chytrid and bacteria.

A randomized, replicated, controlled study in 2010 in a laboratory in Virginia, USA (4) found that infection rate of red-backed salamanders *Plethodon cinereus* with chytridiomycosis was significantly lower following exposure to chytrid-inhibiting skin bacteria in the soil. Infection rate was 40% with exposure to the bacteria *Janthinobacterium lividum* compared to 83% without. All salamanders exposed tested positive for the skin bacteria up until day 29, but by day 42 it was no longer detected. Salamanders infected with chytrid had significantly higher densities of bacteria than uninfected individuals. Fifteen randomly selected wild caught salamanders were exposed to skin bacteria in soil followed by chytrid in solution. Twelve were exposed to chytrid alone, six to skin bacteria in soil alone and five were unexposed controls. Each tank received 150 g of soil, which had 1.5 ml of skin bacteria suspension (2.9×10^7 colony-forming units/dry g soil) or pond water. *Janthinobacterium lividum* was isolated from the skin of four-toed salamanders *Hemidactylium scutatum*. Salamanders were tested for chytridiomycosis and the skin bacteria on days 8, 13, 20, 29 and 42.

A randomized, replicated, controlled study in 2007 in a laboratory in Virginia, USA (5) found that survival of mountain yellow-legged frogs *Rana muscosa* naturally infected with chytridiomycosis was not increased by adding chytrid-inhibiting skin bacteria. Survival of frogs treated with bacteria was 50% compared to 39% for infected controls. Infection was not cleared in surviving frogs. However, weight loss was reduced with treatment (0.1 vs 0.4 g/week). Wild-caught frogs were randomly assigned to treatments. Twenty were bathed in water containing bacteria (*Pedobacter cryoconitis*) isolated from mountain yellow-legged frog and 13 control frogs in water alone for 2 hours. Frogs were swabbed and tested at 7 and 13 days after treatment.

A randomized, replicated, controlled study in 2010 in a laboratory in Switzerland (5) found that survival of common toad *Bufo bufo* toadlets was not significantly increased by treatment with antimicrobial skin peptides before or after infection with chytridiomycosis, although treatment may have cured infection in some individuals. Survival of toads treated with peptides immediately before or eight days after infection was not significantly different from chytrid infected controls (12 vs 18%). However, none of the 3 treated toadlets that survived to 35 days were infected with chytridiomycosis, compared to all 3 of the untreated infected controls. Peptide treatment alone did not reduce survival compared to uninfected controls (64% vs 58%). Captive toadlets were randomly assigned to treatments. Seventeen were infected with chytridiomycosis alone. Seventeen were treated with skin peptides from edible frogs *Pelophylax esculentus* (2 minute bath in 400 μg/ml peptide solution) immediately before infection and 17 on day 8 following infection. Twenty four were uninfected controls, 12 of which were bathed with peptides. Swabs were taken and tested for the chytrid fungus on day 35.

(1)   Harris R.N., Brucker R.M., Walke J.B., Becker M.H., Schwantes C.R., Flaherty D.C., Lam B.A., Woodhams D.C., Briggs C.J., Vredenburg V.T. & Minbiole K.P.C. (2009a) Skin microbes on frogs prevent morbidity and mortality caused by a lethal skin fungus. *The ISME Journal*, 3, 818–824.
(2)   Harris R.N., Lauer A., Simon M.A., Banning J.L. & Alford R.A. (2009b) Addition of antifungal skin bacteria to salamanders ameliorates the effects of chytridiomycosis. *Diseases of Aquatic Organisms*, 83, 11–16.
(3)   Becker M.H., Harris R.N., Minbiole K.P.C., Schwantes C.R., Rollins-Smith L.A., Reinert L.K., Brucker R.M., Domangue R. J. & Gratwicke B. (2011) Towards a better understanding of the use of probiotics for preventing chytridiomycosis in Panamanian golden frogs. *EcoHealth*, 8, 501–506.
(4)   Muletz C.R., Myers J.M., Domangue R.J., Herrick J.B. & Harris R.N. (2012) Soil bioaugmentation with amphibian cutaneous bacteria protects amphibian hosts from infection by *Batrachochytrium dendrobatidis*. *Biological Conservation*, 152, 119–126.
(5)   Woodhams D.C., Geiger C.C., Reinert L.K., Rollins-Smith L.A., Lam B., Harris R.N., Briggs C.J., Vredenburg V.T. & Voyles J. (2012) Treatment of amphibians infected with chytrid fungus: learning from failed treatments with itraconazole, antimicrobial peptides, bacteria, and heat therapy. *Diseases of Aquatic Organisms*, 98, 11–25.

8.21   Use antifungal treatment to reduce infection

- Twelve of 16 studies (including 4 randomized, replicated, controlled studies) in Europe, Australia, Tasmania, Japan and the USA found that antifungal treatment cured[2, 3, 5, 7, 9, 11–14, 16] or increased survival[1, 15] of amphibians with chytridiomycosis. Four studies found that treatments did not cure chytridiomycosis[6], but did reduce infection levels[8, 10] or had mixed results[17].

- Six of the eight studies (including two randomized, replicated, controlled studies) in Japan, Tasmania, the UK and the USA testing treatment with itraconazole found that it was effective at curing amphibians of chytridiomycosis[2, 5, 7, 12, 14, 16]. One study found that it reduced infection levels[10] and one found mixed effects[17].

- Six studies found that specific fungicides caused death or other negative side effects in amphibians[2, 4, 7, 8, 12, 17].

Background

Effective treatments for chytridiomycosis, caused by the fungus *Batrachochytrium dendrobatidis*, are vital to ensure the success of amphibian captive-breeding programmes. Also, to reduce the risk of spreading the disease when animals are moved between breeding facilities, released or translocated between field sites.

A replicated, controlled study of captive amphibians in the USA (1) found that benzalkonium chloride was more effective at reducing chytrid infection (misdiagnosed as *Basidiobolus ranarum* (8)) than copper sulphate or formalin-malachite green in dwarf African clawed frogs

Hymenochirus curtipes. Mortality at day 24 was lower for 2 mg/l benzalkonium chloride (10%), compared to 4 mg/l benzalkonium chloride (16%), 1 mg/l copper sulphate (30%) and formalin (10 mg/l)-malachite green (0.8 mg/l; 25%). In the control group 74% died. Frogs treated with 2 mg/l benzalkonium chloride that survived had only mild infections compared to moderate to severe infections following the other two treatments. A group of 135 frogs from an infected population was bathed in each treatment. Frogs were bathed for 30 minutes on alternate days over 6 days, this was repeated after 8 days. There was an untreated control group of 130 frogs. Five frogs from each group were examined for infection before treatment and on days 1, 3, 5, 10 and 15 after treatment had started. The study ended after 24 days.

A replicated, controlled study in a laboratory (2) found that experimentally infected blue-and-yellow poison dart frogs *Dendrobates tinctorius* treated with miconazole or itraconazole were cured of chytridiomycosis. However, frogs were intolerant to miconazole (possibly due to ethyl alcohol in the solution). Juveniles were experimentally infected with the chytrid fungus. Once excessive skin shedding had started, frogs were treated with miconazole (0.01% solution) or itraconazole (0.1% suspension). Frogs were bathed in the treatments daily for 5 minutes for 8 or 11 days respectively. Controls were untreated. Frogs were then killed humanely and examined.

A replicated study of captive amphibians at the University of California, Berkeley, USA (3) found that western clawed frog *Xenopus tropicalis* treated with formalin-malachite green solution were cured of chytridiomycosis. Five frogs died within the first 48 hours of treatment. However, following the last treatment, all 10 surviving frogs gradually improved in health. The four examined at three weeks, one and two months showed no signs of infection and the remaining six frogs had regained normal body weight within four months. Fifteen naturally infected frogs were treated 4 times with formalin-malachite green solution (25 parts per million formalin and 0.10 mg/L malachite green) at a dilution of 0.007 ml/L of tank water for 24 hours every second day. Following treatment, four were selected at random and killed humanely at either three weeks, one month or two months for examination for infection.

A replicated, controlled study in 2004 at the University of Alexandria, Egypt (4) found that when fluconazole was swallowed by square-marked toads *Bufo regularis* there were significant changes in blood cells, similar to the effects of a carcinogen. White blood cell structure changed in 60% of the toads force-fed with fluconazole and 80% fed with a carcinogen. Controls showed no change. Most white blood cells showed changes such as nuclear abnormalities, vacuolated cytoplasm and reduced organelles. Red blood cells were anaemic with fragmented or degenerated nuclei, long cytoplasmic projections and vacuolated cytoplasm. Fifty adults were force-fed one of the following treatments for 20 weeks: fluconazole daily at a therapeutic dose level (0.26 mg in 0.5 ml saline), a carcinogenic chemical 7,12-dimethylbenz(a) anthracene (0.5 mg in 0.2 ml olive oil) twice/week, a control of 0.2 ml of olive oil or of 0.5 ml saline. Blood samples were obtained from the heart and examined after 20 weeks.

A before-and-after study of an established collection of amphibians in Cheshire, UK (5) found that frogs, axolotls *Ambystoma mexicanum* and Kaup's caecilians *Potymotyphlus kaupii* treated with itraconazole were cured of chytridiomycosis. Approximately 20 individuals had died before treatment (following introduction of new individuals), but once treated there were no further cases of chytridiomycosis for 60 days. The collection was therefore considered disease free. Amphibians were kept in clear plastic boxes at 19–23°C in quarantine (with strict sterilization protocols). Frogs (mainly poison frogs *Dendrobates*, *Epipedobates* and *Phyllobates* spp.) were bathed or soaked daily in itraconazole (10 mg/ml) for 5 minutes over 11 days. Axolotls and caecilians were treated with itraconazole directly in their tank water (concentration 0.01%) for 30 minutes every 5 days for 4 treatments. Following treatment, itraconazole was removed from tanks by filtering.

A replicated, controlled study of captive amphibians in Melbourne, Australia (6) found that although treatment with benzalkonium chloride or fluconazole resulted in increased survival times for juvenile green tree frogs *Litoria caerulea*, mortality rate was still 100%. All treated and untreated frogs died and all uninfected frogs survived. Treatments significantly increased survival time (benzalkonium chloride: 43–44 days, range 21–67; fluconazole: 44 days, range 29–76) compared to untreated frogs (38 days, range 30–67). Time until death did not differ significantly between treatments. Eighteen experimentally infected frogs were sprayed twice a day and kept in a solution with benzalkonium chloride at 1 mg/L and 18 with fluconazole at 25 mg/L. Half were treated for three days and half for seven days. Fourteen were untreated.

A randomized, replicated, controlled study in England, UK (7) found that treatment with itraconazole cured all captive Mallorcan midwife toad *Alytes muletensis* tadpoles of chytridiomycosis, but caused depigmentation. All treated tadpoles tested negative for chytrid infection. However, tadpoles showed significant depigmentation in all treatments and some controls. Fifteen of 17 infected control tadpoles tested positive for infection over 21 days. Tadpoles were infected over two weeks then randomly assigned to treatments. Nine treatment groups of 6 tadpoles were treated with itraconazole baths of 0.5, 1.0 or 1.5 mg/L over 7, 14 or 21 days. Tadpoles were killed humanely one week later. Three control groups of 4–5 infected tadpoles were euthanized at 14, 21 or 28 days post-treatment to test for infection.

A review in 2010 describing a replicated controlled study (8) found that treatment with benzalkonium chloride, fluconazole or methylene blue did not cure great barred frog *Mixophyes fasciolatus* tadpoles of chytridiomycosis. Although they did not cure infections, benzalkonium chloride and fluconazole reduced infection levels. However, at concentrations above 1 mg/L (2–10 mg/L) benzalkonium chloride caused death of tadpoles (over 29%). Methylene blue at concentrations of 12–24 mg/L also caused high mortality. Fifty-six tadpoles were bathed daily in benzalkonium chloride (1 mg/L; 3 hrs) for three days, repeated five days later, or in fluconazole (7 mg/L; 6 hrs) for seven days, or methylene blue (3 or 6 mg/L) for three days. There were 57 controls. Frogs were tested 18 days after treatment. Other studies included in this review have been summarized individually.

A replicated, controlled study of six amphibian species naturally infected with chytridiomycosis in the USA (9) found that treatment with terbinafine hydrochloride in ethanol was effective at curing infection in all animals. All bullfrogs *Rana catesbeiana*, California tiger salamanders *Ambystoma californiense*, foothills yellow-legged frogs *Rana boylii*, black-eyed litter frogs *Leptobrachium nigrops*, Malaysian horned frogs *Megophrys nasuta* and Cranwell's horned frogs *Ceratophrys cranwelli* treated with 0.01% or 0.005% solutions tested negative for chytrid after 3–4 weeks. However, those treated with 0.0005% solution and all control animals remained infected. There were no adverse effects from daily exposure to solution up to 0.01% for up to 15 minutes over 10 days. Amphibians were tested for chytrid before and after treatment. Wild-caught bullfrogs were randomly assigned to four treatments comprising a five minute bath in terbinafine HCl in ethanol: at 0.01% for 5 consecutive days (n = 14), at 0.005% for 6 treatments over 10 days (n = 18), as the previous treatment but kept in a 0.0005% solution between treatments, and a control group. Six or seven individuals of the five other (captive or wild caught) species received five minute baths on five consecutive days of: 0.005%, 0.0005% or distilled water.

A before-and-after study in 2009–2010 of a pond in Mallorca (10) found that treating resident Mallorcan midwife toads *Alytes muletensis* with itraconazole and drying out the pond reduced the prevalence but did not eradicate chytridiomycosis. All samples from tadpoles came back positive for the chytrid fungus the spring after treatment and pond drying. However, the number of spores detected on each swab was lower than the previous year, suggesting a lower level of infection. Healthy-looking toads were seen breeding in the pond following treatment. Over 2,000 toad tadpoles were removed from the pond in March–August 2009. They were taken to a laboratory and completed a week-long treatment of daily five minute baths in itraconazole. Tadpoles were held in captivity for up to seven months. The pond was emptied and left to dry over the summer. Once the pond refilled in autumn, tadpoles were released. The following spring tadpoles were swabbed to test for chytridiomycosis.

A replicated, controlled study of captive amphibians in Europe (11) found that Iberian midwife toads *Alytes cisternasii* and poison dart frogs (Dendrobatidae) sprayed with voriconazole were cured of chytridiomicosis. All five infected poison dart frogs treated were cured. Infection was eliminated from all but one midwife toadlet sprayed with voriconazole at 1.3 mg/L, but only four of seven sprayed at 0.13 mg/L. The one toad treated with 1.3 mg/L that was not cured was sprayed five (rather than one) months after infection. All toadlets housed on tissue soaked in voriconazole remained infected. No toxic side effects were seen. One week after experimental infection with the chytrid fungus, 14 toadlets were sprayed daily with voriconazole (1.3 or 0.13 mg/L water) and 5 were kept on paper towels soaked in voriconazole (1.3 mg/L) for 7 days. Six animals were controls. Five months after experimental infection a further 20 toadlets were sprayed with voriconazole (1.3 mg/L) for 7 days. Animals were tested weekly for infection. A colony of 52 poison dart frogs, 5 positive for chytridiomycosis, was sprayed daily with voriconazole (1.3 mg/L) for 7 days. Frogs containers were sterilized by heating to 45°C for 3 days.

A randomized, replicated, controlled study in 2011 of captive amphibians in the USA (12) found that Australian green tree frogs *Litoria caerulea* and coastal-plain toads *Incilius nebulifer* treated with itraconazole were cured of chytridiomycosis. Itraconazole at 0.01, 0.005 and 0.003 but not 0.001% cured infection. Survival was highest with 0.003% itraconazole. However, itraconazole caused death, loss of appetite, lethargy and skin discolouration, particularly at 0.01 and 0.005%. Survival did not differ between infected animals treated for 6 or 11 days with 0.003% or 6 days with 0.005% itraconazole and untreated animals. However, treatment with all other concentrations for 11 days resulted in reduced survival (0.01%: 66–100% mortality) compared to infected untreated animals. Nine separately housed green froglets and 9–17 communally housed toadlets were randomly assigned to each treatment: infection with chytrid, infection and itraconazole baths for 5 minutes for 6 or 11 days and an uninfected control. Skin swabs were taken for four weeks after treatment.

A randomized, replicated, controlled study in 2010 of captive amphibians in Tennessee, USA (13) found that southern leopard frog tadpoles *Lithobates sphenocephalus* treated with thiophanate-methyl (TM) were cured of chytridiomycosis. All treated tadpoles tested negative for the infection at day 60, as did controls. All infected untreated tadpoles tested positive. By day 60, treated tadpoles were significantly heavier (TM + chytrid: 2.0; TM: 1.1; controls: 0.8–0.9 g) and longer (TM + chytrid: 22; TM: 18; controls: 17 mm). The same was true for metamorphosis mass (TM + chytrid: 1.1; TM: 0.9; controls: 0.5–0.7 g) and length (TM + chytrid: 23; TM: 22; controls: 18–19 mm). Ten tadpoles were randomly assigned to each treatment: thiophanate-methyl treatment of chytrid infected tadpoles, thiophanate-methyl treatment alone, chytrid infection alone and an uninfected control group. Tadpoles were bathed in thiophanate-methyl (0.6 mg/L) and water was changed every three days. Animals were measured and tested for infection at day 60 and measured on tail resorption.

A replicated study in 2009 of captive amphibians in the USA (14) found that reduced-dose itraconazole was an effective treatment for natural infections of chytridiomycosis in Wyoming toads *Anaxyrus baxteri*, White's tree frogs *Litoria caerulea* and African bullfrogs *Pyxicephalus adspersus*. Although 15 infected toads and 1 tree frog died during treatment, all animals surviving at the end of treatment tested negative for chytrid for 5 or 13 months. Before treatment, 70% of Wyoming toads, 45% of tree frogs and both bullfrogs tested positive for chytridiomycosis. Eighty Wyoming toads were bathed for 5 minutes with itraconazole at 100 mg/L for 3 days, 5 mg/L for 6 days and then 50 mg/L on the last day. Eleven tree frogs and 2 African bullfrogs were treated daily with itraconazole at 50 mg/L for 5 minutes over 10 days. Toads were tested for chytrid monthly for 5 months after treatment and frogs every 2 weeks for 2 months and once at 13 months. Animals were not rinsed following baths.

A replicated, controlled study in a laboratory in Australia (15) found that exposing Peron's tree frogs *Litoria peronii* to low concentrations of sea salt significantly lowered chytrid infection loads and increased survival rates. Infection loads were significantly lower with concentrations of 1–4 parts per trillion (ppt) of sodium chloride compared to 5 ppt or no salt. Frogs exposed to 3 ppt had significantly higher survival rates (100%) than at lower (1 ppt: 37; 2 ppt: 63%) or higher concentrations (4 ppt: 72%; 5 ppt: 54%) or with no salt (37%). Survival and weight gains were not reduced with salt. Concentrations of 0–5 ppt sodium chloride did not reduce chytrid fungus survival, but 4–5 ppt significantly reduced growth (10–12 vs 18–22 developing zoospores) and motility (3–7 vs 27%) compared to controls. Frogs were housed with water containing: 0, 1, 2, 3, 4 or 5 ppt sea salt. Chytrid in solution (1 ml) was added to half of each salt treatment (11 replicates/treatment). After 30 days body mass was measured and at 120 days swabs were tested for chytrid infection. Chytrid culture (100 ml) was added to 10 replicates of 0, 1, 2, 3, 4 or 5 ppt sea salt and incubated at 22°C for 11 days. Fungus survival, growth and motility were assessed.

A small replicated study in Japan (16) found that Japanese giant salamanders *Andrias japonicus* treated with itraconazole were cured of chytridiomycosis. By day five of treatment all four previously infected salamanders tested negative for the disease. Tests remained negative for two weeks. Four naturally infected salamanders were bathed daily in 0.01% itraconazole for 5 minutes over 10 days. Animals were tested for chytrid before treatment, on treatment days 5 and 10 and 7 and 14 days after treatment.

Randomized, replicated, controlled studies in 2007–2009 of amphibians with chytridiomycosis in the USA and Tasmania (17) found that treatment with itraconazole cured northern leopard frogs *Lithobates pipiens*, did not increase survival of mountain yellow-legged frogs *Rana muscosa* and was highly toxic to striped marsh frog *Limnodynastes peronii* metamorphs. All four treated leopard frogs were cured, although one control frog died with signs of toxicity. Eight treated marsh frogs died by the third day of treatment. Although treatment did not increase survival of yellow-legged frogs (treated: 30%; controls: 39%), it reduced weight loss (0.2 vs 0.4 g/week) and cleared infection in surviving frogs. Frogs were randomly assigned to treatments. Ten wild-caught naturally infected yellow-legged frogs, 4 infected leopard frogs and 8 wild-caught naturally infected marsh frogs were bathed with itraconazole (100 mg/L) for 5 minutes daily and then rinsed for 11, 5 or 3 days respectively. There were 13 control yellow-legged frogs, 7 marsh frogs (bathed in water) and 8 leopard frogs. Yellow-legged frogs were tested for infection at 7 and 13 days after treatment and leopard frogs before and 17 days after treatment.

A randomized, replicated, controlled study in 2010 in Switzerland (17) found that common midwife toad *Alytes obstetricans* tadpoles treated with three commercial antifungal treatments were not cured of chytridiomicosis. All but one tadpole treated with PIP Pond

Plus and all those treated with Steriplant N remained infected. Only 3 of 18 treated with Mandipropamid (at 0.1, 1.4 and 1.6 mg/L) were cured. Wild-caught tadpoles were randomly assigned to treatments. Twenty-eight were treated daily with PIP Pond Plus (probiotic bacteria, enzymes and isopropanol) in doses of 0, 25, 50 or 100 µg/ml added to their water for 7 days. Twenty-eight were treated with Steriplant N (water and 0.04% oxidants) on day 0 (control), 1 (5 parts per million), 2 (10 parts per million) or 3 (15 parts per million). Twenty-one tadpoles were treated with Mandipropamid (phenylglycinamides and mandelamides) at 18 different doses from 0.01 to 4 mg/L (in acetone), with 3 controls. Tadpoles were swabbed and tested a week after treatment.

(1)   Groff J.M., Mughannam A., McDowell T.S., Wong A., Dykstra M.J., Frye F.L. & Hedrick R.P. (1991) An epizootic of cutaneous zygomycosis in cultured dwarf African clawed frogs (*Hymenochirus curtipes*) due to Basidiobolus ranarum. *Journal of Medical and Veterinary Mycology*, 29, 215–223.
(2)   Nichols D.K. & Lamirande E.W. (2001) Successful treatment of chytridiomycosis. *Froglog*, 46, 1.
(3)   Parker J.M., Mikaelian I., Hahn N. & Diggs H.E. (2002) Clinical diagnosis and treatment of epidermal chytridiomycosis in African clawed frogs (*Xenopus tropicalis*). *Comparative Medicine*, 52, 265–268.
(4)   Essawya A.E., El-Zoheirya A.H., El-Moftya M.M., Helalb S.F. & El-Bardana E.M. (2005) Pathological changes of the blood cells in fluconazole treated toads. *ScienceAsia*, 31, 43–47.
(5)   Forzán M., Gunn H. & Scott P. (2008) Chytridiomycosis in an aquarium collection of frogs: diagnosis, treatment, and control. *Journal of Zoo and Wildlife Medicine*, 39, 406–411.
(6)   Berger L., Speare R., Marantelli G. & Skerratt L.F. (2009) A zoospore inhibition technique to evaluate the activity of antifungal compounds against *Batrachochytrium dendrobatidis* and unsuccessful treatment of experimentally infected green tree frogs (*Litoria caerulea*) by fluconazole and benzalkonium chloride. *Research in Veterinary Science*, 87, 106–110.
(7)   Garner T., Garcia G., Carroll B. & Fisher M. (2009) Using itraconazole to clear *Batrachochytrium dendrobatidis* infection, and subsequent depigmentation of *Alytes muletensis* tadpoles. *Diseases of Aquatic Organisms*, 83, 257–260.
(8)   Berger L., Speare R., Pessier A., Voyles J. & Skerratt L.F. (2010) Treatment of chytridiomycosis requires urgent clinical trials. *Diseases of Aquatic Organisms*, 92, 165–174.
(9)   Bowerman J., Rombough C., Weinstock S.R. & Padgett-Flohr G.E. (2010) Terbinafine hydrochloride in ethanol effectively clears *Batrachochytrium dendrobatidis* in amphibians. *Journal of Herpetological Medicine and Surgery*, 20, 26–28.
(10)   Lubick N. (2010) Emergency medicine for frogs. *Nature*, 465, 680–681.
(11)   Martel A., Van Rooij P., Vercauteren G., Baert K., Van Waeyenberghe L., Debacker P., Garner T.W., Woeltjes T., Ducatelle R., Haesebrouck F. & Pasmans F. (2011) Developing a safe antifungal treatment protocol to eliminate *Batrachochytrium dendrobatidis* from amphibians. *Medical Mycology*, 49, 143–149.
(12)   Brannelly L.A., Richards-Zawacki C.L. & Pessier A.P. (2012) Clinical trials with itraconazole as a treatment for chytrid fungal infections in amphibians. *Diseases of Aquatic Organisms*, 101, 95–104.
(13)   Hanlon S.M., Kerby J.L. & Parris M.J. (2012) Unlikely remedy: fungicide clears infection from pathogenic fungus in larval southern leopard frogs (*Lithobates sphenocephalus*). *PLoS ONE*, 7, e43573.
(14)   Jones M.E.B., Paddock D., Bender L., Allen J.L., Schrenzel M.S. & Pessie A.P. (2012) Treatment of chytridiomycosis with reduced-dose itraconazole. *Diseases of Aquatic Organisms*, 99, 243–249.
(15)   Stockwell M.P., Clulow J. & Mahony M.J. (2012) Sodium chloride inhibits the growth and infective capacity of the amphibian chytrid fungus and increases host survival rates. *PLOS One*, 7, e36942.
(16)   Une Y., Matsui K., Tamukai K. & Goka K. (2012) Eradication of the chytrid fungus Batrachochytrium dendrobatidis in the Japanese giant salamander *Andrias japonicus*. *Diseases of Aquatic Organisms*, 98, 243–247.
(17)   Woodhams D.C., Geiger C.C., Reinert L.K., Rollins-Smith L.A., Lam B., Harris R.N., Briggs C.J., Vredenburg V.T. & Voyles J. (2012) Treatment of amphibians infected with chytrid fungus: learning from failed treatments with itraconazole, antimicrobial peptides, bacteria, and heat therapy. *Diseases of Aquatic Organisms*, 98, 11–25.

8.22   Use antibacterial treatment to reduce infection

• Two studies (including one randomized, replicated, controlled study) in New Zealand and Australia found that treatment with chloramphenicol antibiotic ointment[2] or solution, with other interventions in some cases[3], cured green tree frogs and one Archey's frog of chytridiomycosis.

- One replicated, controlled study[1] found that treatment with trimethoprim-sulfadiazine increased survival time but did not cure blue-and-yellow poison dart frogs of chytridiomycosis.

Background

Effective treatments for chytridiomycosis, caused by the fungus *Batrachochytrium dendrobatidis*, are vital to ensure the success of amphibian captive-breeding programmes. They are also required to reduce the risk of spreading the disease when animals are moved between breeding facilities, released or translocated between field sites.

A replicated, controlled study in a laboratory (1) found that treatment of blue-and-yellow poison dart frogs *Dendrobates tinctorius* with trimethoprim-sulfadiazine survived longer but were not cured of the chytrid infection. Frogs treated with trimethoprim-sulfadiazine survived longer than untreated frogs. Juveniles were experimentally infected with the chytrid fungus *Batrachochytrium dendrobatidis*. Once excessive skin shedding had started, frogs were treated with trimethoprim-sulfadiazine (0.1% solution). Frogs were immersed in the treatment for 5 minutes each day for 11 consecutive days. Controls were untreated. Frogs were then killed humanely and examined.

A study in a laboratory in New Zealand (2) found that treatment of one Archey's frog *Leiopelma archeyi* with an antibiotic ointment cured it of chytridiomycosis. At the end of five days' treatment with chloramphenicol ointment, the infection was significantly reduced (zoospore equivalents: 176–217 to 7). Over the following three months the frog tested negative for chytridiomycosis in five tests. Chloramphenicol treatment did not appear to have any effect on weight, behaviour or health. The frog had 5 mg of chloramphenicol ointment applied to its back for 5 days. Four other wild caught frogs had chloramphenicol in water (10 mg/L) added to their containers. Containers were disinfected with 70% ethanol and the treatment solution changed daily for five days. They were tested for the chytrid fungus on arrival, at 2, 4, 8, 14 and 19 weeks and at the end of the trial. Behaviour, food consumption and weight gain was monitored daily.

A randomized, replicated, controlled study in 2011 in Queensland, Australia (3) found that treatment of captive green tree frogs *Litora caerulea* with chloramphenicol solution cured terminal and pre-symptom chytridiomycosis infections. The three terminally infected frogs also received electrolyte fluids and increased ambient temperature from 22 to 28°C. All 18 infected frogs bathed in chloramphenicol solution were clinically normal within 4–5 days and cured by day 13–17. All five terminally infected frogs that did not receive treatment died within 24–48 hours. Treated controls remained uninfected and clinically normal. Frogs were collected from the wild and randomly assigned to treatments. Seventeen frogs experimentally infected with chytridiomycosis and one naturally infected frog received treatment and five infected (one naturally) were controls. Eighteen uninfected frogs were also treated. Treatment was continuous immersion in 20 mg/L chloramphenicol solution for 14 (n = 3) or 28 (n= 15) days. Solutions were changed daily. Three terminally infected frogs also received electrolyte fluids under the skin every eight hours for six days and increased ambient temperature (from 22 to 28°C). Frogs were swabbed for testing every 7 days for 34 days and at 102 days.

(1) Nichols D.K. & Lamirande E.W. (2001) Successful treatment of chytridiomycosis. *Froglog*, 46, 1.
(2) Bishop P.J., Speare R., Poulter R., Butler M., Speare B.J., Hyatt A., Olsen V. & Haigh A. (2009) Elimination of the amphibian chytrid fungus *Batrachochytrium dendrobatidis* by Archey's frog Leiopelma archeyi. *Diseases of Aquatic Organisms*, 84, 9–15.

(3)   Young S., Speare R., Berger L. & Skerratt L.F. (2012) Chloramphenicol with fluid and electrolyte therapy cures terminally ill green tree frogs (*Litoria caerulea*) with chytridiomycosis. *Journal of Zoo and Wildlife Medicine*, 43, 330–337.

8.23   Use temperature treatment to reduce infection

• Four of five studies (including four replicated, controlled studies) in Australia, Switzerland and the USA[1–5] found that increasing enclosure or water temperature to 30–37°C for over 16 hours cured frogs and toads of chytridiomycosis. One found that heat treatment at 30–35°C for 36 hours did not cure northern leopard frogs[5].

Background

Treatment of chytridiomycosis is vital to ensure the success of amphibian captive-breeding programmes. Also to reduce the risk of spreading the disease when animals are moved between captive or wild populations.

The chytrid fungus *Batrachochytrium dendrobatidis* is very sensitive to temperatures above 32°C. At 37°C the fungus is killed within 4 hours and at 47°C within 30 minutes (Young *et al*. 2007). A study found that the probability of infection by chytrid in the wild decreased strongly with increasing time spent with body temperatures above 25°C in three frog species (Rowley & Alford 2013). A study in captivity also found that fewer frogs became infected and died when exposed to the chytrid fungus if they were housed at 27°C rather than 17°C or 23°C (50 vs 100% mortality; Berger *et al*. 2004). Increasing temperatures within amphibian housing may therefore provide a treatment for chytridiomycosis.

Berger L., Speare R., Hines H.B., Marantelli G., Hyatt A.D., McDonald K.R., Skerratt L.F., Olsen V., Clarke J.M., Gillespie G., Mahony M., Sheppard N., Williams C. & Tyler M.J. (2004) Effect of season and temperature on mortality in amphibians due to chytridiomycosis. *Australian Veterinary Journal*, 82, 434–438.
Rowley J.J.L. & Alford R.A. (2013) Hot bodies protect amphibians against chytrid infection in nature. *Scientific Reports*, 3, 1515.
Young S., Berger L. & Speare R. (2007) Amphibian chytridiomycosis: strategies for captive management and conservation. *International Zoo Yearbook*, 41, 85–95

A replicated, controlled study in a laboratory at James Cook University, Australia (1) found that heat treatment at 37°C cured red-eyed tree frogs *Litoria chloris* of chytridiomycosis. There was a significant difference in survival between temperature treatments. All infected frogs in the treatment with two 8-hour periods at 37°C tested negative for chytrid after 94 days and survived for at least another 5 months. Infected frogs at a constant 20°C survived for the shortest period (55 days), while survival was intermediate in the treatments with naturally fluctuating temperatures (14–23°C; 83 days) and two 8-hour periods at 8°C (1 frog survived over 94 days). All frogs in these treatments were heavily infected. All but one uninfected frog survived. Eighty juvenile frogs were divided equally into the four temperature regimes. Half in each treatment were infected with chytrid fungus and half with sterile medium as a control. Survival was examined over 94 days and infection level determined at post-mortem.

A replicated, controlled study in 2004 in a laboratory in the USA (2) found that heat treatment at 32°C cured western chorus frogs *Pseudacris triseriata* of chytridiomycosis. Three infected frogs died during treatment, but the remaining four tested negative for chytrid following treatment. All infected frogs kept at room temperature remained infected and four died. No uninfected frogs died with or without treatment. Weight gain in cured frogs was significantly greater than infected frogs (1.1–1.4 vs 0.7–0.9 g). Frogs were raised from eggs collected from the wild and were experimentally infected with chytrid. Seven infected and five uninfected frogs were placed in an incubator for five days at 32°C. Nine infected and 15 uninfected frogs were kept at room temperature (20°C). Frogs were weighed at days 172 and 257 and sampled for chytrid on day 172.

A replicated study in 2010 of captive amphibians in Louisiana, USA (3) found that temperature treatment at 30°C cured northern cricket frogs *Acris crepitans* and bullfrogs *Rana catesbeiana* of chytridiomycosis. All bullfrogs and all but one northern cricket frog (96%) tested negative for chytrid following treatment. Animals were randomly assigned to acclimatization at 23 or 26°C for one month. Sixteen northern cricket frogs (7 at 23°C, 9 at 26°C) and 12 bullfrogs (10 at 23°C, 2 at 26°C) naturally infected with the chytrid fungus were then housed individually at 30°C for 10 consecutive days. Frogs were returned to 23 or 26°C and tested again for infection 6 days later.

A replicated, controlled study in a laboratory at the University of Zürich, Switzerland (4) found that heat treatment over 26°C cured the majority of common midwife toad *Alytes obstetricans* tadpoles of chytridiomycosis. The percentage of tadpoles cured increased significantly with temperature (21°C: 20%; 26°C: 63%; 30°C: 88%). Tadpoles were wild caught and were tested for chytridiomycosis before and 6–10 days after treatment. Ten infected tadpoles were randomly assigned to each treatment: water temperature at 21°C or 26°C for 5 days, or water at room temperature and 30°C for 59 hours. After the experiment, toads were treated using itraconazole fungicide and released at the capture site.

A small, replicated, controlled study in 2007 of captive amphibians (5) found that short-term heat treatment at 30–35°C did not cure northern leopard frogs *Lithobates pipiens* of chytridiomicosis. None of the four infected frogs treated were cured of their infection. Five of six uninfected frogs remained uninfected during treatment, but all control frogs kept in group enclosures were infected by the end of the experiment. Naturally infected frogs were placed in an incubator at 30°C overnight and then 35°C for 24 hours. Control groups of 3–4 frogs were kept at room temperature (23°C).

(1) Woodhams D.C., Alford R.A. & Marantelli G. (2003) Emerging disease of amphibians cured by elevated body temperature. *Diseases of Aquatic Organisms*, 55, 65–67.
(2) Retallick R.W.R. & Miera V. (2007) Strain differences in the amphibian chytrid *Batrachochytrium dendrobatidis* and non-permanent, sub-leathal effects of infection. *Diseases of Aquatic Organisms*, 75, 201–207.
(3) Chatfield M.W.H. & Richards-Zawacki C.L. (2011) Elevated temperature as a treatment for *Batrachochytrium dendrobatidis* infection in captive frogs. *Diseases of Aquatic Organisms*, 94, 235–238.
(4) Geiger C.C., Küpfer E., Schär S., Wolf S. & Schmidt B.R. (2011) Elevated temperature clears chytrid fungus infections from tadpoles of the midwife toad, *Alytes obstetricans*. *Amphibia-Reptilia*, 32, 276–280.
(5) Woodhams D.C., Geiger C.C., Reinert L.K., Rollins-Smith L.A., Lam B., Harris R.N., Briggs C.J., Vredenburg V.T. & Voyles J. (2012) Treatment of amphibians infected with chytrid fungus: learning from failed treatments with itraconazole, antimicrobial peptides, bacteria, and heat therapy. *Diseases of Aquatic Organisms*, 98, 11–25.

8.24   Treat amphibians in the wild or pre-release

- One before-and-after study in Mallorca[1] found that treating wild midwife toads with fungicide, along with pond drying, reduced infection levels but did not eradicate chytridiomycosis.

Background
Studies investigating the effects of treating amphibians in captivity are discussed in 'Use antifungal skin bacteria or peptides to reduce infection', 'Use antifungal treatment to reduce infection', 'Use antibacterial treatment to reduce infection' and 'Use temperature treatment to reduce infection'.

A before-and-after study in 2009–2010 in a pond in Mallorca (1) found that treating wild midwife toads *Alytes muletensis* with a fungicide, along with drying out the pond, reduced the prevalence but did not eradicate chytridiomycosis. All samples from tadpoles came back positive for the chytrid fungus the spring after treatment and pond drying. However, the number of spores detected on each swab was lower than the previous year, suggesting a lower level of infection. Healthy-looking toads were seen breeding in the pond following treatment. Over 2,000 toad tadpoles were removed from the pond in March–August 2009. The pond was emptied and left to dry over the summer. Tadpoles were taken to a laboratory and given daily five minute baths in the fungicide itraconazole for one week. They were held in captivity for up to seven months. Once the pond refilled in autumn, tadpoles were released. The following spring tadpoles were swabbed to test for chytridiomycosis.

(1) Lubick N. (2010) Emergency medicine for frogs. *Nature*, 465, 680–681.

8.25   Immunize amphibians against infection

- One randomized, replicated, controlled study in the USA[1] found that vaccinating mountain yellow-legged frogs with formalin-killed chytrid fungus did not significantly reduce chytridiomycosis infection rate or mortality.

Background
Chytridiomycosis infection often spreads rapidly once it has been introduced to amphibian populations, causing mass mortality and population declines. However, some species of amphibians appear to be resistant to developing the disease if they have previously been exposed to the chytrid fungus *Batrachochytrium dendrobatidis* (Hanselmann *et al.* 2004). This suggests that it may be possible to reduce infection by injecting animals with dead chytrid fungus to stimulate a protective immune response.

Hanselmann R., Rodríguez A., Lampo M., Fajardo-Ramos L., Aguirre A.A., Kilpatrick A.M., Rodríguez J. & Daszak P. (2004) Presence of an emerging pathogen in introduced bullfrogs Rana catesbeiana in Venezuela. *Biological Conservation*, 120, 115–119.

A randomized, replicated, controlled study in a laboratory at the University of California, USA (1) found that vaccinating mountain yellow-legged frogs *Rana muscosa* with formalin-killed chytrid fungus did not significantly reduce infection rate with chytridiomycosis or mortality. The proportion of frogs that became infected (chytrid/adjuvant: 0.8; adjuvant only: 0.9; control: 0.8) and died (chytrid/adjuvant: 0.4; adjuvant: 0.4; control: 0.2) were similar to controls. Following vaccination, there was no significant difference in the time to infection, rate of increase in chytrid zoospores in animals (chytrid/adjuvant: 0.08; adjuvant: 0.08; control: 0.09) or the maximum number of zoospores per frog (chytrid/adjuvant: 53,990; adjuvant: 17,831; control: 5,106). Frogs were randomly assigned into 3 groups of 19–20 individuals. Controls received an injection of saline. One group received a 1:1 vaccination of formalin-killed chytrid fungus in Freund's complete adjuvant (to increase effectiveness) and one month later formalin-killed chytrid in Freund's incomplete adjuvant. Another group received saline with Freund's complete adjuvant and one month later saline with Freund's incomplete adjuvant. Injections comprised 0.05 cm^3 into the dorsal lymph sac. Frogs were exposed to live chytrid (10^5 zoospores) one month after treatments. Individuals were monitored weekly for chytridiomycosis using swabs of the ventral surface.

(1)   Stice M.J. & Briggs C.J. (2010) Immunization is ineffective at preventing infection and mortality due to the amphibian chytrid fungus *Batrachochytrium dendrobatidis*. *Journal of Wildlife Diseases*, 46, 70–77.

Reduce parasitism and disease – ranaviruses

8.26   Sterilize equipment to prevent ranavirus

- We found no evidence for the effects of sterilizing equipment to prevent ranavirus on the spread of disease between amphibian individuals or populations.

Background
Ranavirus, sometimes known as 'red-leg', causes two forms of disease in amphibians, skin ulcers and internal bleeding. In some populations the virus causes mass mortality followed by population recovery, in others the disease is recurrent with long-term population declines of up to 80% (Teacher *et al.* 2010). Survival time of the virus outside a host is unknown and so equipment should be disinfected to prevent the spread of the disease.

There is additional literature examining the effectiveness of using a range of disinfectants to kill ranavirus. For example, a study found that chlorhexidine, household bleach and Virkon S, but not potassium permanganate, were effective at inactivating ranavirus when used at certain concentrations (Bryan *et al.* 2009).

Studies investigating prevention of the spread of chytridiomycosis are discussed in 'Sterilize equipment when moving between amphibian sites' and 'Use gloves to handle amphibians'.

Bryan L.K., Baldwin C.A., Gray M.J. & Miller D.L. (2009) Efficacy of select disinfectants at inactivating Ranavirus. *Diseases of Aquatic Organisms*, 84, 89–94.
Teacher, A.G.F., Cunningham, A.A. & Garner, T.W.J. (2010) Assessing the long-term impact of Ranavirus infection in wild common frog populations. *Animal Conservation*, 13, 514–522.

9 Threat: Pollution

Pollution, from many sources, has direct and indirect impacts on amphibians. Amphibian skin is highly permeable to allow gas, water and electrolyte exchange with the environment making them particularly susceptible to pollutants. For example, uptake of three heavily used herbicides was found to be up to 300 times faster through the skin of amphibians than through the skin of mammals (Quaranta et al. 2009). Amphibians also have life stages in water and on land and so can be exposed to toxicants in both environments. As well as direct effects, water-borne pollutants can have significant impacts on aquatic habitats. Little is known of the long-term effects of many pollutants, including those that persist and accumulate in the environment.

Quaranta, A., Bellantuono, V., Cassano, G. & Lippe, C. (2009) Why amphibians are more sensitive than mammals to xenobiotics. *Plos One* 4, e7699.

Key messages – agricultural pollution

Plant riparian buffer strips
One replicated, controlled study in the USA found that planting buffer strips along streams did not increase amphibian abundance or numbers of species.

Prevent pollution from agricultural lands or sewage treatment facilities entering watercourses
We captured no evidence for the effects of preventing pollution from agricultural lands or sewage treatment facilities entering watercourses on amphibian populations.

Create walls or barriers to exclude pollutants
One controlled study in Mexico found that installing filters across canals to improve water quality and exclude fish increased weight gain in axolotls.

Reduce pesticide, herbicide or fertilizer use
One study in Taiwan found that halting pesticide use, along with habitat management, increased a population of frogs.

Key messages – industrial pollution

Add limestone to water bodies to reduce acidification
Five before-and-after studies, including one controlled, replicated study, in the Netherlands and the UK found that adding limestone to ponds resulted in establishment of one of three translocated amphibian populations, a temporary increase in breeding and metamorphosis by natterjack toads and increased egg and larval survival of frogs. One replicated, site comparison study in the UK found that habitat management that included adding limestone to ponds increased natterjack toad populations. However, two before-and-after studies,

including one controlled study, in the UK found that adding limestone to ponds resulted in increased numbers of abnormal eggs, high tadpole mortality and pond abandonment.

Augment ponds with ground water to reduce acidification

We captured no evidence for the effects of augmenting ponds with ground water to reduce acidification effects on amphibian populations.

Agricultural pollution

9.1   Plant riparian buffer strips

- One replicated, controlled study in the USA[1] found that planting buffer strips along streams did not increase amphibian abundance, numbers of species, or the ratio of adults to tadpoles.

> **Background**
> Uncultivated strips of vegetation at the edge of waterways are often used to help reduce pollution entering the water within agricultural and forestry systems. These buffer strips therefore help to protect aquatic and semi-aquatic species.
>
> Studies that investigated retaining riparian buffers are discussed in 'Threat: Biological resource use – Logging and wood harvesting – Retain riparian buffer strips during timber harvest', 'Habitat protection – Retain buffer zones around core habitat' and 'Threat: Agriculture – Exclude domestic animals or wild hogs from ponds by fencing'.

A replicated, controlled study in 2006–2009 of channelized agricultural streams in Ohio, USA (1) found that planting buffer strips along streams had no significant effect on amphibian communities. There was no significant difference in species richness, diversity, abundance or ratio of adult frogs to tadpoles between sites with and without buffer strips. Amphibians were monitored in three streams with planted non-woody buffer strips (<15 m) and three without. Two 125 m long sections were established along each stream (average 743 m apart). Six permanent transects (25 m apart) were sampled along each section in spring, summer and autumn each year.

(1)   Smiley P.C., King K.W. & Fausey N.R. (2011) Influence of herbaceous riparian buffers on physical habitat, water chemistry, and stream communities within channelized agricultural headwater streams. *Ecological Engineering*, 37, 1314–1323.

9.2   Prevent pollution from agricultural lands or sewage treatment facilities entering watercourses

- We found no evidence for the effects of preventing pollution from agricultural lands or sewage treatment facilities entering watercourses on amphibian populations.

> **Background**
> Agricultural intensification has resulted in increased use of chemicals such as fertilizers and pesticides. Studies have found a relationship between proximity to agricultural lands with high levels of pesticide and fertilizer use and amphibian malformations and

population declines (Ouellet *et al.* 1997; Bishop *et al.* 1999; Davidson 2004; Taylor *et al.* 2005). As well as direct effects, pollutants such as fertilizers can stimulate algal growth which can have significant effects on aquatic habitats.

Other studies that investigated methods to reduce pollution levels in aquatic habitats in agricultural landscapes are described in 'Create walls or barriers to exclude pollutants' and 'Reduce pesticide, herbicide or fertilizer use'.

Bishop C.A., Mahony N.A., Struger J., Ng P. & Pettit K.E. (1999) Anuran development, density and diversity in relation to agricultural activity in the Holland River watershed, Ontario, Canada (1990–1992). *Environmental Monitoring and Assessment*, 57, 21–43.
Davidson C. (2004) Declining downwind: amphibian population declines in California and historical pesticide use. *Ecological Applications*, 14, 1892–1902.
Ouellet M., Bonin J., Rodrigue J., Desgranges J.L. & Lair S. (1997) Hindlimb deformities (*ectromelia, ectrodactyly*) in free-living anurans from agricultural habitats. *Journal of Wildlife Diseases*, 33, 95–104.
Taylor B., Skelly D., Demarchis L.K., Slade M.D., Galusha D. & Rabinowitz P.M. (2005) Proximity to pollution sources and risk of amphibian limb malformation. *Environmental Health Perspectives*, 113, 1497–1501.

9.3   Create walls or barriers to exclude pollutants

• One controlled study in Mexico[1] found that installing filters across canals to improve water quality and exclude fish increased weight gain in axolotls.

Background

In some situations it may be possible to install barriers to prevent pollutants reaching habitat that supports amphibians.

A controlled study in 2009 of canals within agricultural land in Xochimilco, Mexico (1) found that installing filters to improve water quality and exclude competitive fish increased weight gain in axolotls *Ambystoma mexicanum*. Only 4 of 12 previously marked axolotls were recaptured; however, their weight had increased by 16%. Weight gain was greater than that of axolotls in control colonies over the same period. After four months, water was significantly lower in ammonia (77%), nitrates (87%) and turbidity (15%) compared to control canals. Working with farmers in 2009, a canal used as a refuge by axolotls was isolated from the main system using filters made of wood to exclude fish and improve water quality. Farmers benefited from improved farm products with the improved water quality and the protection of traditional agriculture.

(1)   Valiente E., Tovar A., Gonzalez H., Eslava-Sandoval D. & Zambrano L. (2010) Creating refuges for the axolotl (*Ambystoma mexicanum*). *Ecological Restoration*, 28, 257–259.

9.4   Reduce pesticide, herbicide or fertilizer use

• One study in Taiwan[1] found that halting pesticide use along with habitat management increased a population of Taipei frogs.

Background

Agricultural land often receives high chemical inputs to control pests, weeds and fungal infections. These chemicals also enter water bodies through spray drift or run-off. With both aquatic and terrestrial life stages, amphibians can be exposed to toxicants in two environments. These pollutants can have significant effects on populations. For example, atrazine is one of the most commonly used herbicides in the world. One study found that atrazine caused 10–92% of wild male leopard frogs *Rana pipiens* tested across the USA to have abnormal reproductive organs including slowed development and development of egg cells within their testes (Hayes *et al.* 2002).

Hayes T., Haston K., Tsui M., Hoang A., Haeffele C. & Vonk A. (2002) Herbicides: feminization of male frogs in the wild. *Nature*, 419, 895–896.

A study in 1999–2006 of a water lily paddy field in Taipei County, Taiwan (1) found that stopping using pesticides along with habitat-improvement work doubled a population of Taipei frogs *Rana taipehensis*. In 2002, a farmer stopped using herbicides and pesticides on his field, which was at the centre of the frogs' breeding habitat. By August 2003, the Taipei frog population in the field had more than doubled (from 28 to 85) and the farmer fully adopted organic-farming practices. Pollution from river construction work resulted in a drastic decline in the population in 2004–2005 (20 to 4), but by 2006 the population appeared to be recovering (19). Habitat-improvement work included cutting weeds in the field.

(1) Lin H.-C., Cheng L.-Y., Chen P.-C. & Chang M.-H. (2008) Involving local communities in amphibian conservation: Taipei frog *Rana taipehensis* as an example. *International Zoo Yearbook*, 42, 90–98.

Industrial pollution

9.5   Add limestone to water bodies to reduce acidification

- One before-and-after study in the UK[4] found that adding limestone to ponds resulted in establishment of one of three translocated populations of natterjack toads. One replicated, site comparison study in the UK[5] found that species-specific habitat management that included adding limestone to ponds increased natterjack toad populations.

- One before-and-after study in the UK[2] found that adding limestone to ponds temporarily increased breeding by natterjack toads. Three before-and-after studies (including one controlled, replicated study) in the Netherlands and UK found that adding limestone increased larval and/or egg survival of moor frogs[1] and common frogs[3] and resulted in metamorphosis of natterjack toads at two of three sites[4].

- Two before-and-after studies (including one controlled study) in the UK found that adding limestone to ponds resulted in high tadpole mortality and pond abandonment by natterjack toads[2] and higher numbers of abnormal common frog eggs[3].

Background

Water bodies can become acidified from atmospheric pollution that includes sulphur dioxide and nitrogen oxides, which fall as acid rain. Polluted water, from mines for example, can also enter water bodies changing the acidity. This can have significant effects on amphibian abundance, diversity and reproduction (Leuven *et al.* 1986). Adding lime (calcium and magnesium-rich minerals) to water bodies can help reduce acidity.

Leuven, R.S.E.W., den Hartog C., Christiaans, M.M.C. & Heijligers, W.H.C. (1986). Effects of water acidification on the distribution pattern and the reproductive success of amphibians. *Experientia*, 42, 495–503.

A replicated, controlled, before-and-after study in 1987–1989 of eight acidic moorland ponds in a Nature Reserve in central Netherlands (1) found that adding limestone decreased fungal infection of moor frog *Rana arvalis* eggs. Fungal infection rate decreased from 75–100% pre-treatment and in unlimed ponds to 0–25% in limed ponds. No differences were found between temporary and permanent ponds. Removal of *Sphagnum* moss had no effect on infection rate. Ponds were 1–3 m^2 in size and 13–43 cm deep. In March 1988, powdered limestone was added to four ponds (15–48 kg; grain <3 mm). *Sphagnum* moss and most organic sediment had previously been removed from two of the ponds. An additional two ponds were controls and two just had *Sphagnum* removed. Each treatment had a permanent and temporary (re-limed annually) pond. Fungal infection of eggs was estimated for entire ponds every two weeks in March–May.

A before-and-after study in 1983–1989 of a heathland pond in Hampshire, UK (2) found that adding limestone temporarily increased breeding by natterjack toads *Bufo calamita*. The pond was used for breeding more frequently while it was being limed (1–9 vs. 0–3 spawn strings), but tadpole mortality was high and metamorphic success low and toads abandoned the pond before liming ceased. A naturally acid pond (735 m^2) had 25 kg of powdered limestone added annually in April (1983–1989). Toads were monitored before, during and after the intervention, once every 10 days in March and August each year.

A controlled before-and-after study in 1985–1989 of two acidic upland ponds in England, UK (3) found that adding limestone resulted in a significant increase in egg and larval survival of common frogs *Rana temporaria*. Egg survival increased from 0–22 to 69–93% the season after liming, but decreased the following year (93 to 79%). The treated pond had significantly higher egg survival, but also significantly higher numbers of abnormal eggs at day 14 than the control (3.0 vs 2.4%). At least 2% of eggs in limed ponds produced metamorphs. In 1988 and 1989, 20 egg clumps were removed from a pond and each halved. Half were returned to the original pond, which had powdered limestone spread over its surface (250 g/m^2; 70 m^2). Half were placed in a control pond (160 m^2), where frogs had not spawned since 1975. Both ponds received limestone in 1989 (333 mg/L). Eggs were removed at days 7, 14 and 19 and reared in the lab or a container in the ponds. Larvae were counted in ponds in July–August.

A before-and-after study in 1972–1995 of ponds at three heathland sites in England, UK (4) found that adding limestone to ponds resulted in the establishment of a translocated population of natterjack toads *Bufo calamita* at one site, metamorphosis at a second, but no population increase at the third site. The translocated population was dependent on limed ponds at the one site. At the second site, metamorphosis occurred at several previously acidic ponds. However, at the third site, the population did not increase (see (2)). Three sites received minimal powdered limestone (to raise pH to 7) in early spring. At one site, silt (with accumulated sulphate) was removed during the summer (pH increase: 4.5 to 5.5). Pond creation, vegetation clearance and establishment of livestock grazing were also undertaken at some sites. Ponds were monitored by counting spawn strings and estimating toadlet production.

A replicated, site comparison study in 1985–2006 of 20 sites in the UK (5) found that natterjack toad *Bufo calamita* populations increased with species-specific habitat management including adding limestone to ponds. In contrast, long-term trends showed population declines at unmanaged sites. Individual types of habitat management (aquatic, terrestrial or common toad *Bufo bufo* management) did not significantly affect trends, but length of management did. Overall, 5 of the 20 sites showed positive population trends, 5 showed negative trends and 10 trends did not differ significantly from zero. Data on populations (egg string counts) and management activities over 11–21 years were obtained from the Natterjack Toad Site Register. Habitat management for toads was undertaken at seven sites. Management varied between sites, but included pond creation, adding limestone to acidic ponds, maintaining water levels, vegetation clearance and implementing grazing schemes. Translocations were also undertaken at 7 of the 20 sites using wild-sourced (including head-starting) or captive-bred toads.

(1)   Bellemakers M.J.S. & van Dam H. (1992) Improvement of breeding success of the moor frog (*Rana arvalis*) by liming of acid moorland pools and the consequences of liming for water chemistry and diatoms. *Environmental Pollution*, 78, 165–171.
(2)   Banks B., Beebee T.J.C. & Denton J.S. (1993) Long-term management of a natterjack toad (*Bufo calamita*) population in southern Britain. *Amphibia-Reptilia*, 14, 155–168.
(3)   Beattie R.C., Aston R.J. & Milner A.G.P. (1993) Embryonic and larval survival of the common frog (*Rana temporaria*) with particular reference to acidic and limed ponds. *Herpetological Journal*, 3, 43–48.
(4)   Denton J.S., Hitchings S. P., Beebee T.J.C. & Gent A. (1997) A recovery program for the natterjack toad (*Bufo calamita*) in Britain. *Conservation Biology*, 11, 1329–1338.
(5)   McGrath A.L. & Lorenzen K. (2010) Management history and climate as key factors driving natterjack toad population trends in Britain. *Animal Conservation*, 13, 483–494.

9.6   Augment ponds with ground water to reduce acidification

• We found no evidence for the effects of augmenting ponds with ground water to reduce acidification effects on amphibian populations.

Background
Disturbance of soils during land clearing or development can result in release of salts and therefore water with a low pH. This acidic water can end up in water bodies and have significant effects on aquatic biodiversity including amphibians. Adding uncontaminated ground water to ponds can help to regulate the pH of the water.

Studies that investigated regulating water levels of ponds are discussed in 'Threat: Natural system modifications – Regulate water levels'.

10 Threat: Climate change and severe weather

Climate change and extreme weather are very large-scale threats. Therefore, most interventions used in response to them are general conservation interventions such as creating additional breeding sites, captive breeding and translocations, which are discussed in: 'Habitat restoration and creation' and 'Species management'.

Key messages

Use irrigation systems for amphibian sites
One study investigating the effect of applying water to an amphibian site is discussed in 'Threat: Energy production and mining'.

Maintain ephemeral ponds
Studies investigating the effects of regulating water levels or deepening ponds are discussed in 'Threat: Natural system modifications – Regulate water levels' and 'Habitat restoration and creation – Deepen, de-silt or re-profile ponds'.

Deepen ponds to prevent desiccation
Studies investigating the effects of deepening ponds are discussed in 'Habitat restoration and creation – Deepen, de-silt or re-profile ponds'.

Provide shelter habitat
We captured no evidence for the effects of providing shelter habitat on amphibian populations.

Artificially shade ponds to prevent desiccation
We captured no evidence for the effects of artificially shading ponds to prevent desiccation on amphibian populations.

Create microclimate and microhabitat refuges
Studies investigating the effects of creating refuges are discussed in 'Habitat restoration and creation' and 'Biological resource use – Leave coarse woody debris in forests'.

Protect habitat along elevational gradients
We captured no evidence for the effects of protecting habitat along elevational gradients on amphibian populations.

10.1   Use irrigation systems for amphibian sites
One study investigating the effect of applying water to an amphibian site is outlined in 'Threat: Energy production and mining'.

Background

Conservation of some species may require intensive management such as the redistribution of water resources. This could be achieved by using irrigation systems. Irrigation sprayers have been used to manipulate water at toadlet breeding sites (Mitchell 2001).

Mitchell, N.J. (2001) Males call more from wetter nests: effects of substrate water potential on reproductive behaviours of terrestrial toadlets. *Proceedings of the Royal Society B: Biological Sciences*, 268, 87–93.

10.2   Maintain ephemeral ponds

Studies investigating the effects of regulating water levels or deepening ponds are discussed in 'Threat: Natural system modifications – Regulate water levels' and 'Habitat restoration and creation – Deepen, de-silt or re-profile ponds'.

Background

Drying out of amphibian breeding sites, either permanently or before terrestrial life stages have developed, can have significant detrimental effects on populations. It may be possible to maintain temporary ponds by deepening them or by artificially increasing water levels. Grazing may also help to maintain ponds as a study found that temporary pools that had not been grazed for three years dried 50 days earlier than grazed pools (Pyke & Marty 2005). Studies investigating the effects of grazing on amphibians are discussed in 'Threat: Agriculture – Manage grazing regime'.

Pyke C.R. & Marty J. (2005) Cattle grazing mediates climate change impacts on ephemeral wetlands. *Conservation Biology*, 19, 1619–1625.

10.3   Deepen ponds to prevent desiccation

Studies investigating the effects of deepening ponds are discussed in 'Habitat restoration and creation – Deepen, de-silt or re-profile ponds'.

Background

If ponds dry out, breeding habitat may be lost which could have significant effects on amphibian populations. It may be possible to maintain ponds by deepening them.

10.4   Provide shelter habitat

- We found no evidence for the effects of providing shelter habitat on amphibian populations.

Background

Planting trees or scrub can act as a windbreak and create a warm microclimate around breeding sites. Alternatively, management such as increasing tree canopies can help to provide shade from the sun to reduce temperatures where they become too high.

10.5   Artificially shade ponds to prevent desiccation

• We found no evidence for the effects of artificially shading ponds to prevent desiccation on amphibian populations.

Background
Where it is not possible to shade ponds naturally with tree canopy cover for example, artificial shades could help to reduce temperatures and prevent drying.

10.6   Create microclimate and microhabitat refuges

Studies investigating the effects of creating refuges are discussed in 'Habitat restoration and creation' and 'Biological resource use – Leave coarse woody debris in forests'.

Background
Refuge habitats can provide amphibians with microclimates that they require to keep them at the correct temperature and humidity, protecting them from extreme temperatures and dehydration. Refuges, such as log piles, vegetation cover or ponds can be created for amphibians to help to reduce exposure to stressful conditions as a result of climate change.

10.7   Protect habitat along elevational gradients

• We found no evidence for the effects of protecting habitat along elevational gradients on amphibian populations.

Background
Where suitable habitat remains, it may be possible to protect large enough areas that would allow amphibians to migrate with climatic changes and any changes in habitat that result.

11 Habitat protection

Habitat destruction is the largest single threat to biodiversity and habitat fragmentation and degradation often reduces the quality of remaining habitat. Habitat protection is therefore one of the most frequently used conservation interventions, particularly in the tropics and in other areas with large patches of surviving natural vegetation.

Habitat protection can be through the designation of legally protected areas, using national or local legislation. It can also be through the designation of community conservation areas or similar schemes, which do not provide formal protection but may increase the profile of a site and make its destruction less likely. Alternatively protection can be of entire habitat types, for example through the European Union's Habitats Directive. On a smaller scale, habitat protection may involve ensuring areas of important habitat are retained during detrimental activities.

It can be difficult to measure the effectiveness of legally protected areas as there may be no suitable controls. Monitoring often only begins with the designation of the protected areas and they are often in areas that would be less likely to be cleared even if they were not protected.

Key messages

Protect habitats for amphibians
One replicated, site comparison study in the UK found that statutory level habitat protection helped protect natterjack toad populations. One before-and-after study in the UK found that protecting a pond during development had mixed effects on populations of amphibians.

Retain connectivity between habitat patches
One before-and-after study in Australia found that retaining native vegetation corridors maintained populations of frogs over 20 years.

Retain buffer zones around core habitat
Two studies, including one replicated, controlled study, in Australia and the USA found that retaining unmown buffers around ponds increased numbers of frog species, but had mixed effects on tadpole mass and survival. One replicated, site comparison study in the USA found that retaining buffers along ridge tops within harvested forest increased salamander abundance, body condition and genetic diversity. However, one replicated study in the USA found that 30 m buffer zones around wetlands were not sufficient to protect marbled salamanders.

11.1   Protect habitats for amphibians

• One replicated, site comparison study in the UK[1] found that populations of natterjack toads were better protected at sites with a statutory level of habitat protection than those outside protected areas. One before-and-after study in the UK[2] found that a common frog population increased but common toads decreased following the protection of a pond during development.

Background

The effectiveness of protecting areas for amphibian populations is rarely assessed. For example, since 2005, Conservation International and the Amphibian Specialist Group have partnered in the creation of 14 new protected areas in Africa, Asia and Latin America. These cover 22,000 ha and support 55 threatened or endemic amphibian species (Moore 2011). However, the effectiveness of these protected areas has not been monitored.

A modelling study of the distribution of endemic Mexican amphibians found that 65% of species may have less than 20% of their range protected and 20% are not protected by governmental Protected Areas (Ochoa-Ochoa *et al.* 2009). Private and community conservation areas were also found to play a role in protecting endemic Mexican species, with 73% of species represented within those areas (Ochoa-Ochoa *et al.* 2009). Another modelling study reported that the range of 50% of threatened amphibian species in Australia was not considered to be adequately covered by the protected area system (Watson *et al.* 2010).

Moore R. (2011) *Protecting the Smaller Majority Amphibian Conservation Case Studies*. Conservation International and the IUCN/SSC Amphibian Specialist Group.

Ochoa-Ochoa L., Urbina-Cardona J.N., Vázquez L.-B., Flores-Villela O. & Bezaury-Creel J. (2009) The effects of governmental protected areas and social initiatives for land protection on the conservation of Mexican amphibians. *PLoS One*, 4, e6878.

Watson J.E.M., Evans M. C., Carwardine J., Fuller R.A., Joseph L.N., Segan D.B., Taylor M.F.J., Fensham R.J. & Possingham H.P. (2011) The capacity of Australia's protected-area system to represent threatened species. *Conservation Biology*, 25, 324–332.

A replicated, site comparison study in 1970–1989 of natterjack toads *Bufo calamita* in the UK (1) found that populations at sites with a statutory level of habitat protection were better protected than those outside protected areas. Populations within Sites of Special Scientific Interest or National Nature Reserves were better protected from damaging activities (before 1980: 40%; 1989: 100% of threats defended) than those outside (0–29%). Protection for natterjacks in the wider countryside did not improve following the Wildlife and Countryside Act of 1981 (1970–1979: 0–20%; 1980–1989: 0–29%). Populations that were not 'protected' were either lost, damaged or had a planning decision made against their conservation interest. 'Damaging activities' included direct development such as caravan parks or intensification of agriculture. Surveys of known and new populations were undertaken annually.

A before-and-after study in 1986–1999 of a pond within a housing development in Cambridgeshire, UK (2) found that pond protection during development did not prevent a significant decrease in common toads *Bufo bufo*, but resulted in an increase in common frogs *Rana temporaria* during the following ten years. Toad day counts decreased from 145–262 in 1990–1991 to 63 in 1999. Night counts showed a similar trend (240–434 to 59). However, numbers of frog egg masses increased significantly from 12 in 1990 to 96 in 1999. Development was undertaken in 1988–1989 and part of the largest of three breeding ponds was protected. The pond section was 375 m², with a terrestrial margin of 5 m. Each spring, day and night pond counts were undertaken.

(1) Banks B., Beebee T.J.C. & Cooke K.S. (1994) Conservation of the natterjack toad *Bufo calamita* in Britain over the period 1976–1990 in relation to site protection and other factors. *Biological Conservation*, 67, 11–118.

(2) Cooke A.S. (2000) Monitoring a population of common toads (*Bufo bufo*) in a housing development. *Herpetological Bulletin*, 74, 12–15.

11.2   Retain connectivity between habitat patches

- One before-and-after study in Australia[1] found that retaining native vegetation corridors maintained populations of 8 of 13 frog species over 20 years.

Background

Habitat destruction and fragmentation are important factors in the decline of amphibian populations. Small patches of habitat support smaller populations and if individuals are unable to move to other suitable areas of habitat, populations become isolated. This can make them more vulnerable to extinction. On a smaller scale, amphibians often occupy two distinct types of habitat, aquatic and terrestrial. They therefore require suitable habitat to enable them to migrate between these different areas. A study found that as the amount of connecting habitat decreased, so did the diversity of amphibian species with aquatic larvae (Becker *et al.* 2007). Retaining corridors of native vegetation between suitable habitat patches may help to maintain amphibian populations.

Studies investigating the effect of restoring connectivity are discussed in 'Habitat restoration and creation – Restore habitat connectivity' and 'Threat: Transportation and service corridors – Install culverts or tunnels as road crossings'.

Becker C.G., Fonseca C.R., Haddad C.F.B., Batista R.F. & Prado P.I. (2007) Habitat split and the global decline of amphibians. *Science*, 318, 1775–1777.

A before-and-after study in 1998–1999 of pine plantations and surrounding native forest in New South Wales, Australia (1) found that retaining native vegetation corridors helped maintain populations of 8 of 13 frog species over 20 years. Eight of the species that had been present in 1980–1984 were recorded within native forest remnants and plantations in 1998–1999. Five species were not found, but two new species were observed. Numbers of species or individuals captured did not increase significantly with corridor width or distance to continuous native vegetation. Species diversity and abundance did not differ between sites that bordered pine or were surrounded by pine (>450 m from native forest). Following a wildfire in 1983, pines were replanted and native vegetation strips (20 m to over 100 m wide) regenerated. Strips were originally retained along drainlines linking native forest remnants. Twenty-four breeding sites within and around the forest were surveyed four times between November 1998 and December 1999. Call and visual surveys were undertaken.

(1)   Lemckert F.L., Brassil T.E. & Towerton A. (2005) Native vegetation corridors in exotic pine plantations provide long-term habitat for frogs. *Ecological Management Restoration*, 6, 132–134.

11.3   Retain buffer zones around core habitat

- One before-and-after study in Australia[2] found that grassland restoration that included leaving unmown buffers around ponds increased numbers of frog species.

- One replicated, site comparison study in the USA[3] found that retaining buffers along ridge tops within harvested forest increased Red Hills salamander abundance, body condition and genetic diversity. One replicated, controlled study in the USA[4] found that retaining unmown buffers around ponds had mixed effects on tadpole survival and mass depending on species and site.

- One replicated study in the USA[1] found that 30 m buffer zones around wetlands were not sufficient to protect marbled salamanders.

Background

Retaining areas of natural or semi-natural vegetation around core habitats can help to protect the habitat and wildlife that it supports from the detrimental effects of habitat loss or disturbance.

Studies that investigated retaining buffer zones within harvested forests are discussed in 'Threat: Biological resource use – Logging and wood harvesting – Retain riparian buffer strips during timber harvest'. Buffer zones provided by excluding livestock from water bodies are discussed in 'Threat: Agriculture – Exclude domestic animals or wild hogs from ponds by fencing' and those created by planting are described in 'Threat: Pollution – Agricultural pollution – Plant riparian buffer strips'.

A replicated study in 1999–2003 of seasonal ponds within hardwood forest in Massachusetts, USA (1) found that most migrating marbled salamanders *Ambystoma opacum* originated further than 30 m from breeding sites indicating that regulation 30 m buffer zones around wetlands were not sufficient. Of the 366 breeding adults immigrating to the ponds, 84–96% captured at 3 m from ponds were first captured at 30 m. Of animals emigrating from the ponds, 58–85% of newly emerging juveniles (n= 2,282 captures) and 60–79% of adults captured at 3 m were subsequently captured at 30 m. Juveniles were captured 111–1,230 m from breeding ponds (n = 284). Standard wetland buffer zones that extended 30 m from the pond edge were simulated using drift-fencing. Fencing with pitfalls every 10 m on both sides was installed around 3 ponds (up to 0.35 ha) at 3 m and 30 m from the pond margin. Traps were checked daily in May–November 1999–2000 and individuals marked. Juveniles were captured emigrating from 10 ponds in 1999–2003.

A before-and-after study in 1997–2004 of a golf course with degraded woodland and grassland in Sydney, Australia (2) found that restoration that included leaving unmown buffers around ponds increased frog species over two years. Frogs increased from 7 to 10 species in the first year and then remained stable for the following 6 years. A total of 18 species of frogs were predicted in the area and so 56% were present following restoration. The golf course was developed in 1993 and restoration undertaken in 1997–2001. The mowing regime was changed to develop grasslands and a narrow band of herb vegetation retained around ponds as a buffer zone. In addition, native shrubs and trees were planted, non-native weeds were removed and coarse woody debris was reintroduced onto the woodland floor. Pond perimeters were walked to record frog calls in 1996–2004.

A replicated, site comparison study in 2010–2011 of 15 sites in commercial forest in Alabama, USA (3) found that where buffers along ridge tops were retained, Red Hills salamanders *Phaeognathus hubrichti* had greater abundance, body condition and genetic diversity compared to unbuffered sites. Burrow density was significantly higher in buffered habitat (0.7 vs $0.4/m^2$) and individuals maintained a better body condition (mass/length: 0.09 vs 0.08) with no difference between sexes. In terms of genetic diversity, allelic richness was significantly higher in buffered compared to unbuffered woodland (82 vs 70 alleles). However, heterozygosity and inbreeding coefficients did not differ between sites. Burrows were more clumped in buffered habitat. Seven sites with ridge top buffers and eight unbuffered sites were selected. Transects were walked to estimate burrow density and distribution.

Salamanders were caught, measured and tissue samples taken from 110 animals from 10 sites for genetic analysis.

A replicated, controlled study in 2008–2009 of golf course ponds in Ohio, USA (4) found that unmown buffers around ponds had mixed effects on tadpole survival and mass depending on frog species and site. When reared in ponds with buffers, Blanchard's cricket frog *Acris blanchardi* tadpoles had significantly greater survival (0.5 vs 0.2) and mass (0.4 vs 0.3 g) at one site of three sites and green frog *Rana clamitans* tadpoles had significantly lower survival (0.1 vs 0.4–0.7) and mass (3.0–4.0 vs 1.5–2.5 g) at two sites. Mass was significantly lower at one buffered site for green frogs (0.5 vs 2.5 g). Rate of development did not differ in buffered and unbuffered ponds. In 2008, 40 green or cricket frog tadpoles were placed in 10 enclosures in 2 ponds at 3 sites, 1 pond with and without a 1 m unmown terrestrial buffer zone. Enclosures were monitored daily for metamorphs.

(1) Gamble L.R., McGarigal K., Jenkins C.L. & Timm B.C. (2006) Limitations of regulated 'buffer zones' for the conservation of marbled salamanders. *Wetlands*, 26, 298–306.
(2) Burgin S. & Wotherspoon D. (2009) The potential for golf courses to support restoration of biodiversity for biobanking offsets. *Urban Ecosystems*, 12, 145–155.
(3) Godwin J. & Apodaca J.J. (2012) Comparison of Red Hills salamander (*Phaeognathus hubrichti*) populations between undisturbed and disturbed sites. Alabama Department of Conservation and Natural Resources Report.
(4) Puglis H.J. & Boone M.D. (2012) Effects of terrestrial buffer zones on amphibians on golf courses. *PLoS One*, 7, e39590.

12 Habitat restoration and creation

Habitat destruction is one of the largest threats to amphibian species and populations and habitat protection remains one of the most important and frequently used conservation interventions. However, in many parts of the world, restoring damaged habitats or creating new habitat patches may also be possible.

Habitat restoration or creation is often required by law as a response to mining or other activities that destroy large areas of natural habitats. Activities may include planting vegetation, removing invasive species or creating breeding or shelter habitats for example.

Studies describing the effects of interventions that involve restoration through processes such as fire and water management are discussed in the section 'Threat: Natural system modifications'.

Key messages – terrestrial habitat

Replant vegetation
Four studies, including one replicated study, in Australia, Spain and the USA found that amphibians colonized replanted forest, reseeded grassland and seeded and transplanted upland habitat. Three of four studies, including two replicated studies, in Australia, Canada, Spain and the USA found that areas planted with trees or grass had similar amphibian abundance or community composition to natural sites and one found similar or lower abundance compared to naturally regenerated forest. One found that wetlands within reseeded grasslands were used less than those in natural grasslands. One before-and-after study in Australia found that numbers of frog species increased following restoration that included planting shrubs and trees.

Clear vegetation
Seven studies, including four replicated studies, in Australia, Estonia and the UK found that vegetation clearance, along with other habitat management and in some cases release of amphibians, increased or maintained amphibian populations or increased numbers of frog species. However, great crested newt populations were only maintained for six years, but not in the longer term.

Change mowing regime
One before-and-after study in Australia found that restoration that included reduced mowing increased numbers of frog species.

Create refuges
Two replicated, controlled studies, one of which was randomized, in the USA and Indonesia found that adding coarse woody debris to forest floors had no effect on the number of amphibian species or overall abundance, but had mixed effects on abundance of individual species. One before-and-after study in Australia found that restoration that included reintroducing coarse woody debris to the forest floor increased frog species. Three studies, including two

replicated studies, in New Zealand, the UK and the USA found that artificial refuges were used by amphibians and, along with other interventions, maintained newt populations.

Create artificial hibernacula or aestivation sites
Two replicated studies in the UK found that artificial hibernacula were used by two of three amphibian species and along with other terrestrial habitat management maintained populations of great crested newts.

Restore habitat connectivity
One before-and-after study in Italy found that restoring habitat connectivity by raising a road on a viaduct significantly decreased amphibian deaths.

Create habitat connectivity
We captured no evidence for the effects of creating habitat connectivity on amphibian populations.

Key messages – aquatic habitat

Create ponds
Sixty-five studies investigated the colonization of created ponds by amphibians. Fifty-five of 56 studies, including 3 reviews, in Australia, Canada, China, Europe and the USA found that amphibians used, reproduced or established breeding populations in some or all created ponds. One found that captive-bred frogs did not establish populations. Sixteen of the studies found that created ponds were colonized by up to 15 naturally colonizing species, up to 10 breeding species and some captive-bred amphibians. Five of nine of the studies found that numbers of amphibian species were similar or higher in created compared to natural ponds. Four found that species composition differed and abundance, reproductive success and growth differed depending on species. One found that numbers of species were similar or lower and one found that populations in created ponds were less stable. Fourteen studies in Europe and the USA found that pond creation, along with other interventions, maintained or increased amphibian populations, or in one case increased numbers of species. One systematic review in the UK found that habitat management, which often included pond creation, did not result in self-sustaining great crested newt populations.

Add nutrients to new ponds as larvae food source
We captured no evidence for the effects of adding nutrients such as zooplankton to new ponds on amphibian populations.

Create wetlands
Fifteen studies, including one review and seven replicated studies, in Australia, Kenya and the USA, investigated the effectiveness of creating wetlands for amphibians. Six studies found that created wetlands had similar amphibian abundance, numbers of species or communities as natural wetlands or in one case adjacent forest. Two of those studies found that created wetlands had fewer amphibians, amphibian species and different communities compared to natural wetlands. One global review and two other studies combined created and restored wetlands and found that amphibian abundance and numbers of species were similar or higher compared to natural wetlands. Five of the studies found that up to 15 amphibian species used created wetlands. One study found that captive-bred frogs did not establish in a created wetland.

Restore ponds
Fifteen studies investigated the effectiveness of pond restoration for amphibians. Three studies, including one replicated, controlled, before-and-after study in Denmark, the UK and the USA found that pond restoration did not increase or had mixed effects on population numbers and hatching success. One replicated, before-and-after study in the UK found that restoration increased pond use. One replicated study in Sweden found that only 10% of restored ponds were used for breeding. Three before-and-after studies, including one replicated, controlled study, in Denmark and Italy found that restored and created ponds were colonized by up to seven species. Eight of nine studies, including one systematic review, in Denmark, Estonia, Italy and the UK found that pond restoration, along with other habitat management, maintained or increased populations, increased numbers of amphibian species, pond occupancy or ponds with breeding success. One found that numbers of species did not increase and one found that great crested newt populations did not establish.

Restore wetlands
Seventeen studies, including one review and eleven replicated studies, in Canada, Taiwan and the USA, investigated the effectiveness of wetland restoration for amphibians. Seven of ten studies found that amphibian abundance, numbers of species and species composition were similar in restored and natural wetlands. Two found that abundance or numbers of species were lower and species composition different to natural wetlands. One found mixed results. One global review found that in 89% of cases, restored and created wetlands had similar or higher amphibian abundance or numbers of species to natural wetlands. Seven of nine studies found that wetland restoration increased numbers of amphibian species, with breeding populations establishing in some cases, and maintained or increased abundance of individual species. Three found that amphibian abundance or numbers of species did not increase with restoration. Three of the studies found that restored wetlands were colonized by up to eight amphibian species.

Deepen, de-silt or re-profile ponds
Four studies, including one replicated, controlled study, in France, Denmark and the UK found that pond deepening and enlarging or re-profiling resulted in establishment or increased populations of amphibians. Four before-and-after studies in Denmark and the UK found that pond deepening, along with other interventions, maintained newt or increased toad populations.

Create refuge areas in aquatic habitats
We captured no evidence for the effects of creating refuge areas in aquatic habitats on amphibian populations.

Add woody debris to ponds
We captured no evidence for the effects of adding woody debris to ponds on amphibian populations.

Remove specific aquatic plants
Studies investigating the effects of removing specific aquatic plants are discussed in 'Threat: Invasive alien and other problematic species – Control invasive plants'.

Add specific plants to aquatic habitats
We captured no evidence for the effects of adding specific plants, such as emergent vegetation, to aquatic habitats on amphibian populations.

Remove tree canopy to reduce pond shading
One before-and-after study in the USA found that canopy removal did not increase hatching success of spotted salamanders. One before-and-after study in Denmark found that following pond restoration that included canopy removal, translocated toads established breeding populations.

Terrestrial habitat

12.1   Replant vegetation
- Three studies (including two replicated studies) in Australia, Canada and Spain found that amphibian abundance[4] or community composition was similar to natural sites following tree planting[1], or became more similar with time since grassland reseeding[2]. One before-and-after study in Australia[5] found that numbers of frog species increased following restoration that included planting shrubs and trees.

- One replicated, site comparison study in Canada[4] found that following logging, amphibian abundance was lower or similar in forests that were planted and had herbicide treatment compared to those left to regenerate naturally, depending on species and forest age.

- Four studies (including one replicated study) in Australia, Spain and the USA found that amphibians colonized replanted forest[1], reseeded grassland[2, 6] and seeded and transplanted upland habitat[3]. Three of the studies investigated restoration following mining[1, 2, 6]. One site comparison study in the USA[7] found that wetlands within reseeded grasslands were used more frequently than those within farmland, but less than those in natural grasslands.

> **Background**
> Vegetation can be replanted to replace habitat that has been lost.
> Studies investigating the effect of replanting as one of a combination of interventions during the restoration of wetlands are discussed in 'Restore wetlands'.

A site comparison study in 1978–1984 of restored sites within bauxite mined areas in Western Australia (1) found that six frog species were recorded in replanted sites compared to eight in the surrounding unmined forest. Community composition comparisons indicated high degrees of similarity between some rehabilitated sites and high quality forests. Species use of revegetated sites depended largely on suitable microhabitats being present. Restoration included just planting native eucalypt species or adding topsoil soil, planting with 50% eucalypts and a native understory and fertilizing the area. Amphibians were monitored monthly in a wide range of restored areas and in surrounding unmined forest. More detailed studies were conducted between December 1980 and February 1981 in three rehabilitated areas and four unmined forests. Surveys involved pitfall trapping, live-traps and hand-collecting.

A replicated, before-and-after, site comparison study in 1988–1994 of spoil benches of a lignite mine in northwest Spain (2) found that reseeded benches were colonized by nine amphibian species. Species richness increased steadily with time since seeding. Species composition was most similar to that in control plots in the oldest restored plots (10-years-old). Common midwife toad *Alytes obstetricans* and Perez's Frog *Rana perezi* were the first species to colonize, in the second year. Spoil benches (60 ha) were created, planted with a slurry of pasture mix seeds and mulch and were fertilized in 1984–1994. Subsequent management was minimal. Monitoring was undertaken annually on a single 2 ha plot over the 6 years following seeding and in 1994 on 10 randomly selected 2 ha plots seeded 0–10 years previously. Three randomly selected undisturbed control plots close to the mine were also monitored in 1994. Surveys involved a total of 30 hours of visual searches in February–November.

A before-and-after study in 1999–2003 of retired agricultural land in California, USA (3) found that upland habitat restored by seeding and transplanting native plant species was colonized by western toads *Bufo boreas*. The species was recorded annually from 2000 and was the only amphibian observed. In 1999, native plants were introduced to 20 plots (4 ha) in randomized blocks by either seeding or transplanting, with or without surface contouring. Visual encounter surveys (circular plots and transects) and artificial coverboard surveys (4/plot) were undertaken four times annually.

A replicated, site comparison study in 2001–2002 of boreal forest stands in Ontario, Canada (4) found that amphibian abundance was not higher following planting and herbicide treatment after logging compared to stands left to regenerate naturally. Wood frogs *Rana sylvatica* were significantly less abundant in 20–30-year-old stands that had been managed by planting and herbicide treatment with or without scarification (0.06 captures/trap night) compared to those that had been left to regenerate naturally (0.09). However, capture rates in 32–50-year-old managed stands (0.07) did not differ significantly from naturally regenerated (0.12) and unharvested stands (0.06). For American toads *Bufo americanus*, there was no significant difference in capture rates between treatments or ages of stands (managed: 0.02–0.04; natural regeneration: 0.02–0.03; unharvested: 0.03). Nineteen stands that had received each treatment and five unharvested stands were selected. Drift-fencing with pitfall traps were used for monitoring in August–September 2001–2002.

A before-and-after study in 1997–2004 of a golf course with degraded woodland and grassland in Sydney, Australia (5) found that restoration that included planting increased frog species over two years. Frogs increased from 7 to 10 species in the first year and then remained stable for the following 6 years. A total of 18 species of frogs were predicted in the area and so 56% were present following restoration. The golf course was developed in 1993 and restoration undertaken in 1997–2001. Endemic shrubs and trees were planted, non-native weeds were removed and coarse woody debris was reintroduced onto the woodland floor. The mowing regime was changed to develop grasslands and a narrow band of herb vegetation retained around ponds as a buffer zone. Pond perimeters were walked to record frog calls in 1996–2004.

A before-and-after study in 2009 of a coal spoil prairie in Indiana, USA (6) found that 4 species of salamanders and 9 species of frogs and toads colonized habitat restored by planting, over 27 years. Two species recorded were species of conservation concern. Each of the four study wetlands had different species compositions. As a comparison, another restoration site (in a former prairie area) had one species of salamander and eight species of frogs and toads. Abundances varied from 6–2739 captures/species. Once extraction was completed in 1982, the area was graded to the approximate original contours, topsoil was added (15–38 cm) and the area was re-vegetated. Planting was initially of non-native tall fescue *Festuca arundinacea*, but since 1999 was replaced with native prairie grasses and forbs. Drift-fences with pitfall traps were installed (920 m) around four seasonal or semi-permanent wetlands and were sampled daily in March–August 2009. Visual encounters were also recorded.

A site comparison study in 2005–2006 of four restored wetlands in restored grasslands in the Prairie Pothole Region, USA (7) found that wetlands within restored grasslands were used more frequently by amphibians than those within farmland, but not as much as those within native prairie grasslands. This was true for two frog, one toad and one salamander species. Four wetlands from each category were selected: farmed (drained with ditches), conservation grasslands (wetland hydrology restored, area reseeded with perennial grassland ≤ 10 years previously) and native prairie grasslands (natural). Call surveys, aquatic funnel traps and visual encounter surveys were undertaken biweekly in May–June 2005–2006.

(1)   Nichols O.G. & Bamford M.J. (1985) Reptile and frog utilisation of rehabilitated bauxite minesites and dieback-affected sites in Western Australia's Jarrah *Eucalyptus marginata* forest. *Biological Conservation*, 34, 227–249.
(2)   Galán P. (1997) Colonization of spoil benches of an opencast lignite mine in northwest Spain by amphibians and reptiles. *Biological Conservation*, 79, 187–195.
(3)   Uptain C.E., Garcia K.R., Ritter N.P., Basso G., Newman D.P. & Hurlbert S.H. (2005) Results of a habitat restoration study on retired agricultural lands in the San Joaquin Valley, California. In *Land Retirement Demonstration Project Five Year Report*. US Department of the Interior, Interagency Land Retirement Team, Fresno, California. pp. 107–175.
(4)   Thompson I.D., Baker J.A., Jastrebski C., Dacosta J., Fryxell J. & Corbett D. (2008) Effects of post-harvest silviculture on use of boreal forest stands by amphibians and marten in Ontario. *Forestry Chronicle*, 84, 741–747.
(5)   Burgin S. & Wotherspoon D. (2009) The potential for golf courses to support restoration of biodiversity for biobanking offsets. *Urban Ecosystems*, 12, 145–155.
(6)   Lannoo M.J., Kinney V.C., Heemeyer J.L., Engbrecht N.J., Gallant A.L. & Klaver R.W. (2009) Mine spoil prairies expand critical habitat for endangered and threatened amphibian and reptile species. *Diversity*, 1, 118–132.
(7)   Balas C.J., Euliss Jr. N.H. & Mushet D.M. (2012) Influence of conservation programs on amphibians using seasonal wetlands in the Prairie Pothole region. *Wetlands*, 32, 333–345.

12.2   Clear vegetation

- Six studies (including four replicated studies) in Australia, Estonia and the UK found that vegetation clearance, along with other habitat management and in some cases release of animals, increased numbers of frog species[7], or increased[2, 3, 5, 8], stabilized[4, 5] or maintained[2, 8] populations of natterjack toads. One before-and-after study in the UK[6, 9] found that vegetation clearance, along with other habitat management, maintained a population of great crested newts for the first six years, but not in the longer term.

- One before-and-after study in England[1] found that vegetation clearance, resulted in increased occupancy by natterjack toads.

Background
Vegetation can be removed to prevent natural succession where specific habitat types are desired, or where invasive species are out-competing native species for example.

Studies that used fire or removed vegetation in forest are discussed in 'Threat: Natural system modifications – Use prescribed fire or modifications to burning regime', 'Use herbicides to control mid-storey or ground vegetation' and 'Mechanically remove mid-storey or ground vegetation'.

Studies that investigated the effect of removing aquatic plants are discussed in 'Habitat restoration and creation – Remove specific aquatic plants', 'Threat: Invasive alien and other problematic species – Control invasive plants' and as one of a combiniation of restoration management actions in 'Habitat restoration and creation – Restore wetlands' and 'Restore ponds'.

A before-and-after study in 1972–1991 of heathland in Hampshire, UK (1) found that vegetation clearance resulted in increased occupancy by natterjack toads *Bufo calamita* (see also (3)). At least six toads, including four juveniles, took up residence and established home ranges at the site within a year of vegetation clearance. However, within two years rank vegetation covered 90% of the ground and no toads remained in the area. Scrub was cleared from 40 ha by cutting and uprooting and bracken was treated with herbicide over 12 ha. Toads were monitored once every 10 days in March and August each year.

A replicated, before-and-after study in 1972–1995 of 10 dune and heathland sites in England, UK (2) found that extensive vegetation clearance, along with other terrestrial and aquatic habitat management, increased or maintained natterjack toad *Bufo calamita* populations. Abundance and range increased at 4 of the 10 sites. Success at another four sites was not yet clear, although toads persisted. At two sites, where vegetation clearance was less complete, populations continued to decline. At six sites where scrub removal was needed but not possible, populations remained low or were maintained by artificial methods (e.g. common toad removal). Clearance of invasive scrub and woodland and rotovation to clear patches of ground was undertaken. Low-density sheep or cattle grazing (<1 animal/3 ha) was also established at some sites (all/part year) to control succession. New ponds were created at most sites. Small numbers of ponds were treated with limestone. Translocations were made to some restored habitats. Ponds were monitored by counting spawn strings and estimating toadlet production.

A replicated, before-and-after study in 1972–1999 at 2 sites in England, UK (3) found that vegetation clearance, along with pond creation and restoration and release of captive-reared toadlets increased natterjack toad *Bufo calamita* populations over 20 years. The continuation of a study in 1972–1991 (1) until 1999 indicated that there was a doubling of the population. Spawn string counts (female population) increased from 15 in 1972 to 32 in 1999, with a maximum number of 48 in 1989. At a second site, spawn string counts increased from 1 in 1973 to 8 in 1999, with a maximum number of 29 in 1997. Ponds were created and restored by excavation, scrub and bracken was cleared and captive-reared toadlets raised from spawn and released. Toads were monitored annually.

A before-and-after study in 1994–2004 of a coastal meadow on an islet in Estonia (4) found that vegetation clearance, along with other terrestrial and aquatic habitat restoration, resulted in a stable population of natterjack toads *Bufo calamita*. A total of 17 natterjacks were counted in 1992 and 7 in 2004, with numbers in the range 1–17/year. It is considered by the author that without management the natterjack population might have declined further or become extinct. Common toad *Bufo bufo* counts were 8 in 1992 and 4 in 2004 and ranged from 3–40/year. Restoration involved reed and scrub removal, mowing (cuttings removed) and implementation of sheep grazing. Toads were counted along a 1 km transect.

A replicated, before-and-after study in 2001–2004 of three coastal meadows in Estonia (5) found that vegetation clearance, along with other terrestrial and aquatic habitat restoration increased numbers of natterjack toads *Bufo calamita* on one island and halted the decline on the other two islands. In 2001–2004, habitats were restored on three coastal meadows where the species still occurred. Restoration included reed and scrub removal, mowing (cuttings removed) and implementation of grazing where it had ceased. Sixty-six breeding ponds and natural depressions were cleaned, deepened and restored.

A before-and-after study in 2005 of a mitigation site in England, UK (6) found that vegetation clearance, along with other terrestrial and aquatic habitat management, maintained a great crested newt *Triturus cristatus* population (see also (9)). The population was classified as 'large' (peak count: 167) 6 years after habitat management. Management included tree felling, clearance of both terrestrial and aquatic vegetation and the re-profiling of ponds. Artificial hibernacula and refugia were also created in 1999. Monitoring was undertaken in March–May 2005 using egg searches, torch surveys and bottle trapping.

A before-and-after study in 1997–2004 of a golf course with degraded woodland and grassland in Sydney, Australia (7) found that restoration that included removing non-native weeds resulted in an increase in frog species over two years. Frogs increased from 7 to 10 species in the first year and then remained stable for the following 6 years. A total of 18 species of frogs were predicted in the area and so 56% were present following restoration. The golf course was developed in 1993 and restoration undertaken in 1997–2001. Non-native weeds were removed, endemic shrubs and trees were planted and coarse woody debris was reintroduced to the woodland floor. The mowing regime was changed to develop grasslands and a narrow band of herb vegetation retained around ponds as a buffer zone. Pond perimeters were walked to record frog calls in 1996–2004.

A replicated, site comparison study in 1985–2006 of 20 sites in the UK (8) found that natterjack toad *Bufo calamita* populations increased with species-specific habitat management that included vegetation clearance, in some cases with translocations. In contrast, long-term trends showed population declines at unmanaged sites. Individual types of habitat management (aquatic, terrestrial or common toad *Bufo bufo* management) did not significantly affect trends, but length of management did. Five of the 20 sites showed positive population trends, 5 showed negative trends and 10 trends were not significantly different from zero. Data on populations (egg string counts) and management activities over 11–21 years were obtained from the Natterjack Toad Site Register. Habitat management for toads was undertaken at seven sites. Management varied between sites, but included vegetation clearance, pond creation, adding lime to acidic ponds, maintaining water levels and implementing grazing schemes. Translocations were also undertaken at 7 of the 20 sites using wild-sourced (including head-starting) or captive-bred toads.

A continuation of a study (6) in 2006–2010 of a mitigation site in England, UK (9) found that although vegetation clearance, along with other terrestrial and aquatic habitat management, initially maintained a great crested newt *Triturus cristatus* population, numbers then declined. The number of newts recorded declined from 167 in 2005, 6 years after management, to just 10 in 2010. Management included tree felling, clearance of both terrestrial and aquatic vegetation and the re-profiling of ponds. Artificial hibernacula and refugia were also created in 1999. Monitoring was undertaken in March–May using egg searches, torch surveys and bottle trapping.

(1)   Banks B., Beebee T.J.C. & Denton J.S. (1993) Long-term management of a natterjack toad (*Bufo calamita*) population in southern Britain. *Amphibia-Reptilia*, 14, 155–168.
(2)   Denton J.S., Hitchings S.P., Beebee T.J.C. & Gent A. (1997) A recovery program for the natterjack toad (*Bufo calamita*) in Britain. *Conservation Biology*, 11, 1329–1338.
(3)   Buckley J. & Beebee T.J.C. (2004) Monitoring the conservation status of an endangered amphibian: the natterjack toad *Bufo calamita* in Britain. *Animal Conservation*, 7, 221–228.
(4)   Lepik I. (2004) Coastal meadow management on Kumari Islet, Matsalu Nature Reserve. In R. Rannap, L. Briggs, K. Lotman, I. Lepik & V. Rannap (eds) *Coastal Meadow Management – Best Practice Guidelines*, Ministry of the Environment of the Republic of Estonia, Tallinn. pp. 86–89.
(5)   Rannap R. (2004) Boreal Baltic coastal meadow management for *Bufo calamita*. In R. Rannap, L. Briggs, K. Lotman, I. Lepik & V. Rannap (eds) *Coastal Meadow Management – Best Practice Guidelines*, Ministry of the Environment of the Republic of Estonia, Tallinn. pp. 26–33.

(6) Lewis B., Griffiths R.A. & Barrios Y. (2007) Field assessment of great crested newt *Triturus cristatus* mitigation projects in England. Natural England Report. Research Report NERR001.
(7) Burgin S. & Wotherspoon D. (2009) The potential for golf courses to support restoration of biodiversity for biobanking offsets. *Urban Ecosystems*, 12, 145–155.
(8) McGrath A.L. & Lorenzen K. (2010) Management history and climate as key factors driving natter-jack toad population trends in Britain. *Animal Conservation*, 13, 483–494.
(9) Lewis B. (2012) An evaluation of mitigation actions for great crested newts at development sites. PhD thesis. The Durrell Institute of Conservation and Ecology, University of Kent.

12.3   Change mowing regime

• One before-and-after study in Australia[1] found that restoration that included reduced mowing increased numbers of frog species.

Background

Many amphibians require damp terrestrial habitat once they move out of water. If vegetation surrounding water bodies is mown very short, it will not retain sufficient humidity or provide cover for amphibians during their terrestrial stages. Cutting can also disturb amphibians.

A before-and-after study in 1996–2004 of a golf course with degraded woodland and grass-land in Sydney, Australia (1) found that restoration that included changing the mowing regime resulted in an increase in frog species over two years. Frogs increased from seven to ten species in the first year and then remained stable for the following six years. A total of 18 species of frogs were predicted in the area of which 56% were present following restoration. The golf course was developed in 1993 and restoration undertaken in 1997–2001. The mowing regime was changed to maintain taller areas of rough grass. In addition, during mowing a narrow band of herb vegetation was retained around ponds as a buffer zone for amphibians. Endemic shrubs and trees were planted, non-native weeds were removed and coarse woody debris was reintroduced onto the woodland floor. Pond perimeters were walked to record frog calls in 1996–2004.

(1) Burgin S. & Wotherspoon D. (2009) The potential for golf courses to support restoration of biodiversity for biobanking offsets. *Urban Ecosystems*, 12, 145–155.

12.4   Create refuges

• Two replicated, controlled studies, including one randomized study, in the USA and Indonesia found that adding coarse woody debris to forest floors had no effect on the number of amphibian species or overall abundance[4, 6], but had mixed effects on abundance of individual species[6]. One before-and-after study in Australia[5] found that restoration that included reintroducing coarse woody debris to the forest floor increased frog species.

• One replicated, before-and-after study in the UK[3, 8] found that creating refugia for great crested newts, along with other interventions, maintained four populations.

• Two studies (including one replicated study) in New Zealand and the USA found that artificial refugia were used by translocated Hamilton's frogs[1, 2] and hellbenders, although few were used for breeding[7].

Background

Refuge habitats can provide amphibians with microclimates to keep them at the correct temperature and prevent them from dehydrating and can protect them from predation. Many amphibians seek shelter in rocks, logs or other refuges created by tree falls and other disturbances. Refuges can be created for amphibians where natural shelter habitat is limited, or to replace these habitats where they have been lost.

Other studies investigating the creation of shelter habitat are discussed in 'Create artificial hibernacula or aestivation sites' and 'Threat: Biological resource use – Leave coarse woody debris in forests'.

A before-and-after study in 1990–2000 of 12 endangered Hamilton's frog *Leiopelma hamiltoni* on Stephens Island, New Zealand (1, 2) found that 3 frogs survived in a created refuge within an predator-proof enclosure for at least 8 years. There was no evidence of breeding by 1992 and only one juvenile was ever recorded, in 1996. Eight frogs survived the first year in the rock-filled pit and were recaptured 61 times by 2000. Two of three frogs that were not recorded at the release site after 1994 were found back at their original habitat (76–89 m). After eight years, 42% of translocated frogs had been recaptured compared to 47% marked at the original site. In May 1992, frogs were translocated 40 m to a new rock-filled pit (72 m^2) in a forest remnant. A predator-proof fence was built around the new habitat to exclude tuatara *Sphenodon punctatus* and the area was 'seeded' with invertebrate prey. Frogs were surveyed regularly from November 1990 to May 1992 (90 visits), intermittently in 1992–1996 and at least 4 times annually (over 6 days) in 1997–2000.

A replicated, before-and-after study in 2005 of four mitigation projects in England, UK (3) found that providing refugia and artificial hibernacula for great crested newts *Triturus cristatus* helped maintain populations (see also (8)). Populations persisted at all four sites following development, although numbers were lower than pre-development at two sites. Three populations were classified as 'medium' (peak count: 19–86) and the other as 'large' (167) after 3 or more years. Mitigation projects during development work had been carried out at least three years previously. Artificial hibernacula and refugia were created at sites in 1992–1999. Terrestrial habitat management was also undertaken at the sites and two sites received 37–73 translocated newts. Monitoring was undertaken in March–May 2005 using egg searches, torch surveys, bottle trapping and mark-recapture.

A randomized, replicated, controlled study in 1998–2005 of pine stands in South Carolina, USA (4) found that adding coarse woody debris to forest did not effect amphibian abundance, species richness or diversity. Plots with added downed woody debris did not differ significantly from controls in terms of amphibian abundance (1–2 vs 2), species richness (6–7 vs 7) or diversity (17 vs 19). One species, the southern leopard frog *Rana sphenocephala,* had lower capture rates with the addition compared to removal of woody debris (0.02 vs 0.11/night). Treatments were randomly assigned to 9 ha plots within 3 forest blocks. The first set of treatments was undertaken in 1996–2001 and a second set in 2002–2005. Woody debris was increased five-fold. Control plots had no manipulation of woody debris. Five drift-fence arrays with pitfall traps/plot were used for sampling in 1998–2005.

A before-and-after study in 1997–2004 of a golf course with degraded woodland and grassland in Sydney, Australia (5) found that restoration that included reintroducing coarse woody debris to the woodland floor increased frog species over two years. Frogs increased from 7 to 10 species in the first year and then remained stable for the following 6 years. A total of 18 species of frogs were predicted in the area and so 56% were present following restoration. The golf course was developed in 1993 and restoration was undertaken in 1997–2001. Coarse woody debris was reintroduced onto the woodland floor, endemic shrubs and trees were planted and non-native weeds were removed. The mowing regime was changed to develop grasslands and a narrow band of herb vegetation retained around ponds as a buffer zone. Pond perimeters were walked to record frog calls in 1996–2004.

A replicated, controlled, before-and-after study in 2007–2008 of a cacao plantation in Sulawesi, Indonesia (6) found that adding woody debris and/or leaf litter to plots had no effect on overall amphibian abundance or species richness. However, following addition of woody debris plus leaf litter, *Hylarana celebensis* abundance increased and Asian toad *Duttaphrynus melanostictus* decreased. Forty-two plots (40 x 40 m^2) were divided into four treatments: addition of woody debris (trunks and branch piles), addition of leaf litter, addition of woody debris plus leaf litter and an unmanipulated control. Monitoring was undertaken twice 26 days before and twice 26 days after habitat manipulation. Visual surveys were undertaken along both plot diagonals (transects 113 x 3 m).

A replicated study in 2007–2011 in Missouri, USA (7) found that artificial shelters were used by hellbenders *Cryptobranchus alleganiensis* in the wild and captivity, but breeding was limited. Six hellbenders used five of the seven shelters in the wild in 2010–2011. One clutch of 182 eggs was found being guarded within 1 shelter. In captivity, many shelters were used by hellbenders, but only one clutch of eggs was recorded. Artificial shelters were constructed from chicken wire covered with concrete (chamber: 41 x 37 cm). Six prototype shelters were installed in a riverbed in winter 2007–2008. A couple of these attracted females but no eggs were laid. Following modifications, 7 L-shaped shelters were installed in a river in June 2010 and 20 in a captive enclosure in August 2011. 'Wild' shelters were checked in July and November 2010 and October 2011 and captive shelters were checked weekly.

A continuation of a study (3) in 2006–2010 of four mitigation projects in England, UK (8) found that providing refugia and artificial hibernacula, along with other management for great crested newts *Triturus cristatus* helped to maintain populations. Numbers decreased initially at 2 sites (over 100 to 19; 42 to 31 in 2005), but had increased to 60 at both sites by 2009 and 2010 respectively. Populations decreased between 2005 and 2010 at the other 2 sites (167 to 10; 86 to 40). Artificial hibernacula and refugia were created at sites in 1992–1999. Terrestrial habitat management was also undertaken at the sites and one site received 73 translocated newts. Monitoring was undertaken in March–May using egg searches, torch surveys, bottle trapping and mark-recapture.

(1)   Brown D. (1994) Transfer of Hamilton's frog, *Leiopelma hamiltoni*, to a newly created habitat on Stephens Island, New Zealand. *New Zealand Journal of Zoology*, 21, 425–430.
(2)   Tocher M.D. & Brown D. (2004) *Leiopelma hamiltoni* homing. *Herpetological Review*, 35, 259–261.
(3)   Lewis B., Griffiths R.A. & Barrios Y. (2007) Field assessment of great crested newt *Triturus cristatus* mitigation projects in England. Natural England Report. Research Report NERR001.
(4)   Owens A.K., Moseley K.R., McCay T.S., Castleberry S.B., Kilgo J.C. & Ford W.M. (2008) Amphibian and reptile community response to coarse woody debris manipulations in upland loblolly pine (*Pinus taeda*) forests. *Forest Ecology and Management*, 256, 2078–2083.
(5)   Burgin S. & Wotherspoon D. (2009) The potential for golf courses to support restoration of biodiversity for biobanking offsets. *Urban Ecosystems*, 12, 145–155.
(6)   Wanger T.C., Saro A., Iskandar D.T., Brook B.W., Sodhi N.S., Clough Y. & Tscharntke T. (2009) Conservation value of cacao agroforestry for amphibians and reptiles in South-East Asia: combining correlative models with follow-up field experiments. *Journal of Applied Ecology*, 46, 823–832.
(7)   Briggler J.T. & Ackerson J.R. (2012) Construction and use of artificial shelters to supplement habitat for hellbenders (*Cryptobranchus alleganiensis*). *Herpetological Review*, 43, 412–416.
(8)   Lewis B. (2012) An evaluation of mitigation actions for great crested newts at development sites. PhD thesis. The Durrell Institute of Conservation and Ecology, University of Kent.

12.5   Create artificial hibernacula or aestivation sites

- One replicated, before-and-after study in the UK[1, 4] found that providing artificial hibernacula, along with other terrestrial habitat management, maintained populations of great crested newts.

- One replicated study in the UK[2] found that created hibernacula were used by common frog and smooth newts, but not great crested newts. One replicated study in the UK[3] found four amphibian species close to hibernacula at two of three sites.

Background

Amphibians need damp sheltered places for overwintering or aestivating during hot arid summers. Overwintering or aestivating sites, or 'hibernacula', can be created for amphibians where natural sites are limited or where these habitats have been lost, for example at newly restored sites or in gardens.

Other studies investigating the creation of shelter habitat are discussed in 'Create refuges' and 'Biological resource use – Leave coarse woody debris in forests'.

A replicated, before-and-after study in 2005 of four mitigation projects in England, UK (1) found that providing artificial hibernacula and refugia for great crested newts *Triturus cristatus* helped to maintain populations (see also (4)). Populations persisted at all four sites following development, although numbers were lower than pre-development at two sites. After 3 or more years, 3 of the populations were classified as 'medium' sized (peak count: 19–86) and the other as 'large' (167). Mitigation projects during development work had been carried out at least three years previously. Artificial hibernacula and refugia were created at sites in 1992–1999. Terrestrial habitat management was also undertaken at the sites and 2 sites received 37–73 translocated newts. Monitoring was undertaken in March–May 2005 using egg searches, torch surveys, bottle trapping and mark-recapture.

A replicated study in 2004–2005 of three created hibernacula in parkland in Lancashire, UK (2) found that they were used by common frogs *Rana temporaria* and smooth newts *Triturus vulgaris*. Thirty-one frogs and nine smooth newts were captured leaving the hibernacula. Although great crested newt *Triturus cristatus* were recorded breeding in the adjacent pond, none were found to use the three hibernacula. Six hibernacula were created around a pond in 2002. Drift-fencing with four pitfalls were installed around three of the hibernacula in December 2004. Traps were checked in January–March 2005.

A replicated study in 2007–2008 of ten created hibernacula at three sites in Tyne and Wear, UK (3) found four amphibian species near hibernacula at two of the sites. In autumn six common frogs *Rana temporaria* and nine common toads *Bufo bufo* were found under tiles at two of the sites (with six hibernacula). In spring six great crested newts *Triturus cristatus*, six smooth newts *Triturus vulgaris*, seven common toads and two common frogs were caught in pitfall traps near the two hibernacula at one of those sites. Hibernacula were constructed in 2005–2007 by excavating an area 2 x 1 m and 0.5 m deep. This was filled with rubble and covered with tree cuttings and leaves, a permeable geotextile fabric and then soil and grass turf (total height 1 m). In autumn, amphibians were surveyed using six roofing felt tiles (0.5 x 0.5 m) around each hibernaculum. At one site a combination of drift-fencing and pitfall trapping was used to monitor species in spring.

A continuation of a study (1) in 2005–2010 of four mitigation projects in England, UK (4) found that providing artificial hibernacula and refugia, along with other management for great crested newts *Triturus cristatus* helped to maintain populations. Numbers initially decreased at two sites (2005: over 100 to 19; 42 to 31), but increased to 60 newts at both sites by 2009 and 2010 respectively. Populations decreased from 2005 to 2010 at the other 2 sites (167 to 10; 86 to 40). Artificial hibernacula and refugia were created at sites in 1992–1999. Terrestrial habitat management was also undertaken at the sites and 1 site received 73 translocated newts. Monitoring was undertaken in March–May using egg searches, torch surveys, bottle trapping and mark-recapture.

(1)   Lewis B., Griffiths R.A. & Barrios Y. (2007) Field assessment of great crested newt *Triturus cristatus* mitigation projects in England. Natural England Report. Research Report NERR001.
(2)   Neave D.W. & Moffat C. (2007) Evidence of amphibian occupation of artificial hibernacula. *Herpetological Bulletin*, 99, 20–22.
(3)   Latham D. & Knowles M. (2008) Assessing the use of artificial hibernacula by great crested newts *Triturus cristatus* and other amphibians for habitat enhancement, Northumberland, England. *Conservation Evidence*, 5, 74–79.
(4)   Lewis B. (2012) An evaluation of mitigation actions for great crested newts at development sites. PhD thesis. The Durrell Institute of Conservation and Ecology, University of Kent.

12.6   Restore habitat connectivity

- One before-and-after study in Italy[1] found that restoring connectivity between two wetlands by raising a road on a viaduct, significantly decreased deaths of migrating amphibians.

Background

Habitat destruction and fragmentation are important factors in the decline of amphibian populations. Small patches of habitat support smaller populations and if individuals are unable to move to other suitable areas, populations become isolated. This can make them more vulnerable to extinction. Restoring corridors of native vegetation between patches of suitable habitat may help to maintain amphibian populations.

Studies investigating the effects of restoring habitat connectivity with wildlife tunnels are discussed in 'Threat: Transportation and service corridors – Install culverts or tunnels as road crossings'.

A before-and-after study in 1994–2004 of a brackish and freshwater wetland in southern Tuscany, Italy (1) found that restoring connectivity between wetlands, by raising a road on a viaduct, significantly decreased deaths of migrating amphibians. Post-construction, many species were found migrating between wetlands under the viaduct. No remains of amphibians were found on the road post-construction, compared to thousands during some periods pre-construction. For example, after a night rainstorm in July 1997, over 6,500 newly emerged Italian edible frog *Rana hispanica* juveniles were counted on a 100 m stretch of road. A viaduct 215 m long was constructed in 2003 to raise the road. The supports of the viaduct (1.6 m high) were built on a bank 1 m higher than potential flood waters to prevent mixing of wetlands. Drift-fencing was installed for 300 m from each end of the viaduct along both sides of the road. Amphibian road kills were monitored before and after construction.

(1)   Scoccianti C. (2006) Rehabilitation of habitat connectivity between two important marsh areas divided by a major road with heavy traffic. *Acta Herpetologica*, 1, 77–79.

12.7   Create habitat connectivity

- We found no evidence for the effects of creating habitat connectivity on amphibian populations.

Background
Creating corridors of native vegetation between suitable habitat patches may enable amphibians to move between isolated populations and therefore help to maintain viable populations.

Aquatic habitat

12.8   Create ponds

- Twenty-eight studies investigated the colonization of created ponds by amphibians in general (rather than by targeted species, which are discussed below). All of the studies found that amphibians used all or some of the created ponds.

- Nine site comparison studies (including seven replicated studies) in Australia, Canada, Spain, the UK and the USA[8, 11, 16–18, 21, 24, 26, 30, 31] compared amphibian numbers in created and natural ponds. Five found that numbers of species[18, 24, 30, 31] or breeding species[16] were similar or higher in created ponds, and numbers of ponds colonized were similar[8]. Four found that species composition differed[8, 11], and comparisons between abundance of individual species[16, 21], juvenile productivity[17] and size at metamorphosis differed depending on species[11]. One[26] found that numbers of species were similar or lower depending on the permanence of created water bodies. One[24] found that populations in created ponds were less stable.

- One review and two replicated before-and-after studies in Denmark and the USA found that amphibians established stable populations in 50–100% of created ponds[1–3]. Six replicated studies (including one randomized study) in France, the Netherlands, the UK and the USA found that amphibians used 64–100%[7, 23, 25] and reproduced in 64–68% of created ponds[7, 25], or used 8–100%[5, 28] and reproduced in 2–62% depending on species[9]. One review and 15 studies (including 12 replicated studies, one of which was randomized) in Europe and the USA found that created ponds were used or colonized by up to 15 naturally colonizing species[2, 5–7, 9–13, 20, 21, 28, 29], up to 10 species that reproduced[4, 7, 10, 12, 16, 19, 30], as well as by captive-bred amphibians[6]. Five replicated studies (including three site comparison studies) in Denmark, Estonia, France, Italy and the USA found that pond creation, and restoration in three cases, maintained and increased amphibian populations[6, 13, 17, 30] or increased numbers of species[27].

- Seven studies (including one review) in Austria, Denmark, Poland, the Netherlands and the USA found that use or colonization of[1, 7, 10, 19, 24] and reproductive success in created ponds[19, 22, 29] was affected by pond age, permanence, vegetation cover, surrounding landscape, distance to existing ponds and presence of fish.

Background

Many ponds have been lost as land has been converted for agriculture or development, and with the intensification of agriculture, for example. Creation of additional breeding habitat may help to replace some of that lost and therefore help to maintain and increase amphibian populations. Different pond types can be created and some may be beneficial to certain species but not to others. Also new ponds may promote establishment of non-target species, such as the American bullfrog *Lithobates catesbeianus* or predatory fish, which can compromise the aims of the project.

Evidence from studies that monitored colonization of created ponds by amphibians in general is combined. Studies that targeted specific species have been separated into separate sections. Some of the targeted species may have colonized ponds monitored in the general amphibian section and vice versa. Study 'ponds' and 'pools' have been referred to as 'ponds' within this section.

There is additional literature that shows that 40–75% of garden ponds are used as breeding sites by amphibians (Beebee 1979, 1984; Banks and Laverick 1986). Common species were found to have colonized naturally, rather than having been introduced, in 47–77% of cases (Banks and Laverick 1986). There is also some literature describing the presence of amphibians in urban stormwater ponds and Sustainable Urban Drainage systems designed to intercept and process pollutants and storm water (Bray 2001; O'Brian 2001; Powell *et al.* 2001; Scher and Thiery 2005; Simon *et al.* 2009).

Studies investigating the creation of wetlands are discussed in 'Create wetlands'.

Banks B. & Laverick G. (1986) Garden ponds as amphibian breeding sites in a conurbation in the north east of England (Sunderland, Tyne & Wear). *Herpetological Journal*, 1, 44–50.
Beebee T.J.C. (1979) Habitats of the British amphibians (2): suburban parks and gardens. *Biological Conservation*, 15, 241–257.
Beebee, T. (1984) Successes and failures of amphibians in garden ponds. *The Herpetological Society Bulletin*, 9, 21–24
Bray R. (2001) Environmental monitoring of sustainable drainage at Hopwood Park motorway service area M42 junction 2. *Proceedings of the Proceedings of the 1st national conference SUDS*. Coventry University.
O'Brian D. (2001) *Sustainable Drainage System (SuDS) Ponds in Inverness and the Favourable Conservation Status of Amphibians*. MSc Thesis. University of Bath.
Powell A., Biggs J., Williams P., Whitfield M., Logan P. & Fox G. (2001) Biodiversity benefits from SUDS – results and recommendations. *Proceedings of the Proceedings of the 1st National Conference SUDS*. Coventry University.
Scher O. & Thiery A. (2005) Odonata, amphibia and environmental characteristics in motorway retention ponds. *Hydrobiologica*, 551, 237–251.
Simon J.A., Snodgrass J.W., Casey R.E. & Sparling D.W. (2009) Spatial correlates of amphibian use of constructed wetlands in an urban landscape. *Landscape Ecology*, 24, 361–373.

A small, replicated before-and-after study in 1965–1986 of three created ponds in Missouri, USA (1) found that stable amphibian populations were established in all ponds. Between 10 and 12 amphibian species colonized the ponds, some within 11 days of construction. However, fish invaded 2 of the ponds after 9 and 16 years. Only 2 of 11 amphibian species remained after the invasion of 6 fish species in 1 pond. In the other pond, amphibian species did not appear to be significantly affected during the year after invasion by two fish species. The three ponds were created in 1965–70. Eggs of spotted salamanders *Ambystoma maculatum*, wood frogs *Rana sylvatica* and ringed salamander *Ambystoma annulatum* were translocated to two of the ponds in 1965–1980. Ponds were monitored using drift-fencing with pitfall traps until 1986.

A replicated before-and-after study in 1984–1987 of nine created ponds on a new golf course near Aarhus, Denmark (2) found that breeding populations of common frogs *Rana temporaria* were established in six ponds and common newts *Triturus vulgaris* in four ponds. Common toads *Bufo bufo* were also heard calling in one pond from 1986. Common frogs had colonized 8 of the new ponds by 1987, with 14–117 egg masses found in 6 of the ponds. Common newts were first recorded breeding in four ponds in 1987. After the development of a golf course in 1984, nine ponds were created over four years.

A review in 1991 of amphibian translocation programmes in the USA (3) found that four of five amphibian translocations to created ponds resulted in established breeding populations. In one study in Missouri, breeding populations of spotted salamanders *Ambystoma maculatum* and wood frogs *Rana sylvatica* established from translocated eggs in one created pond and ringed salamanders *Ambystoma annulatum* but not wood frogs established in a second created pond. In a study in New Jersey, a breeding population of tiger salamanders *Ambystoma tigrinum* established at a created pond, with returning adults and 18–25 egg masses recorded within 4 years. In Missouri, eggs of spotted salamanders, wood frogs and ringed salamander were translocated to two created ponds in 1965–1980. Both ponds were monitored until 1986. In New Jersey, 1,000 tiger salamander eggs were translocated 20 km to a created pond (0.2 ha) each year in 1982–1985.

A before-and-after study in 1990–1992 of a man-made pond within woodland on the constructed Danube Island, Austria (4) found that at least six amphibian species reproduced in the pond within the first two years. In the year of completion, three species reproduced successfully: green toad *Bufo viridis*, European tree frog *Hyla arborea* and marsh frog *Rana ridibunda*. A further three species reproduced in the pond over the next two years: European fire-bellied toad *Bombina bombina*, common toad *Bufo bufo* and agile frog *Rana dalmatina*. Two smooth newts *Triturus vulgaris* were also recorded. The 2 ha pond was sealed with a layer of clay and was completed in 1990. The slope, shore and planting scheme were created to attract amphibians and dragonflies and to discourage human disturbance. Amphibians were monitored during 30 visits in March–October by visual surveys and hand-netting.

A replicated before-and-after study in 1977–1996 of 13 created ponds on chalkland in England, UK (5) found that five amphibian species colonized ponds. Six of 13 new ponds were occupied by common frogs *Rana temporaria* (46%) and 3 (23%) by common toads *Bufo bufo*, which constituted 67% and 75% of their total 1996 distributions (total ponds: 26). Great crested newts *Triturus cristatus* colonized only one new pond, smooth newts *Triturus vulgaris* colonized four (only 40% of distribution) and palmate newts *Triturus helveticus* did not colonize created ponds. Thirteen ponds had been created since 1977, within a 150 km² area. Ponds were surveyed in spring 1995 or 1996 for species presence by egg counts, torchlight surveys, netting and newt trapping.

A replicated before-and-after study of projects in 1986–1997 that created and restored 3,446 ponds for amphibians in Denmark (6) found that pond management maintained and increased populations. The national population of European tree frog *Hyla arborea* doubled. A total of 387 (42%) created ponds were naturally colonized by rare species and 38 colonized by captive-bred animals. Alpine newt *Triturus alpestris* was the most efficient colonizer (72% of new ponds). Approximately 2,000 ponds were created or restored for rare species, over half of which were for the European tree frog. A questionnaire was sent to all those responsible for pond projects across Denmark to obtain data. Over half of the projects created new ponds. For a pond to be defined as 'colonized' a species had to be present but not breeding.

A randomized, replicated before-and-after study in 1994 of 133 ponds created for amphibians in the Netherlands (7) found that 80% contained amphibians and 68% breeding amphibians. A total of nine species were recorded and each pond supported up to five reproducing species. Amphibians were found in ponds of all ages (> 50% presence in ponds of 1–7 years); however, presence was higher in older ponds. Amphibian presence was affected by pond characteristics such as surrounding topography, vegetation cover and electrical conductivity of the water. A random, stratified sample of 133 of 1,691 created ponds was taken across a number of provinces. Amphibians (eggs, larvae, juveniles and adults) were sampled in spring and autumn using netting and visual observation. Sixteen pond characteristics were recorded.

A replicated, site comparison study of 78 constructed farm ponds in England, UK (8) found that amphibian colonization of constructed and existing ponds was similar, although species composition differed. Amphibians were found in 65% of constructed and 71% of existing ponds, or 26% and 39% respectively once ponds with frogspawn introductions had been removed (16 new; 3 existing). Numbers of species in each type were also similar (3–4). Common toad *Bufo bufo* was found significantly more frequently (40 vs 22%) and great crested newt *Triturus cristatus* (9 vs 20%) and smooth newt *Triturus vulgaris* (23 vs 39%) less frequently in constructed ponds. Common frogs *Rana temporaria* and toads were found significantly more frequently, smooth newts less and great crested newts were never found with fish. Constructed ponds were significantly larger (1,704 vs 409 m²) and had higher proportions of fish (54 vs 20%) and waterfowl (46 vs 14%) than existing ponds. Egg, torch and dip-netting surveys were undertaken at 78 new and 49 existing ponds over 3,000 km². Habitat data were also collected.

A replicated before-and-after study in 1996–1997 of 37 created ponds in forest, farmland, grassland and residential areas in Latah County, Idaho, USA (9) found that up to 7 species of amphibians were present. Three species were present within 24–33 of the ponds and 4 within 3–4 ponds. The proportion of ponds used for breeding varied with species (Pacific tree frog *Hyla regilla*: 54%; Columbia spotted frog *Rana luteiventris*: 35%; eastern long-toed salamander *Ambystomam acrodactylum columbianum*: 62%; American bullfrog *Rana catesbeiana*: 5%; roughskin newt *Taricha granulosa*: 8%). Western toad *Bufo boreas* and blotched tiger salamander *Ambystoma tigrinum melanostictum* reproduced in a single pond. Ponds (25–860 m²) that had been created by excavation and damming areas of high water runoff were surveyed 12–20 times in March–August. Surveys comprised visual encounter searches of the shore, egg searches, dip-netting and call surveys at four locations around ponds. Four to 8 funnel or minnow traps were also set for a minimum of 14 days in February–April.

A before-and-after study in 1998 of constructed ponds and the restructured shoreline of the constructed Danube Island, Austria (10) found that in the first year, 9 of 12 species found on the island colonized and bred in most of the 9 inshore water bodies (see also (12)). There was a significantly higher number of species and number of successfully breeding species at those inshore sites compared to water bodies connected to the Danube River. Up to eight species bred in one pond. Colonization was more likely in ponds closer to older ponds. All but two of the other water bodies provided summer habitat for some species. The 21 km shoreline, which was straight with steep embankments, was restructured by creating shallow water areas, gravel banks, small permanent backwaters and temporary waters. Thirteen newly-created inshore zones and existing artificial water bodies (created 1989–1997) and one natural water body were monitored for amphibian colonization. Monitoring was undertaken during 20–32 visits (day and night) in February–October 1998 by visual surveys, audio strip transects and hand-netting.

A before-and-after site comparison study in 1979–1991 of three created ponds in a Carolina bay wetland in South Carolina, USA (11) found that the permanent created ponds supported a significantly different amphibian community structure compared to the seasonal wetlands they were replacing. Four to 13 frog and toad and 2 salamander species were recorded in created ponds, with 3 other salamanders seen rarely. Juveniles of 10 frog and toad and 2 salamander species metamorphosized and left the ponds. The original wetland had breeding populations of 7–15 frog and toad and 4–5 salamander species. Few frog and toad colonists had been recorded at the original wetland. Mean size at metamorphosis was significantly smaller for two species of frogs and greater for two salamander species at created ponds compared to a reference site. In 1983, three ponds (200 m², <1 m deep) were created (and lined) on the edge of the original wetland. Amphibians were monitored in the original wetland, created ponds and a reference wetland. Drift-fencing with pitfall traps, minnow traps, dip-netting and seine netting was used.

A continuation of a before-and-after study (10) of 9 created inshore zones and ponds and restructured shoreline of the constructed Danube Island, Austria (12) found that in the first 2 years, 10 of 12 species found on the island colonized and bred in most of the inshore water bodies. Eight species were recorded in the first year and an additional two were found in the second year. Common toad *Bufo bufo* disappeared in the second year. Green toad *Bufo viridis* was found at six sites in 1998 but due to successional processes in 1999 it was only found at three sites. Numbers of species and breeding species remained higher at inshore sites compared to those connected to the Danube River. Monitoring was undertaken during 20–32 visits in 1998 and 16–39 visits in 1999 by visual surveys, audio strip transects and hand-netting.

A replicated before-and-after study in 1998–2000 of created and restored ponds in 12 natural parks in mountain and plains areas in the Lombardy District, Italy (13) found that pond creation and restoration increased amphibian populations and resulted in colonization by new species within 2 years. Populations increased at 6 sites (average: 1.5 species; range 1–4). Between 1 and 7 species colonized ponds at each site (average: 1.7 species). Numbers of egg clumps increased in the second year. Ponds were created or restored and were lined with clay or PVC if necessary. Other habitat management was also undertaken at some sites including increasing dead wood, excavating tributary canals and removing fish.

A review of 15 seasonal pond creation projects in New England, USA (15) found that one of the two projects that specifically monitored amphibian eggs, larvae, juveniles and adults was considered successful, the other was not. Overall, amphibians were monitored in 12 of the projects, with 6 monitoring indicator species such as the wood frog *Rana sylvatica* or spotted salamander *Ambystoma maculatum*. Programmes created ponds to provide amphibian habitat. Eight translocated amphibian eggs or adults to ponds.

A replicated, site comparison study in 1996–2002 of 10 constructed ponds within a wetland restoration area in North Carolina, USA (16) found that created ponds were rapidly colonized by amphibians and contained significantly more breeding species than natural ponds (4 vs 3/pond; see also (17, 22)). Overall, 7 species bred in created ponds within the first year and 10 species in 2002. Species richness reached equilibrium within two years. A total of 10 species bred in natural ponds. One species was only recorded in one natural pond. Six species occupied constructed ponds significantly more frequently than natural ponds (33–99 vs 0–78%). Occupancy of pond types by the other four species did not differ significantly (0–99 vs 9–93%). Ten ponds were created in 1995–1996 (average 480 m²). Restoration in 1996–2002 also included restoring original channels, filling ditches, removing fill and planting native trees. Ten natural ponds were surveyed for comparison. Ponds were surveyed using dip-netting and open-bottomed samplers every 1–3 weeks in January–August each year.

In a continuation of a study in North Carolina, USA (16), a replicated, site comparison study (17) found that breeding populations of wood frogs *Rana sylvatica* and spotted salamanders *Ambystoma maculatum* increased from 1997–1998 following pond construction (see also (22)). Numbers then decreased to pre-construction levels in 2002, due to drought and ranavirus. Wood frogs reproduced within 71% and spotted salamanders within 59% of created ponds in the first year. From 1996–2002 juvenile productivity was significantly higher in created than natural ponds for spotted salamanders (47 vs 24%), but similar for wood frogs (34 vs 26%). Juvenile productivity and survival tended to decrease in both types of ponds over time. Numbers of eggs tended to be higher in ponds located where breeding sites existed prior to construction. Egg mass counts were undertaken every 1–3 weeks during the breeding season in 1996–2002.

A replicated, site comparison study of 22 paired pond sites in 1999–2000 in New South Wales, Australia (18) found that constructed farm dams had similar amphibian species diversity to natural ponds. Both ponds types supported an average of 5 species and an overall total of 11–12 species. Only one species, the striped marsh frog *Limnodynastes peronii*, showed a different trend, occurring in only 6 dams compared to 14 natural ponds. Farm ponds (>10 years old) were paired with natural ponds (1–3 km away) with similar stock access and landscape features. Surveys were undertaken on two nights/site in spring and summer 1999–2000. Pond pairs were surveyed on the same night by call counts (50 m transect). Four observation surveys were also undertaken along transects (5 x 2 m) within different microhabitats at each site.

A replicated, site comparison study in 2000–2001 of 30 created ponds within agricultural landscapes in southeastern Minnesota, USA (19) found that 9 amphibian species reproduced in created ponds. Blue-spotted salamander *Ambystoma laterale* only reproduced in one of the natural ponds. Ponds surrounded by crops had similar species richness and reproductive success as natural ponds surrounded by non-grazed pasture. Ponds used for watering livestock tended to have lower amphibian reproductive success, compared to those with no livestock. Species richness was highest in small ponds without fish. Amphibian reproductive success was highest in ponds with less emergent vegetation and no fish. Thirty created and 10 natural ponds were randomly selected. The 30 created ponds were classified based on adjacent land use: crops, grazed and non-grazed grassland. Other habitat characteristics were recorded. Amphibians were monitored in April–August 2000–2001 by direct observations and larval dip-netting surveys.

A replicated before-and-after study in 1987–2003 of 22 created ponds in a grassland and woodland nature reserve in Limberg, the Netherlands (20) found that the majority of ponds were colonized by two to five amphibian species. Common frog *Rana temporaria* showed a peak in the number of colonized ponds after five years. By 2003, a total of 5,200 egg masses were recorded. Smooth newt *Triturus vulgaris* also colonized rapidly and continued to increase. Common toad *Bufo bufo* and edible frog *Rana klepton esculenta* took longer to colonize and maintained small populations. Calling males of the European tree frog ranged from 3–15 over 11 years. From 1987, 22 ponds (20–66 m²) were created for amphibians in the 2 km² reserve. Ponds were monitored in 1988–2003.

A replicated site-comparison study in 2000–2003 of eight created ponds in restored peatland near Québec, Canada (21) found that within a year three of four species found in natural ponds were breeding in the created ponds. Wood frogs *Rana sylvatica* and green frogs *Rana clamitans melanota* were present in 60–88% of created ponds each year. Numbers were 1–5 times greater than in natural ponds for green frogs (tadpoles: 23 vs 2; frogs: 5 vs 1/100 trap nights) and wood frog tadpoles (127 vs 1). Numbers of wood frog adults to juveniles were similar (1 vs 1). Leopard frogs *Rana pipiens* were not found and American toads *Bufo americanus* only found in created ponds. In 2000, tadpole numbers were lower in the four ponds stocked with plants compared to those left to recolonize naturally; however, numbers were similar in 2001–2002. Amphibians were surveyed using minnow traps set for 1–3 consecutive nights/month in May–August, 2000–2003 (24–192 trap nights/pond/year). Vegetation, invertebrates and pH were also monitored. For comparison 10–12 ponds in each of 3 natural (mined) peatlands were also sampled in 1999 and 2000.

In a continuation of a study in North Carolina, USA (16, 17), a replicated site comparison study (22) found that wood frogs stopped using and spotted salamanders reduced their use of constructed ponds for breeding following the introduction of fish. Egg mass numbers decreased by 97% for wood frogs and 69% for spotted salamander the year after fish introduction. Adults appeared to rapidly recolonize if fish disappeared. Where egg masses were deposited, salamander tadpoles were absent from five of six ponds with fish, compared to just one of nine ponds without fish. Hatchling survival decreased by 96% in ponds with fish relative to fish-free ponds. Fish were introduced five to seven years after construction.

A small replicated study in 1998–2007 of two constructed temporary ponds along a new highway bypass in New Hampshire, USA (23) found that during the first two years, a relatively diverse community of amphibians used the ponds. Spotted salamanders *Ambystoma maculatum* were found in one of the two ponds. In existing ponds, spotted salamander breeding was similar in the six years before and two years after highway construction (14–73 vs 28–77 egg masses/pond). However, the highway had not yet opened for traffic. Created ponds were designed to mimic existing ponds and a 60 m upland buffer was preserved around each. Egg mass counts were undertaken.

A replicated site comparison study in 1996–2006 of 10 constructed ponds within a wetland restoration site in North Carolina, USA (24) found that amphibian species richness in constructed ponds was significantly higher than natural ponds until fish were introduced. There was an average of four species in constructed ponds compared to three in natural ponds in 1996–2002, but in 2003–2006 the number in created ponds had decreased to three. The wood frog *Lithobates sylvaticus* population increased rapidly in created ponds between 1998 (400 egg masses) and 2000 (1,750). It then declined rapidly in 2000–2002 (to 600) and at a slower rate until 2006 (to 200) due to ranavirus, pond drying and fish invasions. Spotted salamander *Ambystoma maculatum* fluctuated less, tending to increase from 1997 (891 egg masses) to 2005 (2,931). Populations in natural ponds were more stable (50–300 egg masses). Despite reproductive failures, success in a few ponds allowed populations to persist at high levels. Ten ponds created in 1995–1996 as part of the wetland restoration were compared to ten natural ponds. Monitoring was undertaken every 1–3 weeks in February–August and less frequently from September–January. Egg mass counts, dip-netting and larval sampling was undertaken and presence of fish and ranavirus recorded.

A replicated before-and-after study in 1992–1994 of 22 constructed ponds within two clear-cut areas of the Monongahela National Forest, West Virginia, USA (25) found that 11 ponds in the first year and 14 in the second were used by breeding amphibians. Of the 14 ponds used, 43% were used by more than one species for breeding. Ponds supporting three species were significantly deeper and tended to have higher nitrate concentrations than those supporting fewer. Species included American toad *Bufo americanus*, wood frog *Rana sylvatica*, mountain chorus frog *Pseudacris brachyphona* and Cope's grey tree frog *Hyla chrysoscelis*. Allegheny mountain dusky salamander *Desmognathus ochrophaeus* and spring salamander *Gyrinophilus porphyriticus* were present but not breeding. Ponds up to 28 m² and 37 cm deep were constructed randomly along an abandoned logging road six months after timber harvest. Monitoring was undertaken monthly in April–September 1993–1994. Dip-netting and funnel traps were used along drift-fences around each pond.

A site comparison study in 1999–2001 of created ponds, lakes and streams on golf courses in Georgia and South Carolina, USA (26) found that numbers of amphibian species in created seasonal water bodies were more similar to natural water bodies than created permanent water bodies. Created seasonal water bodies supported 18 species (at least 4 were breeding), compared to 11 in created permanent water bodies and 24 in natural seasonal water bodies. The number of fish species was 15–16 in created and 10 in natural water bodies. Three amphibian species made up 99% of captures on golf courses with only permanent water bodies and 64% on those that had permanent and seasonal wetlands. Five golf courses from 4 to over 25 years old were selected. Three had permanent and two also had seasonal water bodies. Eleven natural seasonal wetlands were also sampled. Monitoring was over four days/three nights at two monthly intervals using minnow and hoop-net traps, dip-netting and visual surveys. Drift-fencing (50 m) with pitfall traps was installed at seasonal water bodies for one year.

A replicated before-and-after, site comparison study of 450 existing ponds, 208 of which were created and 22 restored in 6 protected areas in Estonia (27) found that amphibian species richness was higher in created and restored ponds than unmanaged ponds within 3 years (3 vs 2 species/pond). There was an increase in proportion of ponds occupied by the declining common spadefoot toad *Pelobates fuscus* (2 to 15%) and great crested newt *Triturus cristatus* (24 to 71%) and by the other five species present (15–58% to 41–82%). Breeding also occured in an increasing number of pond clusters each year for great crested newts (39% to 92%) and spadefoot toads (30% to 81%). In autumn 2005–2007, ponds were created and restored in 27 clusters. Six clusters (46 ponds) were designed for great crested newts, 2 (31 ponds) for spadefoot toads and 19 (153 ponds) for both. Depths, sizes, slopes and shapes varied. Restoration included clearing vegetation, extracting mud, levelling banks, pond drying and ditch blocking (to eliminate fish). Before management, 405 ponds were surveyed. After restoration in 2006–2008, each pond was visited for 10 minutes of visual counts and dip-netting.

A replicated before-and-after study in 2007 of 17 created ponds in a coastal forest in Gironde, France (28) found that 8 of 13 amphibian species known in the area colonized the ponds. A number of new species for the region were also recorded including the common midwife toad *Alytes obstetricans*. Between one and five species colonized each pond, with ponds in the dune or forest fringe having more species that those further inside the forest (≥ 4 vs 2 species). Green frogs *Pelophylax* sp. were found in all 13 ponds that contained water. The other seven species were found in one to eight ponds. Seventeen ponds were created in the 1970s within a 10 km^2 area of forest and dunes. Some dried in summer. Call and visual surveys were undertaken in March 2007.

A review of pond creation projects for amphibians in Poland and Denmark (29) found that targeted species colonized ponds. Following the creation of three permanent and four temporary ponds in 1997 in Bialowieza, conservation species such as the European tree frog *Hyla arborea*, common spadefoot toad *Pelobates fuscus* and great crested newt *Triturus cristatus* successfully colonized the ponds. Temporary ponds were more successful for reproduction. Both fire-bellied toads *Bombina bombina* and European tree frogs colonized and reproduced in temporary, but not permanent ponds created for them (*n* = 10) in Wigry National Park. For details of the pond creation and restoration project in Denmark see (14).

A replicated before-and-after site comparison study in 1999–2003 of eight ponds constructed to replace those lost during highway construction in western France (30) found that five of six species observed in the original ponds colonized created ponds within three years. Successful reproduction was observed for four of those species. Species richness did not differ significantly between the original (3.3 species/pond) and constructed ponds (3.6) by 2003. Diversity scores showed a similar pattern (original: 1.9; 2003: 1.8). Recovery differed between species and ponds. There was a significant increase in population size of agile frog *Rana dalmatina* and European toad *Bufo bufo*, and in the proportion of ponds occupied by them. Common midwife toad *Alytes obstetricans* disappeared from the area in 2001. Ponds were built with similar physical characteristics and within 80–120 m of destroyed original ponds. In January–July 1999–2003, ponds were surveyed up to three times per week and daily during the breeding season. Call and visual transect sampling and dip-netting was undertaken at night.

A small replicated site comparison study in 2006–2008 of two created temporary ponds in Spain (31) found that created ponds had similar or higher amphibian species diversity compared to natural local ponds. The constructed pond in the 'high diversity' area had similar adult but higher larval species richness compared to natural ponds (adults: 9 vs 7–8; larvae: 6–8 vs 4). The constructed pond in the 'low diversity' area had higher species richness than natural ponds (adults: 4 vs 2; larvae: 3–4 vs 2). Numbers of adult natterjack toads *Bufo calamita* entering the created pond was higher in the 'high diversity' area, but the number of post-metamorphic individuals leaving was higher at the 'low diversity' site. Ponds less than 0.5 ha and 1 m deep were created in 2006 on arable land. Amphibians were monitored in March–June using drift-fencing with pitfalls surrounding each pond. Larvae were sampled monthly using dip-netting. Five natural wetlands/ponds within 3 km of each constructed pond were sampled in 2006 using dip-netting and transect surveys at night.

(1)   Sexton J. & Phillips C. (1986) A qualitative study of fish-amphibian interactions in 3 Missouri ponds. Transactions of the Missouri *Academy of Science*, 20, 25–35.
(2)   Skriver P. (1988) A pond restoration project and a tree-frog *Hyla arborea* project in the municipality of Aarhus Denmark. *Memoranda Societatis pro Fauna et Flora Fennica*, 64, 146–147.
(3)   Reinert H.K. (1991) Translocation as a conservation strategy for amphibians and reptiles: some comments, concerns, and observations. *Herpetologica*, 47, 357–363.
(4)   Chovanec A. (1994) Man-made wetlands in urban recreational areas – a habitat for endangered species? *Landscape and Urban Planning*, 29, 43–54.
(5)   Beebee T. (1997) Changes in dewpond numbers and amphibian diversity over 20 years on chalk downland in Sussex, England. *Biological Conservation*, 81, 215–219.
(6)   Fog K. (1997) A survey of the results of pond projects for rare amphibians in Denmark. *Memoranda Societatis pro Fauna et Flora Fennica*, 73, 91–100.
(7)   Stumpel A.H.P. & van der Voet H. (1998) Characterizing the suitability of new ponds for amphibians. *Amphibia-Reptilia*, 19, 125–142.
(8)   Baker J.M.R. & Halliday T.R. (1999) Amphibian colonisation of new ponds in an agricultural landscape. *Herpetological Journal*, 9, 55–64.
(9)   Monello R.J. & Wright R.G. (1999) Amphibian habitat preferences among artificial ponds in the Palouse Region of Northern Idaho. *Journal of Herpetology*, 33, 298–303.
(10)   Chovanec A., Schiemer F., Cabela A., Gressler S., Grotzer C., Pascher K., Raab R., Teufl H. & Wimmer R. (2000) Constructed inshore zones as river corridors through urban areas – the Danube in Vienna: preliminary results. *Regulated Rivers-Research & Management*, 16, 175–187.
(11)   Pechmann J.H.K., Estes R.A., Scott D.E. & Gibbons J.W. (2001) Amphibian colonization and use of ponds created for trial mitigation of wetland loss. *Wetlands*, 21, 93–111.
(12)   Chovanec A., Schiemer F., Waidbacher H. & Spolwind R. (2002) Rehabilitation of a heavily modified river section of the Danube in Vienna (Austria): biological assessment of landscape linkages on different scales. *International Review of Hydrobiology*, 87, 183–195.
(13)   Gentilli A., Scali S., Barbieri F. & Bernini F. (2002) A three-year project for the management and the conservation of amphibians in Northern Italy. *Biota*, 3, 27–33.
(14)   Briggs L. (2003) Recovery of the green toad *Bufo viridis* Laurenti, 1768 on coastal meadows and small islands in Funen County, Denmark. *Deutsche Gesellschaft für Herpetologie und Terrarienkunde*, 14, 274–282.
(15)   Lichko L.E. & Calhoun A.J.K. (2003) An evaluation of vernal pool creation projects in New England: project documentation from 1991–2000. *Environmental Management*, 32, 141–151.
(16)   Petranka J.W., Kennedy C.A. & Murray S.S. (2003) Response of amphibians to restoration of a southern Appalachian wetland: a long-term analysis of community dynamics. *Wetlands*, 23, 1030–1042.

(17)   Petranka J.W., Murray S.S. & Kennedy C.A. (2003) Responses of amphibians to restoration of a southern Appalachian wetland: perturbations confound post-restoration assessment. *Wetlands*, 23, 278–290.

(18)   Hazell D., Hero J., Lindenmayer D. & Cunningham R. (2004) A comparison of constructed and natural habitat for frog conservation in an Australian agricultural landscape. *Biological Conservation*, 119, 61–71.

(19)   Knutson M.G., Richardson W.B., Reineke D.M., Gray B.R., Parmelee J.R. & Weick S.E. (2004) Agricultural ponds support amphibian populations. *Ecological Applications*, 14, 669–684.

(20)   van Buggenum H.J.M. (2004) Sixteen years of monitoring amphibians in new ponds at IJzerenbosch. *Natuurhistorisch Maandblad*, 93, 181–183.

(21)   Mazerolle M.J., Poulin M., Lavoie C., Richefort L., Desrochers A. & Drolet B. (2006) Animal and vegetation patterns in natural and man-made bog pools: implications for restoration. *Freshwater Biology*, 51, 333–350.

(22)   Petranka J.W. & Holbrook C.T. (2006) Wetland restoration for amphibians: should local sites be designed to support metapopulations or patchy populations? *Restoration Ecology*, 14, 404–411.

(23)   Merrow J. (2007) *Effectiveness of Amphibian Mitigation Measures Along a New Highway*. Proceedings of the 2007 International Conference on Ecology and Transportation. Center for Transportation and the Environment, North Carolina State University. pp. 370–376.

(24)   Petranka J.W., Harp E.M., Holbrook C.T. & Hamel J.A. (2007) Long-term persistence of amphibian populations in a restored wetland complex. *Biological Conservation*, 138, 371–380.

(25)   Barry D.S., Pauley T.K. & Maerz J.C. (2008) Amphibian use of man-made pools on clear-cuts in the Allegheny Mountains of West Virginia, USA. *Applied Herpetology*, 5, 121–128.

(26)   Scott D.E., Metts B.S. & Whitfield Gibbons J. (2008) Enhancing amphibian biodiversity on golf courses with seasonal wetlands. In J.C. Mitchell, R.E. Jung Brown & B. Bartholomew (eds) *Urban Herpetology*, SSAR, Salt Lake City. pp. 285–292.

(27)   Rannap R., Lõhmus A. & Briggs L. (2009) Restoring ponds for amphibians: a success story. *Hydrobiologia*, 634, 87–95.

(28)   Berroneau M., Miaud C. & Bernaud J.-P. (2010) Digging ponds on grey dune in Gironde: importance for amphibians and new distribution data. *Bulletin de la Societe Herpetologique de France*, 133, 5–16.

(29)   Briggs L. (2010) Creation of temporary ponds for amphibians in northern and central Europe. Report.

(30)   Lesbarreres D., Fowler M.S., Pagano A. & Lode T. (2010) Recovery of anuran community diversity following habitat replacement. *Journal of Applied Ecology*, 47, 148–156.

(31)   Ruhí A., San Sebastián O., Feo C., Franch M., Gascón S., Richter-Boix À., Boix D. & Llorente G.A. (2012) Man-made Mediterranean temporary ponds as a tool for amphibian conservation. *International Journal of Limnology*, 48, 81–93.

12.8.1 Frogs

- Three of five before-and-after studies (including one replicated study) in Australia, Spain, the UK and the USA[2, 4, 6, 10, 11] found that translocated, head-started, captive-bred and naturally colonizing frogs established breeding populations in created ponds. Two found that breeding populations were established at one of four sites by translocated frogs[4], but were not established by captive-bred frogs[10]. One replicated, before-and-after study in Denmark[3] found that frogs colonized created ponds. One before-and-after study in the Netherlands[12] found that pond creation, along with vegetation clearance, increased a breeding population of European tree frogs.

- An additional three of four replicated, before-and-after studies in Italy, the UK and the USA[1, 5, 7, 9] found that naturally colonizing frog species reproduced in 50–75% of created ponds[1, 5, 7]. Two found that translocated frog species reproduced in only 31% of created ponds[9] or colonized but did not reproduce successfully[5]. One replicated study in the USA[8] found that survival of translocated Oregon spotted frogs increased with increasing age of created ponds.

A replicated before-and-after study in 1983–1993 of eight created ponds in a Country Park on restored farmland in England, UK (1) found that common frogs *Rana temporaria* colonized and reproduced in six of the ponds (see also (7)). By 1992, a total of 195 egg clumps were counted (1–70/pond). Numbers declined to 123 egg clumps in 1993 (0–32/pond), which was considered to be due to drought. Ponds of 4–625 m² were created in 1983–1987. Twenty ponds were also restored in the area increasing the total pond area from 2,248 m² in 1983 to 4,965 m² in 1993. Egg clumps were counted, as an index of numbers of breeding females, in created ponds in February–March.

A before-and-after study in 1986–1993 of 13 created ponds in a marsh reserve in Peterborough, England, UK (2) found that translocation resulted in breeding populations of common frog *Rana temporaria*. Froglets emerged in 1986 and 1987 and the first naturally laid eggs were recorded in 1988 for frogs (peak in 1989: 162 clumps). Up to 16–39% of eggs were lost to desiccation each year. In 1985, 13 ponds were excavated. Local frog eggs were introduced to the ponds in spring 1986 (200 clumps), 1987 (150), 1990 (8), 1991 (4) and 1993 (14). Adults and eggs were monitored 1–3 times/week in spring 1986–1993.

A replicated, before-and-after study in 1991–1994 of nine created ponds on the island of Lolland, Denmark (3) found that European tree frogs *Hyla arborea* colonized three of the ponds by 1994. Those colonized were within 500 m of densely populated ponds. The ponds were dug in 1991–1993. Frogs were monitored by call surveys and dip-netting each year.

A before-and-after study in 1974–1995 of seven created forest ponds in Missouri, USA (4) found that one of four translocations of wood frogs *Rana sylvatica* established breeding populations in five ponds. The successful translocation resulted in a stable population between 1987 (311 captured) and 1995 (364). Wood frogs also colonized four additional created ponds (0.9–2.4 km). In 1980, 11 wood frog egg masses were translocated 50 km into 4 created ponds. Monitoring was undertaken using drift-fencing with pitfall traps around ponds and by egg mass counts and call surveys.

A small, replicated, before-and-after study in 1995–2000 of two created ponds in agricultural land and a reserve in Ohio, USA (5) found that translocated gray tree frogs *Hyla versicolor* did not reproduce in created ponds. Gray tree frogs were heard calling at one pond in 2000, but no evidence of breeding was found. Green frogs *Rana clamitans*, northern leopard frogs *Rana pipiens* and American toads *Bufo americanus* colonized both and bred in one pond. Ponds were created in 1995–1997 and were 2–4 m deep. Water, vegetation, plankton and organic matter (from local wetlands) were added. Larvae (0–35) and metamorphs (0–4) were added in spring 1996–1998 and 2000. Amphibians were monitored by drift-fencing and pitfall traps around ponds and by dip-netting and egg counts.

A replicated, before-and-after study in 1998–2003 of 13 created and one restored pond in Gipuzkoa province, Spain (6) found that translocated adult and released head-started and captive-bred juvenile stripeless tree frogs *Hyla meridionalis* established breeding populations in 11 ponds. Translocated adults survived in good numbers and returned to 12 ponds. Mating, eggs and well-developed larvae were observed in 11 ponds and froglets were recorded in some ponds. However, introduced predators, dense vegetation, eutrophication and drying resulted in reduced survival and reproduction in some ponds. In 1999–2000, 13 ponds were created, 1 restored and vegetation was planted. In 1998–2003, a total of 1,405 adults were translocated to the ponds. Eggs were collected and reared in captivity (outdoor ponds) and released as 871 metamorphs and 19,478 tadpoles into 8 ponds. An additional 5,767 captive-bred tadpoles were released.

A continuation of a previous study (1), in this case combining data from 31 ponds in a grass and woodland park in 1983–2004 (7), found that pond creation and restoration significantly increased reproduction by common frog *Rana temporaria*. Numbers of egg masses increased from 40 in 1983 to 1,852 in 2002, but then declined to 1,000 in 2004. Numbers of egg clumps increased with pond size and 8 ponds contained 89% of the egg masses. The numbers of ponds used for breeding each year increased from 1 in 1983 to 20 in 2000. Breeding tended to occur two years after pond creation or restoration. Eggs, tadpoles and frogs were introduced and removed from ponds by the public, particularly in 1984. Colonization may not therefore have been natural.

A replicated study in 2001–2004 in 4 created ponds within a wetland in Oregon, USA (8) found that survival of translocated Oregon spotted frogs *Rana pretiosa* increased with increasing pond age. Nine ponds were created in 2001–2004 using explosives (0.01–0.07 ha; 2 m deep). In spring 2001, 9 spotted frog egg masses and in June–September 2001, 41 frogs were translocated to the 4 largest ponds from a site 2.5 km away. Frogs were tagged.

A replicated, before-and-after, site comparison study in 1999–2006 of 13 created ponds in woodland, wetlands and agricultural land in Lombardy, Northern Italy (9) found that translocated Italian agile frog *Rana latastei* tadpoles reproduced in 4 of 13 created ponds. At least one egg mass (1–14) and/or more than 1 adult calling male (4–8 in two ponds) were recorded in 4 of 13 created and 2 of 5 existing unmanaged ponds; the difference was not statistically significant. Up to four adults were found in three of the ponds. Human disturbance and predator presence had a negative effect and woodland, shore incline and pond permanence a positive effect on success. Ponds were excavated in six Natural Parks in 1999–2001. In 2000 and 2001, tadpoles were released in 13 created and 5 existing unmanaged ponds, which had not recently been used for breeding. Ponds were monitored by visual, torch and call surveys from February to April 2006 during 45 field surveys (average 2.5/pond).

A before-and-after study in 2004–2006 of three created ponds in wetlands in New South Wales, Australia (10) found that captive-bred green and golden bell frog *Litoria aurea* released as tadpoles did not establish a stable population because of death from chytridiomycosis. Tadpole survival was high following release and some metamorphs survived for up to a year. However, numbers declined over the following 13 months and no frogs were recorded from March 2006. Four of 6 dead frogs found in 2005 and 53% of 60 juveniles captured tested positive for chytridiomycosis. In 2005, 850 tadpoles were released into 3 ponds created in 2002 within a restored wetland. A fence was installed surrounding the ponds and grassland (2,700 m^2) to contain the frogs and to attempt to exclude competing species, predators and the chytrid fungus. Visual encounter surveys were carried out two to four times each month. A sample of frogs were captured and tested for chytrid fungus.

A before-and-after study in 1999–2004 of two created ponds in Arncliffe, near Sydney, Australia (11) found that a stable population of green and golden bell frogs *Litoria aurea* was established from released captive-bred, translocated and colonizing individuals. By January 2000, five non-translocated frogs had colonized the ponds. In March 2000, 8 adults, eggs, metamorphs and 20 juveniles were recorded, along with other species. The following spring, 14 adults, including 10 first year adults, were recorded in the ponds. The population was estimated at over 50 adults by 2004. Two ponds (25 x 20 m) were created as mitigation for development in 1999. Three frogs were translocated 150 m from the development site to the new ponds in early 2000. Fifty tadpoles were released into the ponds in March 2000 and 150 in February 2001. Frogs were monitored at night.

A before-and-after study in 1978–2011 of 10 created ponds within a nature reserve on historic clay pits and farmland in Limburg, the Netherlands (12) found that pond creation, along with vegetation clearance, increased the breeding population of European tree frogs *Hyla arborea*. Numbers of males increased from 50 to 150–400. Numbers increased with increasing pond area. Ponds (100–450 m^2) were created in 1983, 1985 and 1993. Vegetation removal was also undertaken. Calling males were surveyed two to four times in April–May each year.

(1) Williams L.R. & Green M. (1993) Pond restoration and common frog populations at Fryent Country Park, Middlesex, 1983–1993. *London Naturalist*, 72, 15–24.
(2) Cooke A.S. & Oldham R.S. (1995) Establishment of populations of the common frog *Rana temporaria* and common toad *Bufo bufo* in a newly created reserve following translocation. *Herpetological Journal*, 5, 173–180.
(3) Hels T. & Fog K. (1995) Does it help to restore ponds? A case of the tree frog (*Hyla arborea*). *Memoranda Societatis pro Fauna et Flora Fennica*, 71, 93–95.
(4) Sexton O.J., Phillips C.A., Bergman T.J., Wattenberg E.W. & Preston R.E. (1998) Abandon not hope: status of repatriated populations of spotted salamanders and wood frogs at the Tyson Research Center, St.Louis County, Mo 1998. In *Status and Conservation of Midwestern Amphibians*, University of Iowa Press, Iowa City, Iowa. pp. 340–344.
(5) Weyrauch S.L. & Amon J.P. (2002) Relocation of amphibians to created seasonal ponds in southwestern Ohio. *Ecological Restoration*, 20, 31–36.
(6) Rubio X. & Etxezarreta J. (2003) Plan de reintroducción y seguimiento de la ranita meridional (*Hyla meridionalis*) en Mendizorrotz (Gipuzkoa, País Vasco) (1998–2003). *Munibe*, 16, 160–177.
(7) Williams L.R. (2005) Restoration of ponds in a landscape and changes in common frog (*Rana temporaria*) populations, 1983–2005. *Herpetological Bulletin*, 94, 22–29.
(8) Chelgren N.D., Pearl C.A., Adams M.J. & Bowerman J. (2008) Demography and movement in a relocated population of Oregon spotted frogs (*Rana pretiosa*): influence of season and gender. *Copeia*, 2008, 742–751.
(9) Pellitteri-Rosa D., Gentilli A., Sacchi R., Scali S., Pupin F., Razzetti E., Bernini F. & Fasola M. (2008) Factors affecting repatriation success of the endangered Italian agile frog (*Rana latastei*). *Amphibia-Reptilia*, 29, 235–244.
(10) Stockwell M.P., Clulow S., Clulow J. & Mahony M. (2008) The impact of the amphibian chytrid fungus Batrachochytrium dendrobatidis on a green and golden bell frog *Litoria aurea* reintroduction program at the Hunter Wetlands Centre Australia in the Hunter region of NSW. *Australian Zoologist*, 34, 379–386.
(11) White A.W. & Pyke G.H. (2008) Frogs on the hop: translocations of green and golden bell frogs *Litoria aurea* in Greater Sydney. *Australian Zoologist*, 34, 249–260.
(12) van Buggenum H.J.M. & Vergoossen W.G. (2012) Habitat management and global warming positively affect long-term (1987–2011) chorus counts in a population of the European tree frog (*Hyla arborea*). *Herpetological Journal*, 22, 163–171.

12.8.2 Toads

- Four before-and-after studies (including one replicated study) in Germany, the UK and USA[1–4, 7] found that translocated and naturally colonizing toads established breeding populations in created ponds, or in one case 33% of created ponds[7]. Two before-and-after studies (including one replicated study) in Denmark and Switzerland[6, 8] found that common toads and midwife toads naturally colonized 29–100% of created ponds, whereas captive-bred garlic toads did not colonize. One before-and-after study in Denmark[5] found that creating and restoring ponds, along with head-starting, increased populations of European fire-bellied toads.

- One replicated, before-and-after study in Switzerland[8] found that midwife toads reproduced in 16% of created ponds.

Background

As there is a larger literature for natterjack toads *Bufo calamita* and green toads *Bufo viridis* than other species, evidence is considered in separate sections below.

A before-and-after study in 1986–1992 of a pond created to prevent amphibians migrating across a road between Hollenbeck and Ahlerstedt, Germany (1, 2) found that common toads *Bufo bufo* established a breeding population in the pond and the number migrating across the road decreased significantly. Breeding took place every year from 1986. In 1987, 29% of migrating toads chose the created pond. By 1988 the proportion was 75% and by 1992 it was

99%. Marked individuals indicated that 83% of the population used the new pond (91% of males; 67% of females). Population size did not differ significantly before and after resettlement (522 vs 590). Common frogs *Rana temporaria* migrated to and bred in the pond from 1986. The pond (53 x 20 m) was constructed on wet pasture in 1982. A temporary mesh fence around the pond allowed toads to reach but not leave the pond in spring 1986–1990. An amphibian fence was installed along 400 m of the road. Animals captured in pitfall traps along the fence were placed in the created pond. All animals were tagged.

A before-and-after study in 1986–1993 of 13 created ponds in a marsh reserve in Peterborough, England, UK (3) found that translocation resulted in breeding populations of common toad *Bufo bufo*. Toadlets emerged in 1986 and 1987 and the first naturally laid eggs were recorded in 1987. In 1988, 64% of male and 89% of female toads captured were marked, suggesting that most breeding adults were introduced rather than natural colonizers. The proportion dropped to 15% in 1990 suggesting a 64% loss of males in the first year, reducing to 39% in the second and 42% in the third year. The toad population was estimated at 200–300 adults in 1993. Up to 16–39% of eggs were lost to desiccation each year. In 1985, 13 ponds were excavated. Half a million toad eggs were introduced in spring 1986 and 5,911 marked adults in 1987. Adults and eggs were monitored 1–3 times/week in spring 1986–1993.

A before-and-after study in 1986–1994 of a created forest pond in Gifhorn, Germany (4) found that translocated common spadefoot toads *Pelobates fuscus* established a breeding population in the pond. Monitoring indicated that 33% of translocated toads and 31% of naturally colonizing toads reproduced in the created pond. A total of 152 juveniles were recorded in the pond in 1990. Mortality rate of translocated toads was high, with only 19% of toads recaptured in 1993–1994. A pond (700 m²) was created for amphibians in 1988. From 1989, toads were captured using drift-fencing with pitfall traps along the opposite side of the road. Toads were marked and translocated across the road to the pond. Monitoring was undertaken using drift-fencing with pitfall traps either side of the road and around the pond.

A before-and-after study in 1986–1997 of 69 created and restored ponds at six sites in Funen County, Denmark (5) found that creating and restoring ponds, along with headstarting, increased the population of European fire-bellied toads *Bombina bombina*. Numbers increased from 82 in 1986–1988 to 542 in 1995–1997 (from 1–30 to 8–170/site). Numbers of ponds occupied by adults increased from 8 to 62 and by tadpoles from 1 to 18 over the same period. The population declined at only one site that was flooded with salt water. Ponds were restored by dredging or created. Wild-caught toads were paired in separate nest cages in ponds and eggs collected and reared in aquaria. Metamorphs and one-year-olds were released into ponds. Ponds were monitored for calling males and breeding success (capture-recapture estimate) annually in 1987–1997.

A before-and-after study in 1994–1997 of two created ponds in Jutland, Denmark (6) found that after three years, released captive-bred garlic toads *Pelobates fuscus* had not colonized, but common toads *Bufo bufo* and common frogs *Rana temporaria* had colonized naturally. Authors considered that garlic toads may not have colonized due to predation because of the lack of vegetation and introduction of sticklebacks *Pungitius pungitius*. Common toads and common frogs colonized different ponds. Ponds were created in 1994–1995. One thousand captive-bred garlic toad tadpoles were released at different stages before metamorphosis into one of the ponds in 1994. Monitoring was by tadpole and call surveys.

A replicated, before-and-after study in 1997–2004 of six created ponds in pine forest in Oregon, USA (7) found that western toads *Bufo boreas* established stable breeding populations in two of the ponds. Toads bred in all ponds in the first year after construction (within 2–9 months). At two sites large numbers of juveniles were recruited in the first year

(1,000s–10,000s) and breeding continued in future years. However, breeding effort was small in the other four ponds, with less than three clutches and little or no recruitment of juveniles (<100 observed). With the exception of breeding in the second year at one of those ponds, there was no breeding in following years. Colonization events were estimated to be between three to over 20 pairs/pond. Ponds were created in 1997–2002. Five were <500 m² in area and all were 0.1–4.8 km from natural breeding sites. Eggs, larvae and adults were monitored.

A replicated, before-and-after study in 2010 of 38 created ponds in an area of forest and agricultural land in Bernese Emmental, Switzerland (8) found that midwife toads *Alytes obstetricans* colonized 29% of ponds and only reproduced in 54% of those. Tadpoles were only recorded at 6 of the 38 ponds. The number of ponds at a site was positively related to numbers of ponds colonized and toad abundance. Pond age was positively related to colonization and reproduction and the proportion of forest negatively related to reproduction. In 1985–2009, 38 ponds were created for toads over a 2,800 km² site. Call and visual surveys were undertaken three times and dip-net surveys twice at each pond in April–June 2010.

(1) Schlupp M., Kietz R., Podloucky R. & Stolz F.M. (1989) *Pilot Project Bracken: Preliminary Results from the Resettlement of Adult Toads to a Substitute Breeding Site.* Proceedings of the Amphibians and Roads: Toad Tunnel Conference. Rendsburg, Federal Republic of Germany. pp. 127–135.
(2) Schlupp I. & Podloucky R. (1994) Changes in breeding site fidelity: a combined study of conservation and behaviour in the common toad *Bufo bufo. Biological Conservation*, 69, 285–291.
(3) Cooke A.S. & Oldham R.S. (1995) Establishment of populations of the common frog *Rana temporaria* and common toad *Bufo bufo* in a newly created reserve following translocation. *Herpetological Journal*, 5, 173–180.
(4) Baumann K. (1997) The population ecology of the common spadefoot toad (*Pelobates fuscus*) near Leiferde (district Gifhorn, Germany) with special regard to the effect of its artificial relocation into a new breeding-pond. *Braunschweiger Naturkundliche Schriften*, 5, 249–267.
(5) Briggs L. (1997) Recovery of *Bombina bombina* in Funen County, Denmark. *Memoranda Societatis pro Fauna et Flora Fennica*, 73, 101–104.
(6) Jensen B.H. (1997) Relocation of a garlic toad (*Pelobates fuscus*) population. *Memoranda Societatis pro Fauna et Flora Fennica*, 73, 111–113.
(7) Pearl C.A. & Bowerman J. (2006) Observations of rapid colonization of constructed ponds by western toads (*Bufo boreas*) in Oregon, USA. *Western North American Naturalist*, 66, 397–401.
(8) Kroepfli M. (2011) Factors influencing colonization of created habitats by an endangered amphibian species. MSc thesis. University of Bern.

12.8.3 Natterjack toads

- Five before-and-after studies (including three replicated and one controlled study) in the UK and Denmark found that pond creation, along with other interventions, significantly increased natterjack toad populations[2, 4, 5], or in two cases maintained or increased populations at 75% of sites[3, 6].

- One replicated, site comparison study in the UK[1] found that compared to natural ponds, created ponds had lower natterjack toad tadpole mortality from desiccation, but higher mortality from predation by invertebrates.

A replicated, site comparison study in 1982–1984 of created and natural ponds at a dune and heathland site in England, UK (1) found that artificial ponds had lower natterjack toad *Bufo calamita* tadpole mortality from desiccation, but higher mortality due to predation by invertebrates compared to natural ponds. Invertebrate predator numbers tended to increase with pond permanence. Artificial ponds had been designed to be deeper than most natural ponds (62–95 vs 9–45 cm). Six artificial scrapes were made in 1980 at the dune site, which had 100 natural freshwater ponds (23–63 monitored/year). At the heathland site, natterjacks used 2–3 natural ponds and three artificial scrapes created in the 1970s. Natterjacks were monitored in April–July, 1982–1984 by counting egg strings, netting and undertaking a mark-recapture study of tadpoles.

A before-and-after study in 1972–1991 on heathland in Hampshire, England, UK (2) found that pond creation, along with other interventions resulted in a three-fold increase in a natterjack toad *Bufo calamita* population (see also (5)). Egg string counts (female population) increased from 15 to 43, with a maximum number of 48 in 1989. Ponds tended to be used for breeding within a year of construction. Nine small ponds (< 1,000 m^2) were created and four restored by excavation. Scrub, bracken and swamp stonecrop *Crassula helmsii* were controlled. Captive-reared toadlets raised from eggs were released in 1975 (8,800), 1979, 1980 and 1981 (1,000 each). Limestone was added to one naturally acid pond (735 m^2) in April 1983–1989. Toads were monitored annually, once every 10 days in March and August.

A replicated before-and-after study in 1972–1995 at 26 dune, heathland and salt marsh sites in England, UK (3) found that pond creation, along with terrestrial habitat management and translocations, maintained or increased natterjack toad *Bufo calamita* populations at the majority of sites. At 46% of sites, new ponds were considered by the authors to have prevented extinction. At an additional 19% of sites, populations increased following pond creation. There was no effect of new ponds at 27% of sites. At all 26 sites, at least 1 and usually most new ponds were used by toads within 1–2 years of creation. Over 200 ponds were created by excavation over 25 years. Some were lined with concrete. Vegetation clearance was undertaken at 10 sites. Low-density sheep/cattle grazing was established at seven sites and a small number of acidic ponds were treated with limestone. Twenty translocations to restored habitat were also undertaken. Ponds were monitored by counting egg strings and estimating toadlet production.

A controlled, before-and-after study in 1986–2004 of coastal meadows in Funen County, Denmark (4) found that natterjack toad *Bufo calamita* populations increased significantly after pond creation and restoration and reintroduction of grazing. On 10 islands, natterjack toads increased from 3,106 in 1998–1990 to 4,892 adults in 1997. Numbers of ponds with successful breeding remained similar (28 to 34). Numbers declined on four islands with no restoration (270 to 170). From 1986 to 1991, 8 ponds were created and 6 restored for natterjacks on 16 islands. Cattle grazing was reintroduced on six and continued on ten islands. Four populations were monitored annually and others less frequently during 2–3 call and visual surveys and dip-netting.

A replicated, before-and-after study in 1972–1999 at 2 sites in England, UK (5) found that pond creation and restoration, along with other interventions, increased natterjack toad *Bufo calamita* populations over 20 years. The continuation of a study in Hampshire, UK in 1972–1991 (2) until 1999 indicated that there was a doubling of the population. Spawn string counts (female population) increased from 15 in 1972 to 32 in 1999, with a maximum number of 48 in 1989. At a second site, egg string counts increased from 1 in 1973 to 8 in 1999, with a maximum number of 29 in 1997. Ponds were created and restored by excavation, scrub and bracken was cleared and head-started toadlets were released. Toads were monitored annually.

A replicated, before-and-after, site comparison study in 1985–2006 of 20 dune, heathland and salt marsh sites in the UK (6) found that natterjack toad *Bufo calamita* populations tended to increase or be maintained with species specific habitat management including pond creation. In contrast, long-term trends showed population declines at unmanaged sites. Individual types of habitat management (aquatic, terrestrial or common toad *Bufo bufo* management) did not significantly affect trends, but duration of management did. Overall, 5 of the 20 sites showed positive population trends, 5 showed negative trends and 10 trends were not significantly different from zero. Data on populations (egg string counts) and management activities over 11–21 years were obtained from the Natterjack Toad Site Register. Habitat management for toads was undertaken at seven sites. Management varied between sites, but included pond creation, adding lime to acidic ponds, maintaining water levels, vegetation clearance and implementing grazing. Translocations were also undertaken at 7 of the 20 sites using wild-sourced (including head-starting) or captive-bred toads.

(1) Banks B. & Beebee T.J.C. (1988) Reproductive success of natterjack toads *Bufo calamita* in two contrasting habitats. *Journal of Animal Ecology*, 57, 475–492.
(2) Banks B., Beebee T.J.C. & Denton J.S. (1993) Long-term management of a natterjack toad (*Bufo calamita*) population in southern Britain. *Amphibia-Reptilia*, 14, 155–168.
(3) Denton J.S., Hitchings S.P., Beebee T.J.C. & Gent A. (1997) A recovery program for the natterjack toad (*Bufo calamita*) in Britain. *Conservation Biology*, 11, 1329–1338.
(4) Briggs L. (2004) Restoration of breeding sites for threatened toads on coastal meadows. In R. Rannap, L. Briggs, K. Lotman, I. Lepik & V. Rannap (eds) *Coastal Meadow Management – Best Practice Guidelines*, Ministry of the Environment of the Republic of Estonia, Tallinn. pp. 34–43.
(5) Buckley J. & Beebee T.J.C. (2004) Monitoring the conservation status of an endangered amphibian: the natterjack toad *Bufo calamita* in Britain. *Animal Conservation*, 7, 221–228.
(6) McGrath A.L. & Lorenzen K. (2010) Management history and climate as key factors driving natterjack toad population trends in Britain. *Animal Conservation*, 13, 483–494.

12.8.4 Green toads

- Two before-and-after studies (including one controlled study) in Denmark[2, 3] found that pond creation, along with other interventions, significantly increased green toad populations.

- One replicated, before-and-after study in Sweden[1] found that green toads used 59% and reproduced in 41% of created ponds.

A replicated, before-and-after study in 1986–1993 of 29 created ponds on the island of Samsø, Sweden (1) found that green toads *Bufo viridis* used 17 ponds and bred in 12. Breeding was successful in 10 of the 12 ponds. Toads colonized the ponds over three years. The ponds were created in 1989–1992. Private owners were offered payment by the county to build ponds, provided fish, crayfish and ducks were not introduced and a 10 m pesticide-free zone was maintained around each pond.

A before-and-after study in 1989–1997 of 23 created and 25 restored ponds within coastal meadows on nine islands in Funen County, Denmark (2) found that pond creation and restoration, along with terrestrial habitat management, significantly increased a green toad *Bufo viridis* population. Overall, the population on the islands increased from 1,112 to 3,520 toads over the 7 years. Numbers were similar on islands with just pond creation and restoration (1,020 to 952) and increased on the two where cattle grazing was also reintroduced (92 to 2,568). Overall, pond occupancy increased from 23 to 51 and the number of ponds with successful breeding increased from 9 to 15. In 1989–1997, ponds were created or restored by removing plants and dredging. Cattle grazing was reintroduced to 73 ha of coastal meadows and abandoned fields on 2 islands. Populations were monitored annually in 1990–1997 during 2–3 call and visual surveys and dip-netting surveys. One population was also monitored in 1987–1989.

A controlled, before-and-after study in 1986–2004 of coastal meadows in Funen County, Denmark (3) found that pond creation, along with other interventions, significantly increased populations of green toads *Bufo viridis*. On 10 islands, green toads increased from 1,132 in 1988–1990 to over 10,000 adults in 2004. Numbers remained similar on four islands with no management (512 to 510). Pond occupancy increased from 27 in 1988 to 61 in 1997 and ponds with successful breeding from 11 to 22. From 1986–1991, 23 ponds were created and 25 restored (reed removal) for green toads on 16 islands. Cattle grazing was reintroduced on six and continued on ten islands. Green toad eggs were translocated to one island. Four populations were monitored annually and others less frequently during 2–3 call and visual surveys and dip-netting.

(1) Amtkjær J. (1995) Increasing populations of the green toad *Bufo viridis* due to a pond project on the island of Samsø. *Memoranda Societatis pro Fauna et Flora Fennica*, 71, 77–81.
(2) Briggs L. (2003) Recovery of the green toad *Bufo viridis* Laurenti, 1768 on coastal meadows and small islands in Funen County, Denmark. *Deutsche Gesellschaft für Herpetologie und Terrarienkunde*, 14, 274–282.
(3) Briggs L. (2004) Restoration of breeding sites for threatened toads on coastal meadows. In R. Rannap, L. Briggs, K. Lotman, I. Lepik & V. Rannap (eds) *Coastal Meadow Management – Best Practice Guidelines, Ministry of the Environment of the Republic of Estonia, Tallinn*. pp. 34–43.

12.8.5 Salamanders (including newts)

- Three before-and-after studies (including two replicated studies) in France, Germany and the USA found that naturally colonizing alpine newts[1], captive-bred smooth newts[5] and translocated spotted salamanders[2] established stable breeding populations in 20–100% of created ponds.

- Two replicated, before-and-after study in France and China found that alpine newts[1] and Chinhai salamanders[3] reproduced in 60–100% of created ponds. One small, replicated, before-and-after study in the USA[4] found that translocated spotted salamanders but not tiger salamanders reproduced in created ponds.

Background
As there is a larger literature for great crested newts *Triturus cristatus* than other species, evidence is considered in a separate section below.

A replicated, before-and-after study in 1992–1995 of five created ponds in meadows near Lyon, France (1) found that alpine newts *Triturus alpestris* established a stable breeding population in one of five ponds over the first three years. Breeding occurred in 3 ponds in 1–3 years (3–38 larvae/pond). All ponds were used by newts in the first year, although 4 ponds only had 2–7 animals, the fifth had 40 newts. First year colonizers were biased towards males (38 vs 15) and tended to be 1–2 years old. By the third year one pond was used by 176 newts, 2 by 3–6 and 2 by zero newts. Colonization failed in the two ponds that were colonized by fish, although two of three fish species disappeared within a year. Five ponds were excavated in September 1992. Each was 12 x 5 m and 1.5 m deep with sloping banks. Newts were sampled by netting once a month in March–June 1993–1995. Animals were aged and tagged.

A before-and-after study in 1974–1995 of seven created forest ponds in Missouri, USA (2) found that translocated spotted salamanders *Ambystoma maculatum* established breeding populations in five ponds. Numbers of salamander captures increased from 428 in 1974 to 2,301 in 1995 at the release pond. Salamanders also colonized four additional created ponds (0.9–2.4 km). In 1966, spotted salamander egg masses were translocated 1 km to a newly constructed pond. Another six ponds were constructed at the site in 1965–1979. Monitoring was undertaken using drift-fencing with pitfall traps around ponds and by egg mass counts.

A small, replicated, before-and-after study in 1997–2001 of two ponds created within the range of one of three known populations of the Chinhai salamander *Echinotriton chinhaiensis* in Zhejiang, China (3) found the species breeding in the ponds within two years. By 2001, females and five clutches of eggs were found. By that time, pond banks were 75% covered by vegetation and shrubs were developing. Numbers of female salamanders counted in the area were variable, but similar before (1997: 50; 1998: 88) and after pond construction (1999: 89; 2000: 82; 2001: 58 in 2001). Two or three males were found in 1997–1999, one in 2000 and none in 2001. Two species of frog (*Hylarana latouchii* and *Microhyla mixture*) colonized in the year of construction. In June 1999, ponds were dug 50 m from two existing breeding habitats, within a similar environment. Ponds were 3 x 2 m and 0.4 m deep. Amphibians were monitored from 1997–2001.

A small, replicated, before-and-after study in 1995–2000 of two created ponds in agricultural land and a reserve in Ohio, USA (4) found that translocated spotted salamanders *Ambystoma maculatum*, but not tiger salamanders *Ambystoma tigrinum* reproduced in created ponds. Four adult spotted salamanders and one egg mass were found in one pond in 1997 and three egg masses in the other pond in 2000. Both ponds produced metamorphs in 1996–1998. Tiger salamanders were not recorded following translocation. Ponds were created in 1995–1997 and were 2–4 m deep. Water, vegetation, plankton and organic matter (from local wetlands) were added. Spotted salamander eggs (600–1100), larvae (40–850) and meta-morphs (4–33) and tiger salamander metamorphs (0–25) were added in spring 1996–1998 and 2000. Amphibians were monitored using drift-fencing and pitfall traps around ponds and by dip-netting and egg counts.

A replicated, before-and-after study in 1994–2004 of 14 created ponds in wet meadows in the Luhe valley, Germany (5) found that captive-bred smooth newts *Triturus vulgaris* established stable breeding populations in nine ponds. Fourteen ponds and many small ponds of different designs were created. Some aquatic plants were introduced. Management also included fish removal, hanging wildfowl deterrents, mowing, scrub clearance and creation of hibernacula. From 1994, 90 smooth newts were released into 2 created ponds annually. In 2000–2004, 5–10 adults were also released into the 2 ponds.

(1)   Joly P. & Grolet O. (1996) Colonization dynamics of new ponds, and the age structure of colonizing Alpine newts, *Triturus alpestris. Acta Oecologica-International Journal of Ecology*, 17, 599–608.
(2)   Sexton O.J., Phillips C.A., Bergman T.J., Wattenberg E.W. & Preston R.E. (1998) Abandon not hope: status of repatriated populations of spotted salamanders and wood frogs at the Tyson Research Center, St.Louis County, Mo 1998. In *Status and Conservation of Midwestern Amphibians*, University of Iowa Press, Iowa City, Iowa. pp. 340–344.
(3)   Sparreboom M., Feng X. & Liang F. (2001) Endangered chinhai salamander colonising newly created breeding habitat. *Froglog*, 47, 1–2.
(4)   Weyrauch S.L. & Amon J.P. (2002) Relocation of amphibians to created seasonal ponds in southwestern Ohio. *Ecological Restoration*, 20, 31–36.
(5)   Kinne O. (2004) Successful re-introduction of the newts *Triturus cristatus* and *T. vulgaris. Endangered Species Research*, 1, 25–40.

12.8.6 Great crested newts

- Three before-and-after studies (including two replicated studies) in Germany and the UK found that naturally colonizing[6], captive-bred[2] and translocated[3] great crested newts established breeding populations at 57–75% of created ponds or sites. One systematic review in the UK[6] found that there was no conclusive evidence that mitigation, which often included pond creation, resulted in self-sustaining populations.

- Three replicated, before-and-after studies in the UK found that up to 88% of created ponds were colonized by translocated[3] or by small numbers of naturally colonizing[4, 5] great crested newts. One replicated before-and-after study in the UK[1] found that head-started great crested newts reproduced in 38% of created ponds.

A replicated before-and-after study in 1991–1993 of eight created ponds on restored opencast mining land in England, UK (1) found that head-started great crested newts *Triturus cristatus* returned as adults to five ponds and reproduced in three in the second year. Adults returned to at least five of eight ponds and tadpoles were caught in three of five ponds netted in 1993 (2–5 tadpoles/pond). Sixteen ponds (30 x 20 m) with shelved edges and terrestrial habitat were created on restored opencast land. Ponds were planted with submerged and edge plants. Terrestrial habitat created included scrub, woodland, rough grassland, ditches and hedgerows. Newt eggs were collected and reared to larvae in aquaria. In 1991, 630 larvae were released into 4 ponds and in 1992, 1,366 larvae into 8 ponds (66–243/pond). Ponds were surveyed using dip-netting in July 1993.

A replicated, before-and-after study in 1994–2004 of 14 created ponds in wet meadows in the Luhe valley, Germany (2) found that captive-bred great crested newts *Triturus cristatus* established stable breeding populations in 9 ponds. Fourteen ponds and many small pools of different designs were created. Some aquatic plants were introduced. Management also included fish removal, hanging wildfowl deterrents, mowing, scrub clearance and creation of hibernacula. From 1994, 60 captive-bred great crested newts were released into 2 created ponds annually. In 2000–2004, 5–10 adults were also released into the 2 ponds.

A replicated, before-and-after study in 2005 of seven mitigation projects in England, UK (3) found that translocated great crested newt *Triturus cristatus* established populations at four sites with created ponds. The four populations were classified as 'medium' sized (peak count: 16–86) after 3 or more years. Very low numbers were captured at the other three sites (peak count: 1–2). Newts used 9 of 13 created ponds. Mitigation projects during development work had been carried out at least three years previously. Between one and three ponds were created and 2–164 newts translocated to each site in 1992–2000. Terrestrial habitat management was also undertaken at two sites. Monitoring was undertaken in March–May 2005 using egg searches, torch surveys, bottle trapping and mark-recapture.

A replicated, before-and-after study in 1999–2006 of eight created ponds at a restored steelwork site in North Lanarkshire, Scotland, UK (4) found that small numbers of great crested newts *Triturus cistatus* colonized seven of the ponds within three years. Within one year, three of eight ponds were colonized by breeding newts. Up to six newts used each pond annually in 2000–2003. In 2006, the habitat suitability for newts for five of the created ponds was categorized as 'average' to 'good' (Habitat Suitability Index: 0.6–0.7). Two ponds were dry. There was no significant difference between the habitat suitability of created and existing ponds. Eight ponds were constructed in 1999. Newts were monitored by torchlight sampling, egg counts and metamorph counts at the perimeter fence. Created ponds were compared to seven existing ponds.

A replicated, before-and-after study in 2008–2010 of 13 created ponds in a nature reserve with many existing ponds in England, UK (5) found that some created ponds were colonized by small numbers of great crested newts *Triturus cristatus*. One pond had 6 and another 18 newts in one year. However, the majority of ponds that contained newts had only one or two animals. In winter 2008–2009, 13 new ponds were created. Torchlight surveys were undertaken in March–June 2009–2010.

A before-and-after study in 1998–2011 of eight created ponds in unimproved grassland in Kent, UK (6) found that great crested newts *Triturus cristatus* established a population in the ponds. The population increased by 30% within the first year following construction. The population was 10–14 newts in 2000–2006, increased to 32 in 2008 following draining and relining of ponds and then to 40 following construction of 4 additional ponds. Larvae were recorded in six ponds. There was no significant preference for older or newer ponds (117 vs 134 captures), apart from the first-time breeders that tended to colonize new ponds more than old ponds. Four experimental ponds (2 x 1 m; maximum depth 0.7 m) were created in a row in 1998 and four in 2009. Populations were sampled weekly in March–May using bottle trapping, torch surveys and mark-recapture.

A systematic review in 2011 of the effectiveness of mitigation actions for great crested newts *Triturus cristatus* in the UK (6) found that neither the 11 studies or captured monitoring data from licensed mitigation projects showed conclusive evidence that mitigation, which often included pond creation, resulted in self-sustaining populations or connectivity to populations in the wider countryside. Only 22 of 460 licensed projects provided post-development monitoring data and of those, 16 reported that small populations, 3 medium and 1 large population was sustained. Two reported a loss of the population. A total of 127 (41%) of English and 46 (30%) of Welsh licence files contained licence return (reporting) documents. Of those, only 9% provided post-development monitoring data, a further 7% suggested surveys were undertaken, but no data were provided. The review identified 11 published or unpublished studies and 309 Natural England and 151 Welsh Assembly Government (licensing authorities) mitigation licence files. Mitigation measures were undertaken to reduce the impact of the development and included habitat management such as creating or restoring ponds, as well as actions to reduce deaths including translocations.

(1)   Bray R. (1994) *Case Study: A Programme of Habitat Creation and Great Crested Newt Introduction to Restored Opencast Land for British Coal Opencast.* Proceedings of the Conservation and Management of Great Crested Newts. Kew Gardens, Richmond, Surrey. pp. 113–125.
(2)   Kinne O. (2004) Successful re-introduction of the newts *Triturus cristatus* and *T. vulgaris*. Endangered Species Research. *Endangered Species Research*, 1, 25–40.
(3)   Lewis B., Griffiths R.A. & Barrios Y. (2007) Field assessment of great crested newt *Triturus cristatus* mitigation projects in England. Natural England Report. Research Report NERR001.
(4)   McNeill D.C. (2010) Translocation of a population of great crested newts (*Triturus cristatus*): a Scottish case study. PhD thesis. University of Glasgow.
(5)   Furnborough P., Kirby P., Lambert S., Pankhurst T., Parker P. & Piec D. (2011) The effectiveness and cost efficiency of different pond restoration techniques for bearded stonewort and other aquatic taxa. Report on the Second Life for Ponds project at Hampton Nature Reserve in Peterborough, Cambridgeshire. Froglife Report.
(6)   Lewis B. (2012) An evaluation of mitigation actions for great crested newts at development sites. PhD thesis. The Durrell Institute of Conservation and Ecology, University of Kent.

12.9   Add nutrients to new ponds as larvae food source

- We found no evidence for the effects of adding nutrients, such as zooplankton, to new ponds on amphibian populations.

Background

Providing supplementary food may help to increase survival rates of colonizing amphibians and therefore increase the probability of breeding populations establishing. However, providing supplementary food could also increase predator populations and therefore increase predation pressure.

12.10 Create wetlands

- Fifteen studies investigated the effectiveness of creating wetlands for amphibians.

- Five site comparison studies (including four replicated studies) in the USA compared created to natural wetlands and found that created wetlands had similar numbers of amphibian species[1, 4, 13, 14, 16], amphibian abundance[11] or communities depending on depth[14] as natural wetlands. Two of the studies found that created wetlands had fewer amphibian species[11], lower abundance and different communities[13, 16] compared to natural wetlands. One site comparison study in the USA[5] found that created wetlands had similar numbers of species to adjacent forest. One global review and two site comparison studies (including one replicated study) in the USA combined created and restored wetlands and compared them to natural wetlands and found that numbers of amphibian species and abundance was higher[6] or similar[8], or higher in 54% of studies and similar in 35% of studies reviewed[15] compared to natural wetlands. Three site comparison studies (including one replicated study) in the USA[1, 8, 14] found that certain amphibian species were only found in created or natural wetlands.

- One before-and-after study in Australia[10] found that captive-bred green and golden bell frog tadpoles released into a created wetland did not establish a self-sustaining population.

- Five studies (including two replicated studies) in Kenya and the USA that investigated colonization of created wetlands found that four to 15 amphibian species used[2, 3] or colonized[7, 9, 12] the wetlands. One global review and three studies (including two replicated studies) in the USA found that numbers of amphibian species[7, 8, 15] and amphibian abundance[13, 15, 16] in created wetlands were affected by wetland design, vegetation, water levels, surrounding habitat, fish presence and distance to source wetlands.

Background

Loss and degradation of wetlands are two major factors contributing to the global decline of amphibians. Many wetlands have been drained and altered to allow for agriculture and urban development. It has been estimated that the number of wetlands had declined by 33–90% depending on the region of the world (Mitsch & Gosselink 2007).

Creation of wetlands may help to replace some of the habitat lost and therefore help to maintain and increase amphibian populations. However, wetland dynamics are complex and the success of created wetlands depends on many factors such as vegetation, geomorphology and hydrology.

There is additional literature describing the presence of amphibians in agricultural and mine-drainage treatment wetlands designed to intercept and process pollutants (Lacki *et al.* 1992; Smiley *et al.* 2011). Use by wildlife is not typically considered in the design of treatment wetlands and wildlife may be restricted using exclusion barriers, trapping or other habitat modifications. Therefore, these studies are not included here.

Study 'ponds' and 'pools' have been referred to as 'ponds' within this section. Studies investigating the creation of individual ponds are discussed in 'Create ponds'.

Lacki M.J., Hummer J.W. & Webster H.J. (1992) Mine-drainage treatment wetland as habitat for herpetofaunal wildlife. *Environmental Management*, 16, 513–520.
Mitsch W.J. & Gosselink J.G. (2007) *Wetlands*. Wiley, New York.
Smiley P.C. & Allred B.J. (2011) Differences in aquatic communities between wetlands created by an agricultural water recycling system. *Wetlands Ecology and Management*, 19, 495–505.

A site comparison study in 1994–1996 of a created forested wetland in a Research Refuge in Maryland, USA (1) found that the created wetland supported a similar number of amphibian species to an adjacent natural forested wetland. Ten species were captured in the created wetland (284 individuals) and 11 in the adjacent natural wetland (87 individuals). Spotted salamander *Ambystoma maculatum* was only found in the created site and wood frog *Rana sylvatica* and marbled salamander *Ambystoma opacum* only in the natural wetland. As mitigation for loss of wetland, a 9 ha wetland was constructed in 1994, of which 5.5 ha was forested wetland. Amphibians were captured in pitfall and flannel traps along drift-fencing within the created and adjacent natural forested wetland. Trapping was conducted several times during the year.

A before-and-after study in 1996–1997 of a created wetland in Nairobi, Kenya (2) found that eight species of amphibians used the wetland. Seven species of frog and common toads *Bufo bufo* were recorded in the wetland. In 1996, a 0.5 ha wetland was constructed using a combination of a sub-surface horizontal flow system planted with *Typha*, followed by a series of three pond systems planted with a variety of species including local reeds and ornamental plants. Ponds were shallow near the shore with deep sections in the centre (1.5 m).

A before-and-after study in 1992–1994 of a constructed wetland (32 ha) in Florida, USA (3) found that nine amphibian species used the wetland within the first two years. Seven species were already present as construction was completed in July 1992. Species richness continued to increase throughout the study. Wildlife was monitored quarterly from July 1992 to August 1994. Counts were undertaken on transect and perimeter walks and call counts were undertaken at night.

A replicated, site comparison study in 1999–2000 of nine wetlands in South Dakota, USA (4) found that amphibian and reptile species richness did not differ significantly between created and natural wetlands. A total of 11 amphibian and reptile species were recorded in the wetlands. Four wetlands had been created during the previous 10 years by excavation or enclosing small streams. Five natural wetlands were used for comparison. Monitoring was undertaken using drift-fences and pitfall traps and visual surveys around wetland perimeters in spring and autumn in 1999–2000.

A site comparison study in 1995–1996 of a created wetland in Maryland, USA (5) found that all but one amphibian species present in an adjacent forest were recorded in the created wetland. Spotted salamander *Ambystoma maculatum* was the only species not recorded in the wetland. Nine of the species were recorded in all four wetland terraces created. The 52 ha wetland was constructed in 4 terraces and was surrounded by regenerating forest. Monitoring was undertaken in March–September 1995–1996 using transects, call counts, drift-fencing with pitfall and funnel traps and dip-netting. The adjacent forest was used as a reference site.

A replicated, site comparison study of 11 mitigation wetlands in West Virginia, USA (6) found that amphibian species richness and abundance was significantly higher in created and partially restored wetlands than natural wetlands. Mitigation wetlands had 2.0 species/ point compared to 1.5 in natural wetlands and 4.8 amphibians compared to 4.7 per wetland. Seven species were recorded in both wetland types. Abundance of American bullfrog *Rana catesbeiana*, northern green frog *Rana clamitans* and pickerel frog *Rana palusris* were higher in mitigation than natural wetlands (0.2–7.8 vs 0.1–3.6/wetland). Abundance of northern spring peeper *Pseudacris crucifer*, gray tree frog *Hyla chrysoscelis*, wood frog *Rana sylvatica* and eastern American toad *Bufo americanus* were similar between wetland types (mitigation: 0.4–22.9; natural: 0.1–28.4/wetland). Mitigation wetlands were 3–10 ha, had depths of 5–57 cm and had been constructed 4–21 years previously. The 4 reference wetlands were 7–28 ha, had depths of 5–17 cm and were near mitigation sites. Amphibians were monitored using nocturnal call surveys once a month in April–June 2001–2002.

A replicated, site comparison study in 2000–2001 of 42 wetlands constructed to replace lost wetlands in the Tillplain ecoregion, Ohio, USA (7) found that created wetlands supported 13 species of amphibians and had an average species richness of 4 per site (range 1–7). Occurrence of species varied from 2–76%. Species richness was positively associated with presence of a shallow shoreline (on average two species more than wetlands without) and negatively associated with predatory fish (average one species less). Wetland age, size, emergent vegetation cover and surrounding forest cover did not affect species richness. Wetlands tended to be in areas with little or no forest cover and so amphibian species associated with forested wetlands were rare or absent. Amphibians were sampled three times in March–July 2001–2002 using aquatic funnel traps, dip-netting and visual surveys. Four call surveys were undertaken at the end of each month from March–June.

A site comparison study in 2004–2005 of 8 created and 14 restored wetlands associated with hardwood forests in Louisiana, USA (8) found no significant difference in amphibian species richness between created/restored and natural wetlands. Twelve of 13 species in the area were found within the wetlands, one of which was only found in created/restored wetlands (upland chorus frog *Pseudacris feriarum*). Species richness was higher in 2004 (created/restored: 3.7; natural: 4.2) than 2005 (created/restored: 2.4; natural: 2.2). Richness was positively associated with water depth, canopy cover, flooding, aquatic vegetation and surrounding forest. Temporary and permanent wetlands were 1–174 ha and had been created or restored 1–18 years previously. Restoration had included replanting trees, water management and dredging. Eight natural wetlands within a wildlife refuge were used for comparison. Amphibians were monitored by call surveys (2/season), egg mass counts (1/season) and dip-netting (monthly along a 100 m transect).

A before-and-after study in 2000–2004 of constructed wetlands in southern Illinois, USA (9) found that amphibians rapidly colonized wetlands and restored surrounding terrestrial habitat. A total of 17 species were recorded with one new species each year. There were 12–15 species and 5,216–8,462 animals recorded at each wetland. Wetlands were created in 1999–2000 by enclosing water behind earth dams at the end of valleys. Hardwood tree seedlings were also planted. Wetlands were surveyed in April–June each year. Monitoring was undertaken using drift-fencing (four/wetland and three/adjacent habitat) with funnel traps (4/fence), artificial coverboards (0.7 m^2), visual encounter surveys, dip-netting and frog call surveys.

A before-and-after study in 1998–2004 of created wetland on a golf course in Long Reef, Sydney, Australia (10) found that captive-bred green and golden bell frog *Litoria aurea* tadpoles released into a created wetland did not establish a self-sustaining population. Once releases had stopped, the number of frogs declined to zero. Only 45 adult frogs were recorded. A few males were heard calling, but breeding was not recorded. Releases did not result in any metamorph or immature frogs if they occurred during autumn or involved low numbers of tadpoles, if ponds dried out soon after release or if fish were present. Successive releases into fish-free ponds were less successful in terms of numbers of metamorphs and immatures. Sixteen ponds, 12 of which were inter-connected (20–200 cm), were created in 1996–1997, with aquatic emergent vegetation and shrubs planted. A total of 9,000 captive-bred 3–4 week old tadpoles were released into the ponds over 11 occasions in 1998–2003. Amphibian monitoring was undertaken at 1–4 week intervals using artificial shelters around ponds, dip-netting and visual count surveys.

A replicated, site comparison study in 2007 of a complex of 10 created temporary ponds in central Ohio, USA (11) found that amphibian biomass was similar in the created and a natural complex of temporary ponds, although natural ponds supported more species (7 vs 4). There was no significant difference between created and natural ponds in overall biomass (dip-net: 3 vs 1; funnel trap: 3 vs 6 g/pond) or family biomass (hylidae: 1 vs 1; ranidae: 1 vs 4; ambystomatidae: 1 vs 2 g/pond). Created ponds had higher taxa diversity than the natural ponds (0.95 vs 0.70 Shannon-Weaver index) due to a more even distribution between the three families. Eleven years after construction, significant differences between created and natural ponds were found for hydrology, dissolved oxygen, conductivity and temperature. Wetland construction was completed in 1996. Amphibian larvae were sampled in May–July 2007 using dip-netting (all ponds) and funnel traps (in one of each pond type). Hydrology and physiochemistry were recorded for each pond in April–July. Comparisons were made with six natural temporary ponds in a nature preserve.

A replicated, site comparison study in 2006 of 49 constructed wetlands throughout northern Missouri, USA (12) found that 16 of 22 local amphibian species were recorded in the wetlands. The average number of species per wetland was five (range: 0–10). Cricket frogs *Acris cepitans*, bullfrogs *Lithobates catesbeianus* and leopard frog *Lithobates blairi/sphenocephalus* complex were each found in over 80% of wetlands. Green frogs *Lithobates clamitans* and gray tree frog *Hyla versicolor/chrysoscelis* complex were found in 53–55% of wetlands. Other frog and toad species were recorded in 29–37% and salamanders in 2–18% of wetlands. Species were positively associated with variables such as pond or stream density, grassland, wetland or vegetation cover. Fish were present at 43% of sites. Twenty wetlands were compensatory wetlands for road developments, and many others were farm ponds that were then managed by the agency for wildlife. Wetlands ranged from temporary ponds to large permanent ponds. Amphibians were sampled once or twice in March–August 2006 by visual and call surveys, dip-netting and funnel trapping.

A replicated, site comparison study in 2010 of 14 constructed ridge-top wetlands in a National Forest in Kentucky, USA (13, 16) found that amphibian species richness was similar, communities different and abundance lower in created compared to natural wetlands. Species richness did not differ significantly in constructed (10) and natural wetlands (12). However, captures were lower in constructed (permanent: 650; ephemeral: 407) compared to natural wetlands (1, 315). Amphibian communities differed significantly between constructed and natural wetlands. Larvae of wood frogs *Lithobates sylvaticus* and marbled salamanders *Ambystoma opacum* were almost exclusively found in natural wetlands, whereas large frog species (*Lithobates clamitans, L. catesbeianus, L. palustris*) and eastern newts *Notopthalmus viridescens* tended to be in constructed (particularly permanent) wetlands. Wetland size and depth were positively and aquatic vegetation negatively associated with some species. Captures were not affected by wetland age. Constructed wetlands were either permanent damed wetlands (built 1988–2003; $n = 7$) or ephemeral wetlands with added woody debris (built 2004–2007; $n = 7$). Five natural wetlands were also monitored. Dip-net sampling was undertaken over three days/wetland/month in May–August 2010.

A replicated, site comparison study in 2009–2010 of nine constructed ridge-top wetlands in a National Forest in Kentucky, USA (14) found that amphibian communities in shallow, but not deep, constructed wetlands were similar to natural wetlands. Communities differed significantly in deep constructed and natural wetlands. Species richness was similar in created and natural wetlands (13 vs 12). Constructed wetland communities tended to reflect permanent pond-breeding amphibians, while those in natural wetlands contained temporary pond-breeding species. Abundance of individual species differed between wetlands types and a small number of species were found only in natural or constructed wetlands. Two predatory species American bullfrog *Rana catesbeiana* and eastern newt *Notophthalmus viridescens* were found in higher numbers in constructed wetlands and were considered by the authors to increase predation rates. Five shallow and four deep constructed wetlands and six natural (temporary) wetlands were monitored. Monitoring was undertaken four times/year in March–August and included visual perimeter counts, call surveys, minnow trapping and dip-netting.

A global review in 2012 of studies comparing created and restored wetlands to natural wetlands (15) found that amphibian species richness or abundance at created and restored wetlands tended to be similar to or greater than natural wetlands. Species richness or abundance of some or all species was greater at created or restored wetlands in 54% of studies, similar in 35% of studies and lower than natural wetlands in 11%. Created and restored wetlands tended to be larger, deeper and were wet for more of the year than natural wetlands. Species richness and abundance tended to be positively associated with abundance of emergent vegetation, proximity of source wetlands and the availability of wetlands with varying water levels. They were also influenced by upland habitat and tended to be negatively associated with fish presence. Only peer-reviewed studies were included ($n = 37$; 70% in USA). Only studies that converted existing upland or shallow-water areas to wetland habitat (created; $n = 27$), or restored wetlands ($n = 14$) were included. Wetlands built specifically for water quality improvement were not included. Twenty-six studies had controls, either natural reference wetlands or historic data.

(1)   Perry M.C., Sibrel C.B. & Gough G.A. (1996) Wetlands mitigation: partnership between an electric power company and a federal wildlife refuge. *Environmental Management*, 20, 933–939.
(2)   Nyakang'o J.B. & van Bruggen J.J.A. (1999) Combination of a well functioning constructed wetland with a pleasing landscape design in Nairobi, Kenya. *Water Science and Technology*, 40, 249–256.
(3)   Kent D.M. & Langston M.A. (2000) Wildlife use of a created wetland in central Florida. *Florida Scientist*, 63, 17–19.
(4)   Juni S. & Berry C.R. (2001) A biodiversity assessment of compensatory mitigation wetlands in eastern South Dakota. *Proceedings of the South Dakota Academy of Science*, 80, 185–200.
(5)   Toure T.A. & Middendorf G.A. (2002) Colonization of herpetofauna to a created wetland. *Bulletin of the Maryland Herpetological Society*, 38, 99–117.
(6)   Balcombe C.K., Anderson J.T., Fortney R.H. & Kordek W.S. (2005) Wildlife use of mitigation and reference wetlands in West Virginia. *Ecological Engineering*, 25, 85–99.
(7)   Porej D. & Hetherington T.E. (2005) Designing wetlands for amphibians: the importance of predatory fish and shallow littoral zones in structuring of amphibian communities. *Wetlands Ecology and Management*, 13, 445–455.
(8)   Barlow S.J. (2007) Evaluation of anuran richness in restored wetlands of central Louisiana. MSc thesis. Louisiana State University and Agriculture and Mechanical College.
(9)   Palis J.G. (2007) If you build it, they will come: herpetofaunal colonization of constructed wetlands and adjacent terrestrial habitat in the Cache River drainage of southern Illinois. *Transactions of the Illinois State Academy of Science*, 100, 177–189.
(10)   Pyke G.H., Rowley J., Shoulder J. & White A.W. (2008) Attempted introduction of the endangered green and golden bell frog to Long Reef Golf Course: a step towards recovery? *Australian Zoologist*, 34, 361–372.
(11)   Korfel C.A., Mitsch W.J., Hetherington T.E. & Mack J.J. (2010) Hydrology, physiochemistry, and amphibians in natural and created vernal pool wetlands. *Restoration Ecology*, 18, 843–854.
(12)   Shulse C.D. (2010) Influences of design and landscape placement parameters on amphibian abundance in constructed wetlands. *Wetlands*, 30, 915–928.
(13)   Denton R.D. (2011) Amphibian community similarity between natural ponds and constructed ponds of multiple types in Daniel Boone National Forest, Kentucky. MSc thesis. Eastern Kentucky University.
(14)   Drayer A.N. (2011) Efficacy of constructed wetlands of various depths for natural amphibian community conservation. MSc thesis. Eastern Kentucky University.
(15)   Brown D.J., Street G.M., Nairn R.W. & Forstner M.R.J. (2012) A place to call home: amphibian use of created and restored wetlands. *International Journal of Ecology*, ID 989872.
(16)   Denton R.D. & Richter S.C. (2013) Amphibian communities in natural and constructed ridge top wetlands with implications for wetland construction. *Journal of Wildlife Management*, 77, 886–896.

12.11 Restore ponds

• Fifteen studies investigated the effectiveness of pond restoration for amphibians.

• One replicated, before-and-after study in Denmark[1] found that pond restoration had mixed effects on European tree frog population numbers depending on site. One replicated, controlled, before-and-after study in the UK[14] found that pond restoration did not increase great crested newt populations. Six replicated, before-and-after studies (including one controlled and one site comparison study) in Denmark, Estonia, Italy and the UK found that pond restoration and creation increased numbers of amphibian species[12], maintained[5, 6] or increased populations [5, 6, 8, 9], or increased pond occupancy and ponds with breeding success [4, 8, 9, 12]. One found that numbers of species did not increase[4]. Two before-and-after studies (including one replicated study) in Estonia[10, 11] found that pond restoration, along with terrestrial habitat management, maintained or increased populations of natterjack toads. One systematic review in the UK[15] found that there was no conclusive evidence that mitigation, which often included pond restoration, resulted in self-sustaining great crested newt populations.

• One small, replicated study in the USA[13] found that pond restoration had mixed effects on spotted salamander hatching success depending on restoration method.

• One replicated, before-and-after study in the UK[7] found that restoration increased the number of ponds used by breeding natterjack toads. One replicated study in Sweden[3] found that following restoration green toads only reproduced in 1 of 10 ponds. Three before-and-after studies (including one replicated, controlled study) in Denmark and Italy found that restored and created ponds were colonized by 1–7 species[6,] with 6–65% of ponds colonized and 35% used for breeding[2, 5].

Background
Ponds are often drained, left to dry or degraded during the development of agriculture or expansion of urban areas or other land uses. Although some amphibians are relatively tolerant of poor pond conditions, breeding is likely to be more successful in better quality ponds. Restoration of ponds may therefore help to increase populations of amphibians.
   Studies included here investigated the restoration of ponds using a combination of interventions. Studies that investigated pond restoration using one specific action are discussed under the relevant intervention, i.e. 'Deepen, de-silt or re-profile ponds', 'Add specific plants to aquatic habitats', 'Remove specific aquatic plants' and 'Remove tree canopy to reduce pond shading'. For removal or control of fish populations see 'Threat: Invasive alien and other problematic species – Reduce predation by other species'.
   Studies investigating the restoration of wetlands are discussed in 'Restore wetlands'.

A replicated, before-and-after study in 1983–1985 of 23 restored ponds on Borholm, Denmark (1) considered that 14 were successful and two failed. European tree frogs *Hyla arborea* were found in 10 ponds. There were large tree frog population increases in two ponds, moderate increases in five, no change in one and declines in two ponds post-restoration. One of the failures was due to pollution, another to ducks and restoration had changed the community in a third pond. In 1983–1985, 23 ponds were restored on private land, primarily to improve European tree frog populations. Restoration involved activities such as dredging and tree cutting.

A before-and-after study in 1985–1987 of head-started European tree frogs *Hyla arborea* released into 20 restored and created ponds near Aarhus, Denmark (2) found evidence of breeding a year after release. In 1986, 17–21 males were heard calling in 4 ponds, but no females, eggs or tadpoles were recorded. In 1987, up to 50 males were heard calling in 13 ponds. Four egg masses were found in one pond and tadpoles in six ponds. One hundred and fifty egg masses were collected from the nearest natural population. These were captive-reared in hot houses. Over 6,000 metamorphs were released into 9 created and 11 restored ponds (over 10 km^2) in 1985–1986.

A replicated study in 1987–1993 of 10 ponds on the island of Samsø, Sweden (3) found that restoration only resulted in successful breeding by green toads *Bufo viridis* in one pond. A year after pond cleaning, breeding was recorded in one pond, only males in another and no toads in the third pond. Only one male was seen in one of the seven ponds that were enlarged and had fish removed. In winter 1987–1991, three ponds were cleaned due to eutrophication. Seven ponds had fish removed and were enlarged. Ponds were monitored by call and torch surveys and by counting tadpoles and metamorphs during 4–6 visits in April–September.

A replicated, before-and-after study in 1977–1996 of ponds on chalkland in England, UK (4) found that pond restoration and creation resulted in increased occupancy by amphibians but not species richness/pond. In 1996, 69% of ponds were used compared to 55% in 1977. Species richness was similar in 1977 and 1996 (all ponds: 1.1; used ponds: 1.9 vs 1.6 species). Occupancy increased from 1977 to 1996 for common frogs *Rana temporaria* (4 vs 9 ponds) and toads *Bufo bufo* (2 vs 4). However, occupancy decreased for smooth newts *Triturus vulgaris* (14 vs 10), palmate newts *Triturus helveticus* (6 vs 3) and great crested newts *Triturus cristatus* (9 vs 3). Despite restoration, 17 of 33 original ponds were lost by 1996. However, a higher proportion of surviving ponds (*n* = 26) were in good condition in 1996 (58%) compared with 1977 (24%). Ponds were within a 150 km^2 area. Eleven of 33 ponds had been restored since 1977 and 13 created. Ponds were surveyed in spring 1995 or 1996 for species presence by egg counts, torchlight surveys and netting and trapping for newts.

A replicated, controlled, before-and-after study in 1986–1997 of 3,446 ponds restored and created for amphibians in Denmark (5) found that pond management was effective for maintaining and increasing populations. Populations survived five years after restoration in 92% (74–100%) of cases, compared to just 40% (32–52%) of cases without restoration. A total of 175 (39%) restored ponds were naturally colonized by rare species and 28 colonized by released animals. Approximately 2,000 ponds were restored or created for rare species, over half of which were for the European tree frog *Hyla arborea*. The national population of the species doubled as a result. A questionnaire was sent to all those responsible for pond projects across Denmark to obtain data. Over a third of projects dredged existing ponds and 7% had other types of restoration. For a pond to be defined as 'colonized' a species had to be present but not breeding.

A replicated, before-and-after study in 1998–2000 of restored and created ponds in 12 natural parks in the Lombardy District, Italy (6) found that pond restoration and creation resulted in increases in some existing amphibian species populations and colonization by new species within 2 years. Existing populations increased at six sites (average: 1.5 species; range 1–4). Between one and seven species colonized ponds at each site (average: 1.7 species). Numbers of egg clumps increased in the second year. Ponds were created or were restored by methods such as deepening and were lined with clay or PVC if necessary. Other habitat management was also undertaken at some sites including increasing dead wood, excavating tributary canals and removing and excluding fish.

A replicated, before-and-after study in 1991–1999 in a reserve in Caerlaverock, Scotland, UK (7) found that pond restoration increased the number of ponds used by breeding natterjack toads *Bufo calamita*. Out of 12 ponds restored in 1995–1998, 11 were used for breeding every year until 1999, compared to just 4 before restoration. Overall, breeding occurred one or two years after restoration in eight ponds that had not been used for breeding during the previous two or more years. All ponds used before restoration were still used for breeding and there was little change in use of unmanaged ponds. In 1995–1999, 17 ponds were restored by clearing aquatic vegetation, excavation and redefinition. Electric fences were installed around ponds during the summer to exclude cattle and sheep. Fences were removed after toadlet emergence. Ponds were visited at least four times in May–August 1991–1992 and 1994–1999 to count eggs, tadpoles and toadlets.

A replicated, before-and-after study in 1989–1997 of 25 restored and 23 created ponds on nine islands in Funen County, Denmark (8) found that there was a significant increase in the green toad *Bufo viridis* population. Overall, the population on the islands increased from 1,112 to 3,520 toads over the 7 years. Numbers were similar on islands with just pond creation and restoration (1,020 to 952) and increased on the two where cattle grazing was also reintroduced (92 to 2,568). Overall, pond occupancy increased from 23 to 51 and the number of ponds with breeding success increased from 9 to 15. In 1989–1997, ponds were created or restored by removing plants and dredging. On two of the islands, cattle grazing was also reintroduced to 73 ha of coastal meadows and abandoned fields. Populations were monitored annually in 1990–1997 during two or three call and visual surveys and dip-netting surveys. One population was also monitored in 1987–1989.

A replicated, before-and-after study in 1986–2004 of coastal meadows in Funen County, Denmark (9) found that pond restoration and creation, along with reintroduction of grazing, significantly increased green toad *Bufo viridis* and natterjack toad *Bufo calamita* populations. On 10 islands, green toads increased from 1,132 in 1988–1990 to over 10,000 adults in 2004. Numbers were similar on four islands with no management (512 to 510). Pond occupancy increased from 27 in 1988 to 61 in 1997 and ponds with breeding success from 11 to 22. Natterjack toads increased from 3,106 in 1998–1990 to 4,892 adults in 1997. Numbers of ponds with breeding success was similar (28 to 34). However, in 2000–2004, numbers dropped and small populations were lost due to insufficient grazing. Numbers of natterjacks declined on four islands with no restoration (270 to 170). From 1986–1991, 25 ponds were restored by reed removal and 23 created for green toads and 6 were restored and 8 created for natterjacks on 16 islands. Cattle grazing was reintroduced on six and continued on ten islands. Green toad eggs were translocated to one island. Four populations were monitored annually and others less frequently during 2–3 call and visual surveys and dip-netting.

A before-and-after study in 1992–2004 of a coastal meadow on an islet in Estonia (10) found that pond and terrestrial habitat restoration maintained a population of natterjack toads *Bufo calamita*. A total of 17 natterjacks were counted in 1992 and 7 in 2004, with numbers ranging from 1–17/year. It is considered by the author that without management the natterjack population may have declined or become extinct. Common toad *Bufo bufo* counts were 8 in 1992 and 4 in 2004 and ranged from 3–40/year. Restoration involved reed and scrub removal, mowing (cuttings removed) and reintroduction of sheep grazing. Toads were counted along a 1 km transect.

A replicated, before-and-after study in 2001–2004 of three coastal meadows in Estonia (11) found that restoration of breeding ponds, along with terrestrial habitat management, increased numbers of natterjack toads *Bufo calamita* on one island and stopped a decline on the other two islands. In 2001–2004, habitats were restored on three coastal meadows where the species still occurred. Sixty-six breeding ponds and natural depressions were cleaned, deepened and restored. Restoration also included reed and scrub removal, mowing (cuttings removed) and implementation of grazing where it had ceased.

A replicated, before-and-after, site comparison study of 450 existing ponds, 22 of which were restored, and 208 created ponds in 6 protected areas in Estonia (12) found that within 3 years amphibian species richness was higher in restored and created ponds than unmanaged ponds (3 vs 2 species/pond). The proportion of ponds occupied also increased for common spadefoot toad *Pelobates fuscus* (2 to 15%), great crested newt *Triturus cristatus* (24 to 71%) and the other 5 amphibian species (15–58% to 41–82%). Breeding occurred at increasing numbers of pond clusters from one to three years after restoration and creation for great crested newt (39% to 92%) and spadefoot toad (30% to 81%). Prior to restoration and creation, only 22% of ponds were considered high quality for breeding. In 2005, 405 existing ponds were sampled by dip-netting. In autumn 2005–2007, 22 ponds were restored and 208 created for great crested newts and spadefoot toads in 27 clusters. Restoration included clearing vegetation, extracting mud, levelled banks, pond drying and ditch blocking (for fish elimination). Monitoring was undertaken by visual and dip-netting surveys during one visit in 2006–2008.

A small, replicated study in 2005–2007 of five restored forest ponds in Illinois, USA (13) found that spotted salamander *Ambystoma maculatum* hatching success increased following additional prescribed burning, but not canopy removal. Eggs failed to hatch in three restored ponds. However, hatching success of egg masses increased after a prescribed burn at the one pond (2005: 0%; 2006–2007: 30–54%). This was not the case following canopy thinning at another pond (0%). Restored ponds had similar hatching success to ponds with resident spotted salamanders in 2005–2006 (29 vs 30%), but significantly higher success in 2007 following additional restoration (62 vs 20%). Restoration started in 2000 and included destruction of drainage tiles, clearing of invasive plants and prescribed burning. An egg mass was placed in two mesh enclosures (56 x 36 x 36 cm) in each restored pond. Three enclosures with an egg mass were also placed in each of three ponds with existing spotted salamanders populations (different site). Eggs were monitored every five days.

A replicated, controlled, before-and-after study in 2008–2010 of nine restored ponds in a reserve in England, UK (14) found that dredging and vegetation clearance did not appear to significantly increase great crested newt *Triturus cristatus* numbers in the first two years. Results were difficult to interpret but suggested that complete restoration and partial manual restoration did not significantly change numbers of newts. Data suggested that partial mechanical restoration may have had resulted in slight increases in newts. In winter 2008–2009, three groups of four ponds had sediment and vegetation removed by: partial manual clearance, partial mechanical clearance with an excavator, complete mechanical clearance or no management (controls). Torchlight surveys were undertaken before restoration and in March–June 2009–2010. Survey effort varied between years.

A systematic review in 2011 of the effectiveness of mitigation actions for great crested newts *Triturus cristatus* in the UK (15) found that neither the 11 studies found or monitoring data from licensed mitigation projects showed conclusive evidence that mitigation, which often included pond restoration, resulted in self-sustaining populations or connectivity to populations in the wider countryside. Only 22 of 460 licensed projects provided post-development monitoring data and of those, 16 reported that small, 3 medium and 1 large population was sustained. Two reported a loss of the population. A total of 127 (41%) of English and 46 (30%) of Welsh licence files contained licence return (reporting) documents. Of those, only 9% provided post-development monitoring data and a further 7% suggested surveys were undertaken, but no data were provided. The review identified

11 published or unpublished studies together with 309 Natural England and 151 Welsh Assembly Government (licensing authorities) mitigation licence files. Mitigation measures were undertaken to reduce the impact of the development and included habitat management such as creating or restoring ponds, as well as actions to reduce deaths including translocations.

(1) Fog K. (1988) Pond restoration on Bornholm. *Memoranda Societatis pro Fauna et Flora Fennica*, 64, 143–145.
(2) Skriver P. (1988) A pond restoration project and a tree-frog *Hyla arborea* project in the municipality of Aarhus Denmark. *Memoranda Societatis pro Fauna et Flora Fennica*, 64, 146–147.
(3) Amtkjær J. (1995) Increasing populations of the green toad *Bufo viridis* due to a pond project on the island of Samsø. *Memoranda Societatis pro Fauna et Flora Fennica*, 71, 77–81.
(4) Beebee T. (1997) Changes in dewpond numbers and amphibian diversity over 20 years on chalk downland in Sussex, England. *Biological Conservation*, 81, 215–219.
(5) Fog K. (1997) A survey of the results of pond projects for rare amphibians in Denmark. *Memoranda Societatis pro Fauna et Flora Fennica*, 73, 91–100.
(6) Gentilli A., Scali S., Barbieri F. & Bernini F. (2002) A three-year project for the management and the conservation of amphibians in Northern Italy. *Biota*, 3, 27–33.
(7) Phillips R.A., Patterson D. & Shimmings P. (2002) Increased use of ponds by breeding natterjack toads, *Bufo calamita*, following management. *Herpetological Journal*, 12, 75–78.
(8) Briggs L. (2003) Recovery of the green toad *Bufo viridis* Laurenti, 1768 on coastal meadows and small islands in Funen County, Denmark. *Deutsche Gesellschaft für Herpetologie und Terrarienkunde*, 14, 274–282.
(9) Briggs L. (2004) Restoration of breeding sites for threatened toads on coastal meadows. In R. Rannap, L. Briggs, K. Lotman, I. Lepik & V. Rannap (eds) *Coastal Meadow Management – Best Practice Guidelines*, Ministry of the Environment of the Republic of Estonia, Tallinn. pp. 34–43.
(10) Lepik I. (2004) Coastal meadow management on Kumari Islet, Matsalu Nature Reserve. In R. Rannap, L. Briggs, K. Lotman, I. Lepik & V. Rannap (eds) *Coastal Meadow Management – Best Practice Guidelines*, Ministry of the Environment of the Republic of Estonia, Tallinn. pp. 86–89.
(11) Rannap R. (2004) Boreal Baltic coastal meadow management for *Bufo calamita*. In R. Rannap, L. Briggs, K. Lotman, I. Lepik & V. Rannap (eds) *Coastal Meadow Management – Best practice Guidelines*, Ministry of the Environment of the Republic of Estonia, Tallinn. pp. 26–33.
(12) Rannap R., Lõhmus A. & Briggs L. (2009) Restoring ponds for amphibians: a success story. *Hydrobiologia*, 634, 87–95.
(13) Sacerdote A.B. & King R.B. (2009) Dissolved oxygen requirements for hatching success of two Ambystomatid salamanders in restored ephemeral ponds. *Wetlands*, 29, 1202–1213.
(14) Furnborough P., Kirby P., Lambert S., Pankhurst T., Parker P. & Piec D. (2011) The effectiveness and cost efficiency of different pond restoration techniques for bearded stonewort and other aquatic taxa. Report on the Second Life for Ponds project at Hampton Nature Reserve in Peterborough, Cambridgeshire. Froglife Report.
(15) Lewis B. (2012) An evaluation of mitigation actions for great crested newts at development sites. PhD thesis. The Durrell Institute of Conservation and Ecology, University of Kent.

12.12 Restore wetlands

• Seventeen studies investigated the effectiveness of wetland restoration for amphibians.

• Ten site comparison studies (including eight replicated studies) in Canada and the USA[1–6, 8, 10, 13, 16, 17] compared amphibian numbers in restored and natural wetlands. Eight found that amphibian abundance[6], numbers of species[4, 5, 8, 10, 13, 16] and species composition[3] were similar. Two found that the number of species[1, 2] or abundance[16] was lower and species composition different[16] in restored wetlands. One[17] found that restored wetlands were used more or less depending on the habitat surrounding natural wetlands. One global review[18] found that in 89% of cases, restored and created wetlands had similar or higher amphibian abundance or numbers of species to natural wetlands.

• Seven of nine studies (including six site comparison and/or replicated studies) in Canada, Taiwan and the USA[1, 2, 5, 7–9, 11–12, 15, 16] found that wetland restoration increased numbers of amphibian species, with breeding populations establishing in some cases[5, 9, 12].

Three found that numbers of species[8, 15] or abundance[16] did not increase with restoration. Two[7, 15] found mixed effects, with restoration maintaining or increasing abundance of individual species. Three replicated studies (including two site comparison studies) in the USA found that numbers of species in restored wetlands were affected by wetland size, proximity to source ponds[5] and seasonality[10], but not wetland age[14].

• Three studies (including two replicated, site comparison studies) in Taiwan and the USA[5, 11, 12] found that restored wetlands were colonized by three to eight amphibian species. One before-and-after study in the USA[15] found that three target species did not recolonize restored wetlands.

Background
Wetland habitats are often drained or degraded during the development of agriculture or expansion of urban areas or other land uses. Restoration of these important amphibian habitats can help to increase local amphibian species richness and abundance.

Studies included here tended to investigate the restoration of wetlands using a combination of interventions. Studies investigating the restoration of individual ponds are discussed in 'Restore ponds'. Study 'ponds' and 'pools' have been referred to as 'ponds' within this section.

A before-and-after, site comparison study in 1995–1996 of a degraded forested wetland in South Carolina, USA (1, 2) found that restoration increased numbers of amphibian species over the first four years. Sixteen frog and toad and 13 salamander species were captured in the restoration area. It was assumed that there were no amphibians prior to restoration. Successful reproduction was documented for 16 of the 29 species. However, species diversity was lower in the restored compared to natural site. Planting regimes and treatment (burning or herbicide application) had little effect on species assemblage. Restoration included tree planting in 1993–1995 (549–1078 trees/ha). In some areas herbicide application and prescribed burns were undertaken to control scrub. Approximately 25% of the restoration area was left as unmanaged (control) strips. Amphibians were monitored over 21 months in planted and unplanted areas and in an adjacent natural wetland area.

A before-and-after, site comparison study in 1995–1998 of a wetland restoration site in St. Clair County, Illinois, USA (3) found that by the end of the study, all seven species of amphibians previously found at the site were recorded within the restored area. An eighth species, not present in the adjacent forest, had also colonized the site by 1997. Abundance was higher at the restored site compared to the adjacent forest (5 vs 4 amphibians/man-hour of survey). Restoration of the 95 ha area included removal of low embankments to restore water levels and planting native hardwood trees. Amphibians were monitored at the restoration area and an adjacent forest in May–June 1995–1996. Drift-fences (15 m long) with pitfall and funnel traps at the centre and ends were used. In 1997–1998, visual encounter surveys were carried out twice in March–May.

A replicated, site comparison study in 1999–2000 of 13 wetlands in South Dakota, USA (4) found that combined amphibian and reptile species richness did not differ significantly between restored, enhanced and natural wetlands. Although not significant, there was a trend for higher numbers of species in restored and enhanced wetlands compared to natural wetlands. A total of 11 amphibian and reptile species were recorded. Study sites were four restored, four enhanced and five natural wetlands. Restoration tended to involve plugging drainage ditches or breaking sub-surface drainage tiles. Enhancement included manipulating water levels to increase wetland size or changing vegetation structure. Monitoring was undertaken using drift-fences with pitfall traps and visual surveys around wetland perimeters in spring and autumn in 1999–2000.

A replicated, site comparison study in 1998 of seven restored wetlands in Minnesota, USA (5) found that eight amphibian species rapidly colonized the wetlands and four of those established breeding populations. Natural wetlands supported an additional four species. However, there was no significant difference between average numbers of species in restored and natural wetlands (4 vs 5). Six of the seven restored wetlands supported amphibian populations. Species richness increased with restored wetland size and proximity to source ponds. Wetlands were restored 5–20 months before the study, by destroying drainage tiles or filling ditches to allow flooding. Five natural wetlands were surveyed for comparison. Wetlands were seasonal to semi-permanent and were 0.1–8.6 ha in size. Amphibians were monitored on five visits in April–July 1998 using visual encounter and call surveys along wetland edges. Larval sampling was also carried out using five activity and five minnow traps along pond edges.

A replicated, site comparison study in 1999–2000 of 97 restored wetlands in aspen parkland in Alberta and Saskatchewan, Canada (6) found that restored wetlands had similar amphibian abundance to natural wetlands, suggesting that restored wetlands provide suitable habitat for amphibians. A total of 4,086 wood frogs *Rana sylvatica*, boreal chorus frogs *Pseudacris maculata* and tiger salamanders *Ambystoma tigrinim* were captured over the 2 years. Amphibians, in particular wood frogs, were found in similar numbers in restored and natural wetlands. From 1987, wetlands were restored by installing ditch plugs. Amphibian presence/absence was recorded in 97 restored and 85 natural wetlands. In 1999, 7 and in 2000, 11 restored and natural wetlands were monitored intensively. Pitfall (19,431 trap nights) and minnow traps (6,794 trap nights) were used to compare species between restored and natural ponds.

A replicated, controlled study in 1998–1999 of 22 restored wetlands on Prince Edward Island, Canada (7) found that restored wetlands had significantly higher numbers of amphibian species than non-restored wetlands (2.7 vs 1.8). All five species present on the islands were recorded in both wetland types. Abundance was significantly higher in restored wetlands for spring peeper *Pseudacris crucifer* (2.7 vs 2.1), northern leopard frog *Rana pipiens* (0.5 vs 0.2) and green frog *Rana clamitans* (0.7 vs 0.2). There was no difference in abundance of wood frog *Rana sylvatica* or American toad *Bufo americanus* in restored and non-restored wetlands. Wetlands were 0.3–0.6 ha and had been restored by dredging (30–95% of area) two to seven years before the study. Amphibians were monitored at 22 dredged and 24 undredged wetlands during monthly call surveys in May–July 1998 and/or 1999.

A replicated, site comparison study in 1999 of 11 restored and 29 abandoned wetlands on old mines in southwestern Indiana, USA (8) found that species richness was similar at reclaimed and abandoned wetlands. Restored wetlands supported an average of 3–8 species and abandoned wetlands 4–6 species. Two natural wetlands supported eight of nine local species. Breeding was recorded at both wetland types. The emphasis of reclamation was restoring mined lands to the original land use (e.g. forestry and agriculture). However, standards being followed included actions that enhanced reclaimed wetlands by developing the shoreline and establishing ephemeral wetlands. Wetlands included permanent (average 11 ha), semi-permanent (3 ha) and ephemeral sites (<0.3 ha). Call surveys were undertaken over three hours in February–August 2000. Tadpole surveys were conducted in March–August using dip-nets, minnow traps and seines. Twenty-nine abandoned and two natural wetlands were used as comparisons.

A before-and-after study in 1998–2003 of a wetland landscape restoration project at Kankakee Sands, Indiana, USA (9) found that numbers of amphibian breeding populations increased from 14 to 172 and species richness from 7 to 10, 3 years after restoration began. Prior to restoration in 1998, there were 14 populations of 7 species at 7 breeding sites (> 200 m apart). By 2000, this increased to 33 populations of 7 species at 14 sites and by 2003, 172 populations of 10 species at 44 sites. Average species richness/site increased from 2 in 1998 to 4 in 2003. Species became significantly more common and breeding occurred in every land management unit (vs 50% in 1998). However, apart from in wetter than average years (2002 and 2003), restored wetlands dried before larvae of most species metamorphosis. Restoration began in 1999 and comprised plugging and filling ditches, breaking drainage tiles and recontouring basins. Amphibians were monitored at restored and natural wetlands in April–July 1998, 2000–2003. Call surveys (3/year), seines, dip-nets, minnow traps (each 2/year), terrestrial searches and drift-fences with funnel traps were used.

A replicated, site comparison study in 2003–2004 of two restored wetlands in southwestern Washington, USA (10) found that amphibian species richness was similar and abundance tended to be higher in restored compared to natural wetlands. Abundances were significantly higher at restored and one natural wetland compared to the other three natural wetlands. Restored and natural wetlands had similar species richness (4–5 species). Pacific treefrogs *Pseudacris regilla* were only found in natural wetlands. Abundances of the other five species varied between wetlands. Significantly higher numbers of amphibians emigrated from the restored compared to natural oxbow wetlands (29–58 vs 0.01–0.25/trap night). Abundance was highest in wetlands with intermediate hydroperiods (>7 months) compared to those with temporary or permanent water. Two restored (emergent) and four natural (emergent and oxbow) wetlands were surveyed. Restoration in 1997–1998 involved blocking drainage ditches by constructing water control structures and embankments. Amphibians were monitored using fyke nets and one-way traps in January–June 2003–2004.

A before-and-after study in 2002–2003 of a restored wetland in a tropical forest in Kenting, Taiwan (11) found that 8 of 18 amphibian species known to be in the area colonized the wetland within a year. A total of 1,456 amphibians were recorded (average density: 0.025 m²). Cricket frog *Fejervarya limnocharis* was the most common species (62%), followed by ornate narrow-mouthed frog *Microhyla ornate* and spot-legged treefrog *Polypedates megacephalus*. These three species accounted for 97% of the relative frequency and abundance. Abundance varied with habitat type and within ponds was positively correlated with vegetation cover. From December 2002 to April 2003, a concrete pond was demolished, the hole filled with soil and replanted to restore a 0.5 ha semi-natural permanent wetland. Amphibians were monitored by visual survey within six habitat areas. Surveys were undertaken twice a month from May 2003 to April 2004.

A small, replicated study in 1999–2004 of three seasonal ponds and 200 potholes created at a forested wetland restoration site on Sears Island, Maine, USA (12) found that wood frogs *Rana sylvatica* and spotted salamanders *Ambystoma maculatum* colonized and reproduced in the three ponds and bred in 28% of potholes. Spring peeper *Pseudacris crucifer* colonized and bred in one pond and American toad *Bufo americanus* visited but did not breed in two ponds. Reproductive success varied between ponds for wood frogs (0.2–48.4 juveniles/egg mass) and spotted salamanders (1.8–5.4). Metamorphosis of these species was only completed in one pothole before drying. In 1997, two ponds were excavated within the original wetland (350 and 600 m²) and one dry detention basin was converted to a pond (900 m²). Approximately 200 small potholes (0.3–110 m²) were also created. Amphibians were monitored in the three ponds in March–October using enclosure drift-fencing. Pitfall traps were installed in pairs every 10 m either side of fences (9–18 pairs/pond). Eggs were counted within ponds and 50 potholes.

A replicated, site comparison study in 2004–2005 of 14 restored and 8 created wetlands associated with hardwood forests in Louisiana, USA (13) found no significant difference in amphibian species richness between restored/created and natural wetlands (2.9 vs 2.9). Twelve of 13 species in the area were found within the wetlands, one of which was only found in restored wetlands (upland chorus frog *Pseudacris feriarum*). Species richness was higher in 2004 (restored: 3.7; natural: 4.2) than 2005 (restored: 2.4; natural: 2.2). Species richness was positively associated with water depth, canopy cover, flooding, aquatic vegetation and surrounding forest. Temporary and permanent wetlands were 1–174 ha and had been restored 1–18 years previously. Restoration had included replanting trees, water management and dredging. Eight natural wetlands within a wildlife refuge were used for comparison. Amphibians were monitored by call surveys (2/season), egg mass counts (2/season) and dip-netting (monthly along 100 m transect).

A replicated study in 1992–2004 of 16 restored wetlands in Wisconsin, USA (14) found that amphibian communities stayed the same as wetlands matured over 12 years. The six amphibian species and overall amphibian abundance did not change between 1992 and 2004 (13–14 calls/wetland). Overall wetland species and coefficients of conservation values increased over time (coefficient: 3.6 vs 3.9). Restoration occurred between 1988 and 1991. Amphibians were monitored at eight wetlands. Four amphibian call surveys were undertaken at each in April–July. Plants and birds were also monitored.

A before-and-after study in 2000–2006 of a restored forested wetland in Lake County, Illinois, USA (15) found that restoration did not increase amphibian species diversity or natural recolonization by three target species five years after restoration. There was no natural recolonization by spotted salamander *Ambystoma maculatum*, wood frog *Lithobates sylvaticus* or spring peeper *Pseudacris crucifer*. Species richness was similar before (4–8) and after restoration (4–6); the diversity index tended to increase (0.5 vs 1.2). Post-restoration, the abundance of northern leopard frog *Lithobates pipiens*, American toad *Anaxyrus americanus* and western chorus frog *Pseudacris triseriata* increased. Green frogs *Lithobates clamitans* and bullfrogs *Lithobates catesbeiana* were detected in small numbers, but did not breed. Restoration was undertaken in 2000. Agricultural drainage tiles were removed to restore water levels to previous wetland levels. Non-native European buckthorn *Rhamnus cathartica* and garlic mustard *Alliaria petiolaris* were removed using herbicide, chainsaws, manually and controlled burns. Native trees were also planted. Amphibians were monitored in 2004–2006 using drift-fences with pitfall traps and funnel traps, dip-netting, artificial cover, visual and mark-recapture surveys.

A replicated, controlled, site comparison study in 2008 of 18 sites within a large wetland restoration area in Florida, USA (16) found that restored sites had higher amphibian species richness and abundance than non-restored sites. Species richness was significantly higher in restored compared to non-restored (8 vs 5–6) wetlands and similar to natural (7–8) wetlands. Abundance was significantly higher in restored compared to non-restored wetlands one year (27 vs 10) but not four years after restoration (18 vs 17). Abundance was highest in natural wetlands (28–35). Species assemblages differed between wetland types. Overall, five species of tadpole were found in restored wetlands (mainly 1/wetland) compared to none in non-restored sites. Natural wetlands contained six species (mainly 1/wetland). Restoration within a failed residential development site involved plugging canal systems to restore previous water levels. Permanent ponds and ephemeral wetlands were created. There were two areas, restored either one or four years previously, each with three replicates of three wetland types: restored, non-restored and natural. Amphibians were monitored during six call surveys in May–September and monthly dip-netting in June–August 2008.

A replicated, site comparison study in 2005–2006 of four restored wetlands in restored grasslands in the Prairie Pothole Region, USA (17) found that the restored wetlands were used more frequently by amphibians than wetlands within farmland, but not as much as natural wetlands within native prairie grasslands. This was true for two frog, one toad and one salamander species. Four wetlands from each category were selected: farmed (drained with ditches), conservation grasslands (wetland hydrology restored, area reseeded with perennial grassland ≤ 10 years previously) and native prairie grasslands (natural). Call surveys, aquatic funnel traps and visual encounter surveys were undertaken biweekly in May–June 2005–2006.

A review in 2012 of studies examining restored and created wetlands across the world (18) found that amphibian species richness or abundance at restored and created wetlands tended to be similar or greater than at natural wetlands. Species richness or abundance of some or all species was greater at restored or created wetlands in 54% of studies, similar in 35% of studies and lower than natural wetlands in 11%. Restored and created wetlands tended to be larger, deeper and were wet for more of the year than natural wetlands. Species richness and abundance tended to be positively associated with abundance of emergent vegetation, proximity of source wetlands and the availability of wetlands with varying water levels. They were also influenced by upland habitat and tended to be negatively associated with fish presence. Only peer-reviewed studies were included ($n = 37$; 70% in USA). Only studies that converted existing upland or shallow-water areas to wetland habitat (created; $n = 27$), or restored wetlands ($n = 14$) were included. Wetlands built specifically for water quality improvement were not included. Twenty-six studies had controls, either natural reference wetlands or historic data.

(1)   Barton C., Nelson E.A., Kolka R.K., McLeod K.W., Conner W.H., Lakly M., Martin D., Wigginton J., Trettin C.C. & Wisniewski J. (2000) Restoration of a severely impacted riparian wetland system – the Pen Branch Project. *Ecological Engineering*, 15, S3–S15.
(2)   Bowers C.F., Hanlin H.G., Guynn D.C., McLendon J.P. & Davis J.R. (2000) Herpetofaunal and vegetational characterization of a thermally-impacted stream at the beginning of restoration. *Ecological Engineering*, 15, S101–S114.
(3)   Mierzwa K.S. (2000) Wetland mitigation and amphibians: preliminary observations at a southwestern Illinois bottomland hardwood forest restoration site. *Journal of the Iowa Academy of Science*, 107, 191–194.
(4)   Juni S. & Berry C.R. (2001) *A Biodiversity Assessment of Compensatory Mitigation Wetlands in Eastern South Dakota*. Proceedings of the South Dakota Academy of Science, 80, 185–200.
(5)   Lehtinen R.M. & Galatowitsch S.M. (2001) Colonization of restored wetlands by amphibians in Minnesota. *American Midland Naturalist*, 145, 388–396.
(6)   Paszkowski C.A., Puchniak A.J. & Gray B.T. (2001) Recovery of amphibian assemblages in restored wetlands in Prairie Canada. *Ecological Society of America Annual Meeting Abstracts*, 86, 328.
(7)   Stevens C.E., Diamond A.W. & Gabor T.S. (2002) Anuran call surveys on small wetlands in Prince Edward Island, Canada restored by dredging of sediments. *Wetlands*, 22, 90–99.
(8)   Timm A. & Meretsky V. (2004) *Anuran Habitat Use on Abandoned and Reclaimed Mining Areas of Southwestern Indiana*. Proceedings of the Indiana Academy of Science, 113, 140–146.
(9)   Brodman R., Parrish M., Kraus H. & Cortwright S. (2006) Amphibian biodiversity recovery in a large-scale ecosystem restoration. *Herpetological Conservation and Biology*, 1, 101–108.
(10)   Henning J.A. & Schirato G. (2006) Amphibian use of Chehalis River floodplain wetlands. *Northwestern Naturalist*, 87, 209–214.
(11)   Lee Y.F., Kuo Y.M., Lin Y.H., Chu W.C., Wang H.H. & Wu S.H. (2006) Composition, diversity, and spatial relationships of anurans following wetland restoration in a managed tropical forest. *Zoological Science*, 23, 883–891.
(12)   Vasconcelos D. & Calhoun A.J.K. (2006) Monitoring created seasonal pools for functional success: a six-year case study of amphibian responses, Sears Island, Maine, USA. *Wetlands*, 26, 992–1003.
(13)   Barlow S.J. (2007) Evaluation of anuran richness in restored wetlands of central Louisiana. MSc thesis. Louisiana State University and Agriculture and Mechanical College.
(14)   Nedland T.S., Wolf A. & Reed T. (2007) A reexamination of restored wetlands in Manitowoc County, Wisconsin. *Wetlands*, 27, 999–1015.

(15) Sacerdote A.B. (2009) Reintroduction of extirpated flatwoods amphibians into restored forested wetlands in northern Illinois: feasibility assessment, implementation, habitat restoration and conservation implications. PhD thesis. Northern Illinois University.
(16) Dixon A.D. (2011) Anurans as biological indicators of restoration success in the greater Everglades ecosystem. *Southeastern Naturalist*, 10, 629–646.
(17) Balas C.J., Euliss Jr. N.H. & Mushet D.M. (2012) Influence of conservation programs on amphibians using seasonal wetlands in the Prairie Pothole region. *Wetlands*, 32, 333–345.
(18) Brown D.J., Street G.M., Nairn R.W. & Forstner M.R.J. (2012) A place to call home: amphibian use of created and restored wetlands. *International Journal of Ecology*, ID 989872.

12.13 Deepen, de-silt or re-profile ponds

• Two before-and-after studies in France and Denmark found that pond deepening and enlarging or re-profiling resulted in the establishment of a breeding population of great crested newts[1] and translocated garlic toads[7]. Two studies (including one replicated, controlled study) in the UK and Denmark found that pond deepening and enlarging or dredging increased a population of common frogs[3, 9] or numbers of calling male tree frogs[4].

• Four before-and-after studies in Denmark and the UK found that pond deepening, along with other interventions, maintained newt populations[6] and increased populations of European fire-bellied toads[5] and natterjack toads[2, 8].

Background
If ponds dry out, breeding habitat may be lost which could have a significant impact on amphibian populations. It may be possible to restore ponds by deepening or de-silting them. Re-profiling ponds can also make them more suitable for amphibians. Ponds should ideally contain a range of microhabitats, which can be achieved with a range of depths, an irregular shape and gently sloping sides to encourage a diversity of plants and invertebrates.

Studies that investigated the restoration of ponds using a combination of interventions including deepening, de-silting or re-profiling ponds are discussed in 'Restore ponds'.

A before-and-after study in 1977–1992 of a pond in an abandoned sand-quarry in northwestern France (1) found that pond enlargement for great crested newts *Triturus cristatus* resulted in rapid colonization and fast initial population increase, followed by a dramatic decline. Newts were recorded in the pond the year after enlargement. The population increased to 346 adults within 5 years, but decreased to 16 newts 2 years later. However, by 1992 the population was estimated at 55 adults. Variation in the adult population was largely due to variation in juvenile recruitment. The juvenile cohort was estimated at 300 individuals in 1980, but zero by 1984. Juvenile survival varied from 7 to 45%. Before enlargement, the shallow pond (30 cm) supported a breeding population of natterjack toads *Bufo calamita*, but not great crested newts. In summer 1977, it was enlarged by 7 x 20 m, approximately doubling its area, and to a maximum depth of at least 1.2 m. Newts were monitored in 1979–1984 and in 1992 by torching the shallow part of the pond from dusk to midnight and dip-netting.

A before-and-after study in 1972–1991 of ponds on heathland in Hampshire, UK (2) found that pond restoration by deepening, along with other interventions, tripled a natterjack toad *Bufo calamita* population. Spawn string counts (female population) increased from 15 to 43, with a maximum number of 48 in 1989 (see also (8)). Nine small ponds (< 1,000 m²) were created and four restored by excavation to generate shallow, temporary ponds with gradually shelved margins. Scrub was cleared from 40 ha by cutting and uprooting, bracken was treated with herbicide over 12 ha and swamp stonecrop *Crassula helmsii*, which invaded 6 new ponds, was pulled up and treated with herbicide. Captive-reared toadlets raised from spawn were released in 1975 (8,800), 1979, 1980 and 1981 (1,000 each). Limestone was added to one naturally acid pond (735 m²) annually in April 1983–1989. Toads were monitored annually, once every 10 days in March and August.

A before-and-after study in 1981–1993 of 20 restored ponds in Middlesex, UK (3) found that the population of common frogs *Rana temporaria* increased as the number of ponds restored by deepening increased (see also (9)). Egg clumps increased from 40 in one pond in 1983 to 584 in 1992 (1–370/pond). However, numbers declined to 399 egg clumps in 1993, which was considered by the authors to be due to drought. Many ponds within a country park had dried up and so were restored by deepening and enlarging (4–1,680 m²) in 1981–1993. Eight ponds were also created in the area increasing the total pond area from 2,248 m² in 1983 to 4,965 m² in 1993. Egg clumps were counted in restored ponds in February–March as an index of numbers of breeding females.

A replicated, controlled study in 1991–1994 of 29 restored ponds on the island of Lolland, Denmark (4) found that numbers of calling male tree frogs *Hyla arborea* increased significantly and larvae increased and then decreased after dredging. Numbers of calling males increased significantly in dredged but not undredged ponds from 1991 to 1994. The year after dredging, numbers of larvae were significantly higher in dredged ponds compared to undredged ponds; numbers had been similar before dredging. However, two years after dredging, there was no significant difference between numbers of larvae in dredged and undredged ponds. In 1991–1993, 29 ponds that had at least 3 calling males were restored by dredging. Water was usually pumped out and mud removed from the bottom. Frogs were monitored by call surveys and dip-netting (30 minutes) in 1991–1994.

A before-and-after study in 1986–1997 of 69 restored and created ponds at 6 sites in Funen County, Denmark (5) found that there was an increase in the population of European fire-bellied toads *Bombina bombina*. The total adult population increased from 82 in 1986–1988 to 542 in 1995–1997 (from 1–30 to 8–170/site). Numbers of ponds occupied by adults increased from 8 to 62 and by tadpoles from 1 to 18 over the same period. The population declined at only one site that was flooded with salt water. Ponds were restored by dredging or created. Wild-caught toads were paired in separate nest cages in ponds and eggs collected and reared in aquaria. Metamorphs and one-year-olds were released into ponds. Ponds were monitored for calling males and breeding success (capture-recapture estimate) annually in 1987–1997.

A before-and-after study in 1986–1995 of two ponds within a housing development near Peterborough, UK (6) found that pond deepening, fish removal and regulation of water levels resulted in the maintenance of great crested newt *Triturus cristatus* and smooth newt *Triturus vulgaris* numbers seven years after development. Pre-development numbers were variable for great crested (29–102) and smooth newts (10–18). Adults of both species returned to breed in 1989–1990 following development (crested: 51–67; smooth: 16–42) and until 1995 (crested: 55–123; smooth: 33–125). However, production of metamorphs failed in 1990 due to three-spined sticklebacks *Gasterosteus aculeatus* in one pond and drying of the other. Larval catches increased in 1991 following fish removal (crested: 37; smooth: 13) and maintenance of water level (crested: 62; smooth: 22) and then varied in each pond (crested: 1–15; smooth: 1–27). Development was undertaken in 1987–1989. Ponds (800 m²) were deepened in 1988, fish removed by pond drying in 1990 and water pumped to the pond that dried naturally from 1991. A 1 ha area was retained around ponds. Newts were counted by torch and larvae netted once or twice in 1986–1987 and 3–4 times in March–May 1988–1995.

A before-and-after study in 1994–1997 of two restored ponds in Jutland, Denmark (7) found that translocated garlic toads *Pelobates fuscus* established breeding populations in both ponds. Breeding was recorded in one in 1996 and the other in 1997. Ponds were restored by removing surrounding willows and by levelling the banks of one pond. Forty-three toads were captured from a pond being eliminated by development. Four egg strings were produced and raised in captivity. The 43 adults and 1,000 tadpoles were released into one of the restored ponds in 1994. Toads were monitored by tadpole and call surveys.

A before-and-after study in 1972–1999 of natterjack toads *Bufo calamita* at two sites in England, UK (8) found that pond restoration and creation, vegetation clearance and captive-rearing toadlets resulted in population increases over 20 years. The continuation of a study in Hampshire, UK in 1972–1991 (2) until 1999 indicated that there was a doubling of the population. Egg string counts (female population) increased from 15 in 1972 to 32 in 1999, with a maximum number of 48 in 1989. At a second site, spawn string counts increased from 1 in 1973 to 8 in 1999, with a maximum number of 29 in 1997. Ponds were created and restored by excavation, scrub and bracken was cleared and captive-reared toadlets raised from eggs and released. Toads were monitored annually.

In a continuation of a study (3) a before-and-after study in 1983–2004 of 31 ponds in Middlesex, UK (9) found that pond restoration and creation resulted in a significant increase in total common frog *Rana temporaria* egg masses. Numbers increased from 40 egg masses in 1983 to 1,852 in 2002, although then declined to 1,000 in 2004. Numbers of egg clumps increased with pond size and eight ponds contained 89% of the spawn. The numbers of ponds used for breeding each year increased from 1 in 1983 to 20 in 2000. Breeding tended to occur two years after pond creation or restoration. Egg clumps were counted in restored ponds in February–March as an index of numbers of breeding females. An unmonitored number of eggs, tadpoles and frogs were introduced and removed from ponds by the public, particularly in 1984. Colonization may not therefore have been natural.

(1) Arntzen J.W. & Teunis S.F.M. (1993) A six year study on the populations dynamics of the crested newt (*Triturus cristatus*) following the colonisation of a newly created pond. *Herpetological Journal*, 3, 99–110.
(2) Banks B., Beebee T.J.C. & Denton J.S. (1993) Long-term management of a natterjack toad (*Bufo calamita*) population in southern Britain. *Amphibia-Reptilia*, 14, 155–168.
(3) Williams L.R. & Green M. (1993) Pond restoration and common frog populations at Fryent Country Park, Middlesex, 1983–1993. *London Naturalist*, 72, 15–24.
(4) Hels T. & Fog K. (1995) Does it help to restore ponds? A case of the tree frog (*Hyla arborea*). *Memoranda Societatis pro Fauna et Flora Fennica*, 71, 93–95.
(5) Briggs L. (1997) Recovery of Bombina bombina in Funen County, Denmark. *Memoranda Societatis pro Fauna et Flora Fennica*, 73, 101–104.
(6) Cooke A.S. (1997) Monitoring a breeding population of crested newts (*Triturus cristatus*) in a housing development. *Herpetological Journal*, 7, 37–41.
(7) Jensen B.H. (1997) Relocation of a garlic toad (*Pelobates fuscus*) population. *Memoranda Societatis pro Fauna et Flora Fennica*, 73, 111–113.
(8) Buckley J. & Beebee T.J.C. (2004) Monitoring the conservation status of an endangered amphibian: the natterjack toad *Bufo calamita* in Britain. *Animal Conservation*, 7, 221–228.
(9) Williams L.R. (2005) Restoration of ponds in a landscape and changes in common frog (*Rana temporaria*) populations, 1983–2005. *Herpetological Bulletin*, 94, 22–29.

12.14 Create refuge areas in aquatic habitats

• We found no evidence for the effects of creating refuge areas in aquatic habitats on amphibian populations.

Background
Refuge areas that provide the correct microclimate and some protection from predation can be created for amphibians where natural shelters are limited.

12.15 Add woody debris to ponds

• We found no evidence for the effects of adding woody debris to ponds on amphibian populations.

Background
Woody debris can provide amphibians with refuges and can be added where shelter habitat is limited.

12.16 Remove specific aquatic plants

Studies investigating the effects of removing specific aquatic plants are discussed in 'Threat: Invasive alien and other problematic species – Control invasive plants'.

Background
Non-native plant species can be introduced into or naturally invade waterbodies and out-compete native species altering aquatic habitats. For example, swamp stonecrop *Crassula helmsii* can form thick mats covering whole ponds.

Studies in which aquatic vegetation was removed as one of a combination of interventions for the restoration of ponds or wetlands are discussed in 'Restore ponds' and 'Restore wetlands'.

12.17 Add specific plants to aquatic habitats

• We found no evidence for the effects of adding specific plants, such as emergent vegetation, to aquatic habitats on amphibian populations.

Background
Plants can be added to aquatic habitats to increase shade, cover from predators, or egg laying sites or to attract invertebrates or improve water quality, for example.

12.18 Remove tree canopy to reduce pond shading

- One before-and-after study in Denmark[1] found that translocated garlic toads established breeding populations following pond restoration that included canopy removal.

- One before-and-after study in the USA[2] found that canopy removal did not increase hatching success of spotted salamanders.

Background

Shading of ponds by tree canopies reduces water temperature, affects plant communities and can result in chemical changes resulting from the decomposition of increased leaf litter. Such changes can affect amphibian populations. For example, warm ponds are more favourable for amphibian growth and development.

Studies in which tree canopies were removed as one of a combination of interventions for the restoration of ponds are discussed in 'Restore ponds'.

A before-and-after study in 1994–1997 of two restored ponds in Jutland, Denmark (1) found that translocated garlic toads *Pelobates fuscus* established breeding populations of in both ponds. Breeding was recorded in one in 1996 and the other in 1997. Ponds were restored by removing surrounding willows and by levelling the banks of one pond. Forty-three toads were captured from a pond being eliminated by development. Four egg strings were laid and raised in captivity. The 43 adults and 1,000 tadpoles were released into one of the restored ponds in 1994. Toads were monitored by tadpole and call surveys.

A before-and-after study in 2005–2007 of a restored forest pond in Illinois, USA (2) found that hatching success of spotted salamanders *Ambystoma maculatum* did not increase following canopy removal. Two egg masses failed in 2005 and 2006 before canopy removal and two failed in 2007 after removal. Restoration started in 2000 and included destruction of drainage tiles, clearing of invasive plants and prescribed burning. Canopy thinning was undertaken in winter 2006–2007. An egg mass was placed in two mesh enclosures (56 x 36 x 36 cm) in the pond. Eggs were monitored every five days until hatching was complete.

(1)   Jensen B.H. (1997) Relocation of a garlic toad (*Pelobates fuscus*) population. *Memoranda Societatis pro Fauna et Flora Fennica*, 73, 111–113.
(2)   Sacerdote A.B. & King R.B. (2009) Dissolved oxygen requirements for hatching success of two Ambystomatid salamanders in restored ephemeral ponds. *Wetlands*, 29, 1202–1213.

13 Species management

Most of the chapters in this book are aimed at minimizing threats, but there are also some interventions which aim specifically to increase population numbers by increasing reproductive rates and by introducing individuals.

This chapter describes interventions that can be used to increase population size by translocating wild animals from one area to another, or by breeding or rearing animals in captivity (*ex-situ* conservation) to release amphibians back into the wild.

Key messages – translocate amphibians

Translocate amphibians
Fifty-four studies investigated the effectiveness of translocating amphibians. Three global reviews found that 59% of amphibian translocations that could be assessed resulted in established breeding populations or substantial recruitment to the adult population. Twenty-four of 28 studies, including 3 reviews, in New Zealand, Europe and the USA found that translocating amphibian eggs, tadpoles, juveniles or adults established, or in one case maintained, breeding populations at 25–100% of sites. Four found that breeding populations went extinct within five years, or did not establish. Two studies, including one replicated study, in Denmark and the UK found that translocations, with habitat management in some cases, increased existing populations. One systematic review found that mitigation that included translocations did not result in self-sustaining great crested newt populations. An additional 20 studies, including one review, in Canada, Europe, New Zealand, South Africa and the USA measured aspects of survival or breeding success of translocated amphibians and found mixed results.

Key messages – captive breeding, rearing and releases (*ex-situ* conservation)

Breed amphibians in captivity
Sixty-two studies investigated the success of breeding amphibians in captivity. Forty-four of 60 studies, including 7 reviews, from across the world found that amphibians successfully produced eggs in captivity; 6 studies involved captive-bred females. Twelve found mixed results depending on species, captive population or housing conditions. One found that eggs were only produced by simulating a dry and wet season and three found limited or no breeding. Thirty-three of the studies found that captive-bred amphibians were raised successfully to tadpoles, metamorphs, juveniles or adults in captivity. Five found that survival of captive-bred amphibians was low.

Use hormone treatment to induce sperm and egg release
One review and nine of ten replicated studies, including two randomized, controlled studies, in Austria, Australia, China, Latvia, Russia and the USA found that hormone treatment of male amphibians stimulated or increased sperm production, or resulted in successful breeding. One found that hormone treatment of males and females did not result in breeding. One review and 9 of 14 replicated studies, including 6 randomized

and/or controlled studies, in Australia, Canada, China, Ecuador, Latvia and the USA found that hormone treatment of female amphibians had mixed results, with 30–71% of females producing viable eggs following treatment, or with egg production depending on the combination, amount or number of doses of hormones. Three found that hormone treatment stimulated egg production or successful breeding. Two found that treatment did not stimulate or increase egg production.

Use artificial fertilization in captive breeding

Three replicated studies, including two randomized studies, in Australia and the USA found that the success of artificial fertilization depended on the type and number of doses of hormones used to stimulate egg production. One replicated study in Australia found that 55% of eggs were fertilized artificially, but soon died.

Freeze sperm or eggs for future use

Ten replicated studies, including three controlled studies, in Austria, Australia, Russia, the UK and the USA found that following freezing, viability of amphibian sperm, and in one case eggs, depended on species, cryoprotectant used, storage temperature or method and freezing or thawing rate. One found that sperm could be frozen for up to 58 weeks.

Release captive-bred individuals

Twenty-six studies investigated the success of releasing captive-bred amphibians. Ten of 15 studies, including 3 reviews, in Australia, Europe, Hong Kong and the USA found that captive-bred amphibians released as larvae, juveniles, metamorphs or adults established populations at 38–100% of sites. Five found that leopard frogs, Houston toads and green and golden bell frogs did not establish breeding populations, or only established following one of four release programmes. One review and one before-and-after study in Spain found that 41–79% of release programmes of captive-bred, captive-reared and translocated frogs combined established breeding populations. An additional ten studies, including one review, in Australia, Italy, Puerto Rico, the UK and the USA measured aspects of survival or breeding success of released captive-bred amphibians and found mixed results.

Head-start amphibians for release

Twenty-two studies head-started amphibians from eggs and monitored them after release. A global review and six of ten studies in Europe and the USA found that released head-started tadpoles, metamorphs or juveniles established breeding populations or increased existing populations. Two found mixed results with breeding populations established in 71% of studies reviewed or at 50% of sites. Two found that head-started metamorphs or adults did not establish a breeding population or prevent a population decline. An additional 10 studies in Australia, Canada, Europe and the USA measured aspects of survival or breeding success of released head-started amphibians and found mixed results. Three studies in the USA only provided results for head-starting in captivity. Two of those found that eggs could be reared to tadpoles, but only one successfully reared adults.

Translocate amphibians

13.1   Translocate amphibians

- Overall, three global reviews[1, 3, 4] and one replicated, before-and-after study in the USA[2] found that 35 of 54 (65%) amphibian translocations that could be assessed resulted in established breeding populations or substantial recruitment to the adult population. A further two translocations resulted in breeding and one in survival following release.

- One review[4] found that translocations of over 1,000 animals were more successful, but that success was not related to the source of animals (wild or captive), life-stage, continent or reason for translocation.

Background

Animals are translocated either to reintroduce species to sites that were occupied in the past, introduce species to new sites or to increase population numbers where the species is already present. The strategy can help to rescue populations from threats such as development, maintain or restore a species' historical range or increase the total number of viable populations, to safeguard against loss of other populations. When translocating animals, consideration should be given to within-species diversity and local adaption, particularly when mixing populations in order to ensure that diversity and adaptive potential is not lost (Ficetola & De Bernardi 2005).

The majority of studies translocated amphibians to sites that had been occupied in the past but had no current population or to new sites. A smaller number of studies translocated animals to sites with an existing population.

Ficetola G.F. & De Bernardi F. (2005) Supplementation or in situ conservation? Evidence of local adaptation in the Italian agile frog *Rana latastei* and consequences for the management of populations. *Animal Conservation*, 8, 33–40.

A review of translocation programmes for amphibians (1) found that none of the six programmes identified were considered successful as they did not provide evidence that a stable breeding population had been established. Two of the programmes did result in breeding, in the eastern spadefoot *Pelobates syriacus* (larvae and juveniles translocated) and the banded newt *Triturus vittatus* (juveniles translocated). Translocation of the natterjack toad *Bufo calamita* in England was not considered successful. The release of half a million wild-caught and captive-bred Houston toads *Bufo houstonensis* (adults, juveniles, metamorphs, tadpoles) to 10 sites did not result in establishment of any populations. Success was unknown for the Coeur d'salamander *Plethodon idahoensis* and Puerto Rican crested toad *Peltophryne lemur* (juveniles and adults translocated). Published and unpublished literature was searched.

A replicated, before-and-after study in 1980–1999 of 19 amphibian translocations to 5 upland sites near to New York, USA (2) found that 9 translocations of 4 species resulted in established populations (spring peeper *Pseudacris crucifer*; grey tree frog *Hyla versicolor*; Fowler's toad *Bufo fowleri*; redback salamander *Plethodon cinereus*). Four translocations of four species were likely to have been successful based on persistence of offspring records and one translocation failed. The success of five could not be assessed because of insufficient data. In 1980–1995, nine species of locally caught amphibians of different life stages were translocated to one or more of five sites. Monitoring involved frog call counts, funnel traps, drift-fences with pitfall traps, artificial coverboards and visual searches.

A review of 19 amphibian translocation programmes (3) found that all 7 of the programmes that could be assessed were considered successful. Some programmes may have included head-starting. Six species (1 toad; 5 frog) showed evidence of breeding in the wild for multiple generations (high success) and one toad species only showed evidence of survival following release (low success). The outcome was not known for the other 12 programmes. Species from eight countries were involved in these release programmes, with a bias towards temperate countries. A quarter of the species were classified in the top four highest IUCN threat categories (i.e. vulnerable to extinction in the wild).

A review of 38 global amphibian translocation projects during 1991–2006 (4) found that half were considered successful, with evidence of substantial recruitment to the adult population. Of the 38 translocation projects reviewed (25 species), 52% were successful, 29% failed and long-term success was uncertain for 19%. Projects releasing over 1,000 animals were significantly more successful (success: 65%) than those releasing less than 100 (0%) or 101–1,000 animals (38%). Success was independent of the source of animals (wild, captive, combination), life-stage translocated, continent and motivation for translocation (conservation: 90%; human-wildlife conflict: 8%; research: 3%). Translocations were of eggs, larvae and metamorphs in 71% of cases, adults in 45% and juveniles in 21% of cases. Wild animals were translocated in 76% of projects. The most common reported causes of failure were homing and migration and poor habitat. Success was defined as evidence of substantial recruitment to the adult population during monitoring over a period at least as long as it takes for the species to reach maturity.

(1) Dodd C.K.J. & Seigel R.A. (1991) Relocation, repatriation, and translocation of amphibians and reptiles: are they conservation strategies that work? *Herpetologica*, 47, 336–350.
(2) Cook R.P. (2002) Herpetofaunal community restoration in a post-urban landscape (New York and New Jersey). *Ecological Restoration*, 20, 290–291.
(3) Griffiths R.A. & Pavajeau L. (2008) Captive breeding, reintroduction, and the conservation of amphibians. *Conservation Biology*, 22, 852–861.
(4) Germano J.M. & Bishop P.J. (2009) Suitability of amphibians and reptiles for translocation. *Conservation Biology*, 23, 7–15.

13.1.1 Frogs

- Eight of ten studies (including five replicated studies) in New Zealand, Spain, Sweden, the UK and the USA found that translocating frog eggs, juveniles or adults established breeding populations at 100%[2, 3, 10, 15, 19, 21] or 79% of sites[9]. Two found that breeding populations of two species were initially established but went extinct within five years[13], or did not establish[14].

- Five studies (including one replicated study) in Italy, New Zealand and the USA found that translocated juveniles or adults survived the winter[5], had high survival[10], survived up to two years[20], or up to eight years with predator exclusion[1, 11, 18]. One study in the USA[16] found that survival was lower for Oregon spotted frogs translocated as adults compared to eggs and lower than that of resident frogs. Five studies (including three replicated studies) in Canada, New Zealand and the USA found that translocations of eggs, juveniles or adults resulted in little[1, 7, 11, 18] or no breeding at one[6] or three of four sites[14].

- Two studies (including one before-and-after study) in the USA[4, 8] found that 60–100% of translocated frogs left the release site and 35–73% returned to their original pond within 1–32 days. Two before-and-after studies New Zealand and the USA found that frogs lost weight during the 30 days after translocation[8] or became heavier than animals at the donor site[10].

Background

As there is a larger literature for wood frogs *Rana sylvatica* than other species, evidence is considered in a separate section below.

A study in 1990–2000 on Stephens Island, New Zealand (1, 11) found that 3 of 12 translocated Hamilton's frog *Leiopelma hamiltoni* survived within the new habitat for at least 8 years. Evidence of breeding had not been recorded by 1992. Only one juvenile was ever recorded, in 1996. Eight frogs survived the first year and were recaptured 61 times by 2000. Three were not recorded at the release site after 1994, but two were found back at their original habitat (76–89 m). After eight years, 42% of translocated frogs had been recaptured compared to 47% marked at the original site. Recaptured frogs showed variable weight changes between translocation and 1992 (+23%, −12 to +55%). In May 1992, frogs were translocated 40 m to a new rock-filled pit (72 m²) in a forest remnant. A predator-proof fence was built around the new habitat to exclude tuatara *Sphenodon punctatus* and the area was 'seeded' with invertebrate prey. Frogs were surveyed regularly from November 1990 to May 1992 (90 visits), intermittently in 1992–1996 and at least 4 times annually (over 6 days) in 1997–2000.

A replicated study in 1986–1993 of 13 created ponds in a reserve in England, UK (2) found that translocating common frog *Rana temporaria* eggs established breeding populations. The first naturally laid eggs were recorded in 1988 (92 clumps). The peak count was in 1989 with 162 egg clumps. Numbers of emerged froglets were high in the first year, but low in the second. Up to 12–13% of eggs were lost to collection and 16–39% to desiccation each year. In 1985, 13 ponds were excavated. Local frog spawn was introduced to the ponds in spring 1986 (200 clumps), 1987 (150), 1990 (8), 1991 (4) and 1993 (14). Monitoring was 1–3 times/week in spring 1986–1993.

A replicated, before-and-after study in 1987–1997 in Jersey, UK (3) found that agile frog *Rana dalmatina* breeding populations were established from translocated eggs. Translocated eggs hatched and were successfully reared at all three sites. Populations started breeding within two to three years of release and then bred most years. In 1987, six egg masses were removed from a polluted pond and translocated to a garden pond (1 m²). In 1993, two enclosures in a second pond were stocked with translocated eggs. Surviving frogs were translocated to a third pond in 1994.

A study in 2000 at Guadalupe Dunes, California, USA (4) found that 8 of 11 translocated California red-legged frogs *Rana aurora daytonii* returned to the original pond within a few days. All 7 adults left the release ponds between 24–48 hours after release. Six returned to the original pond in 1–9 days; one was found dead there. The 5 surviving were translocated again and four remained at the release pond for at least 10–17 days. The fifth adult was found back at the original pond within 32 days having travelled 3 km. Two juvenile frogs also returned to the original pond a number of times; the other two were not recaptured. In February 2000, seven adult and four juvenile frogs were marked and translocated 2 km, from a polluted pond to three natural ponds. The original pond was pumped dry at the end of February. Frogs were monitored by radio-tracking for a month.

A replicated study in 1998–2000 in the Lombardy District, Italy (5) found that translocated head-started Italian agile frog *Rana latastei* tadpoles metamorphosed successfully and survived over winter. Metamorphosis occurred in both years. Eggs were collected from sites close to the release sites. Eggs were hatched in semi-natural conditions in captivity. In 2000, a total of 12,000 tadpoles were raised in captivity. Tadpoles with developing hind limbs were released to new and restored ponds and habitat in five natural parks. Tadpoles were released at two sites in 2000 and seven sites in 2001.

A replicated, before-and-after study in 1995–2000 of two created ponds in Ohio, USA (6) found that translocated gray tree frogs *Hyla versicolor* did not reproduce in created ponds. Evidence of reproduction was not recorded, although frogs were heard calling at one pond in 2000. Ponds were created in 1995–1997 and were 2–4 m deep. Vegetation, plankton and organic matter (from local wetlands) were added. Gray tree frog larvae (0–35) and meta-morphs (0–4) were translocated to the pond in spring 1996–1998 and 2000. Monitoring was undertaken using drift-fencing and pitfall traps surrounding ponds, dip-netting and egg counts.

A replicated study in 1999–2002 in Alberta, Canada (7) found limited evidence of breeding by translocated head-started northern leopard frog *Rana pipiens*. Seven released frogs were recaptured, another three were heard calling and one egg mass was observed at the site surveyed. Three to six egg masses were collected from the wild each year and reared to metamorphs in two man-made outdoor ponds. Predators were excluded or removed where possible. Between 1999 and 2002, a total of 6,500 captive-reared frogs were tagged and released at 3 new sites. Surveys were undertaken at one release site in May–July 2002.

A before-and-after study in 1999 on an alpine fell in Kings Canyon National Park, California, USA (8) found that translocated mountain yellow-legged frogs *Rana muscosa* lost weight during the 30 days after translocation. Translocated frogs lost an average 1.2 g in body mass, whereas resident frogs gained 2.5 g over the same period. Seven of the trans-located frogs returned to their original capture site; 5 moved the 206–485 m in 11–30 days. Four frogs moved in the direction of their capture site and nine remained at the transloca-tion site. Twenty frogs with transmitters fitted were translocated 144–630 m to other ponds and lakes that were not typically used. Frogs were monitored intensively for 30 days in August and then surveyed using passive integrated transponder (PIT) tags. Translocated and 18 randomly selected resident frogs were weighed at the start and end of the study.

A replicated, before-and-after study in 1998–2003 of 14 ponds in Gipuzkoa province, Spain (9) found that translocated adults, along with head-started and captive-bred juvenile stripeless tree frogs *Hyla meridionalis* established breeding populations in 11 ponds. Trans-located adults survived in good numbers and returned to 12 of 14 ponds. Mating, eggs and well-developed larvae were observed in 11 of the ponds; froglets were also recorded in some ponds. Introduced predators, dense vegetation, eutrophication and drying resulted in reduced survival and reproduction in some ponds. A small number of additional ponds were colonized by the species. Thirteen ponds were created and one restored, with veget-ation planted in 1999–2000. In 1998–2003, a total of 1,405 adults were translocated to the ponds. Eggs were also collected, reared in captivity (in outdoor ponds) and released as 871 metamorphs and 19,478 tadpoles into 8 of the ponds. An additional 5,767 captive-bred tadpoles were released.

A before-and-after study in 1984–2003 on Maud Island, New Zealand (10) found that translocated Maud Island frogs *Leiopelma pakeka* established a population that remained relat-ively stable. Losses of translocated frogs were offset by local recruitment. Numbers declined initially (survival: 64%), but annual survival rate was then high (97%). Seventy per cent of translocated frogs and 35 young recruits were (re)captured over the 20-years of monitoring. Survival of local recruits was 80%. Most frogs settled within the release site, but a few dispersed up to 26 m. Translocated frogs became significantly heavier (per unit length) than those in the source population; average range size did not differ (12 m²). Frogs were marked and translocated from one forest remnant to one 0.5 km away that had no Maud Island frogs. Forty-three frogs were moved in May 1984 and 57 in May 1985. Monitoring was carried out during 4–5 successive nights over 600 m² at least twice annually until 1994 and then annually until March 2003.

A before-and-after study in 1997–2002 of the translocation of 300 Hamilton's frog *Leiopelma hamiltoni* from Maud Island to Motuara Island, New Zealand (12) found that the population established and stabilized. Losses of translocated frogs were offset by new recruits. High mortality and/or dispersal occurred during the first two months, followed by a constant high survival rate (71–100%). New juveniles were found every breeding season from 1998, just 10 months after the translocation. By August 2002, 155 of the translocated frogs and 42 recruits had been (re)captured. New recruits had survival rates of 29–88%. Frogs were toe-clipped and translocated 25 km to the predator free island in May 1997. Frogs were released into a 10 x 10 m grid with initial densities of 3/m². Frogs were monitored by recapturing within the grid during 2 sessions of 5–10 nights in 1997 and 4 sessions in 1998. The grid and a 100 m² surrounding grid were searched in August 1999–2002.

A replicated, before-and-after study in 2003–2008 of 18 ponds within agricultural landscapes in western Scania, southern Sweden (13) found that although translocation of moor frog *Rana arvalis* and common frog *Rana temporaria* eggs initially resulted in breeding populations, they were extinct within five years. Common frog calling males were found at 2 ponds, eggs in 8 and metamorphs in 12 release ponds. Moor frog calling males were found at one pond, eggs at five and metamorphs at nine ponds. Numbers of egg clumps peaked after two years. However, four years after the translocation, breeding was recorded in only two ponds and one year later those populations were extinct. Eggs were collected from south Scania and introduced into eight ponds in 2003 and ten ponds in 2004 in six areas. Each pond received 20 egg clumps from each species. Ponds were monitored for metamorphs in June–July. Release ponds and other ponds within 750 m were monitored annually.

A replicated, before-and-after study in 1994–1998 at four sites in meadow in Sequoia National Park, California, USA (14) found that following translocation of mountain yellow-legged frogs *Rana muscosa* there was no evidence of reproduction at three sites and insufficient reproduction to maintain a population at the fourth. Survival of all life history stages was high in the first week and metamorphs and adults were present at the end of the first summer. However, nearly all life history stages disappeared within 12 months of translocation. At one site there was recruitment of 28 adults from tadpoles. However, in 1997 all frogs at that site were sick or dead, due, it was thought, to chytridiomycosis and/or pesticides. A total of 22 of 135 frogs were found in nearby ponds. In 1994 and 1995, egg masses (2/site), tadsoles (0–108), sub-adults (0–25) and adults (0–31) were released at 4 previously occupied sites, 30 km from the original population. Release sites were monitored every 1–3 days in summer in 1994–1995, monthly in 1996–1997 and once in 1998. Visual surveys and adult captures were undertaken.

A before-and-after study in 2004 of a pond in parkland in Lancashire, UK (15) found that translocated common frogs *Rana temporaria* established a breeding population. Frogs were translocated to the pond from a nearby building site in 2002 and monitored in spring 2004.

A study in 2001–2004 of created ponds within a wetland in Oregon, USA (16) found that survival was lower for Oregon spotted frogs *Rana pretiosa* translocated as adults compared to those translocated as eggs. Frogs had a significantly lower survival rate during the first year after translocation, compared to the following three years (e.g. large frogs: 28–44% vs 48–74%) and non-relocated frogs. Annual survival rate was significantly higher for large frogs (>53 mm; 48–74%) compared to small frogs (40–53 mm; 5–39%). Survival increased with increasing pond age. Nine ponds were created in 2001–2004 using explosives (0.01–0.07 ha; 2 m deep). In spring 2001, 9 spotted frog egg masses and in June–September 2001, 41 marked frogs were translocated to the 4 largest ponds, from a site 2.5 km away.

A replicated study in 2005–2008 in a restored forested wetland in Lake County, Illinois, USA (17) found that translocated spring peeper *Pseudacris crucifer* tadpoles survived to metamorphosis in enclosures in restored ponds. All tadpoles survived through metamorphosis. In 2008, 12 tadpoles were placed in 2 mesh enclosures (56 x 36 x 36 cm) in 2 restored ponds. Tadpoles were monitored 2–3 times/week until metamorphosis. Tadpoles were moved if ponds dried.

A study in 2006–2009 in Zealandia, Wellington, New Zealand (18) found that survival of Maud Island frogs *Leiopelma pakeka* released in a predator-proof enclosure was high (93%), but in the wild was low. Numbers observed in the wild declined significantly, where house mice *Mus musculus* and little spotted kiwis *Apteryx owenii* were known predators. In the enclosure, two males were recorded breeding in February 2008 and ten nearly metamorphosed young frogs resulted. Sixty frogs from Maud Island were placed in a 2 x 4 m predator-proof mesh enclosure in 2006. In April 2007, 29 were retained in the enclosure and 28 were released into the adjacent forest. Larvae found were moved to incubators to complete metamorphosis and then released into nursery pens.

A before-and-after study in 2004–2011 of 71 Hamilton's frogs *Leiopelma hamiltoni* translocated from Stephens Island to Nukuwaiata Island, New Zealand (19) found that a breeding population was established. Production of juveniles, establishment of the new population and recovery of the donor population was slower than expected. However, by August 2011, repeated breeding had occurred and new recruits were almost at breeding age. There was no evidence of a decline within the donor population. Frogs were translocated between May 2004 and July 2006. Both the translocated frogs and donor population were monitored until 2011.

A study in 2010–2012 in southwest Georgia, USA (20) found that a number of translocated head-started gopher frogs *Lithobates capito* survived. Some froglets released in 2012 were observed later in the year and a large adult female released in 2010 was captured. Portions of egg masses were collected from one of the remaining breeding sites and transferred to partner institutions for rearing to metamorphosis. Tadpoles were reared outdoors in large tanks with plant matter from the egg collection site. Over 4,300 froglets were marked and released onto restored Nature Conservancy land, which lacked a natural population. In 2012, froglets were released directly into burrows as protection from drought. Monitoring began in summer 2012.

A before-and-after study in the UK (21) found that a small population of pool frogs *Pelophylax lessonae* was established from translocations. The frogs were healthy and had good survival rates, but the population did not grow as anticipated. Not all of the ponds were used by the frogs. In 2005, adults, juveniles and tadpoles were collected from Sweden and released at a recently restored site. Releases were repeated three times. Individual frogs were monitored.

(1) Brown D. (1994) Transfer of Hamilton's frog, *Leiopelma hamiltoni*, to a newly created habitat on Stephens Island, New Zealand. *New Zealand Journal of Zoology*, 21, 425–430.
(2) Cooke A.S. & Oldham R.S. (1995) Establishment of populations of the common frog *Rana temporaria* and common toad *Bufo bufo* in a newly created reserve following translocation. *Herpetological Journal*, 5, 173–180.
(3) Gibson R.C. & Freeman M. (1997) Conservation at home: recovery programme for the agile frog *Rana dalmatina* in Jersey. *Dodo*, 33, 91–104.
(4) Rathbun G.B. & Schneider J. (2001) Translocation of California red-legged frogs (*Rana aurora daytonii*). *Wildlife Society Bulletin*, 29, 1300–1303.
(5) Gentilli A., Scali S., Barbieri F. & Bernini F. (2002) A three-year project for the management and the conservation of amphibians in Northern Italy. *Biota*, 3, 27–33.

(6)   Weyrauch S.L. & Amon J.P. (2002) Relocation of amphibians to created seasonal ponds in southwestern Ohio. *Ecological Restoration*, 20, 31–36.

(7)   Kendell K. (2003) Northern leopard frog reintroduction: year 4 (2002). Alberta Sustainable Resource Development & Fish and Wildlife Service Report. Alberta Species at Risk Report.

(8)   Matthews K.R. (2003) Response of mountain yellow-legged frogs, *Rana mucosa*, to short distance translocation. *Journal of Herpetology*, 37, 621–626.

(9)   Rubio X. & Etxezarreta J. (2003) Plan de reintroducción y seguimiento de la ranita meridional (*Hyla meridionalis*) en Mendizorrotz (Gipuzkoa, País Vasco) (1998–2003). *Munibe*, 16, 160–177.

(10)   Bell B.D., Pledger S. & Dewhurst P.L. (2004) The fate of a population of the endemic frog *Leiopelma pakeka* (Anura: Leiopelmatidae) translocated to restored habitat on Maud Island, New Zealand. *New Zealand Journal of Zoology*, 31, 123–131.

(11)   Tocher M.D. & Brown D. (2004) *Leiopelma hamiltoni* homing. *Herpetological Review*, 35, 259–261.

(12)   Tocher M.D. & Pledger S. (2005) The inter-island translocation of the New Zealand frog *Leiopelma hamiltoni*. *Applied Herpetology*, 2, 401–413.

(13)   Loman J. & Lardner B. (2006) Does pond quality limit frogs *Rana arvalis* and *Rana temporaria* in agricultural landscapes? A field experiment. *Journal of Applied Ecology*, 43, 690–700.

(14)   Fellers G.M., Bradford D.F., Pratt D. & Long Wood L. (2007) Demise of translocated populations of mountain yellow-legged frogs (*Rana muscosa*) in Sierra Nevada of California. *Herpetological Conservation and Biology*, 2, 5–21.

(15)   Neave D.W. & Moffat C. (2007) Evidence of amphibian occupation of artificial hibernacula. *Herpetological Bulletin*, 99, 20–22.

(16)   Chelgren N.D., Pearl C.A., Adams M.J. & Bowerman J. (2008) Demography and movement in a relocated population of Oregon spotted frogs (*Rana pretiosa*): influence of season and gender. *Copeia*, 2008, 742–751.

(17)   Sacerdote A.B. (2009) Reintroduction of extirpated flatwoods amphibians into restored forested wetlands in northern Illinois: feasibility assessment, implementation, habitat restoration and conservation implications. PhD thesis. Northern Illinois University.

(18)   Bell B.D., Bishop P.J. & Germano J.M. (2010) Lessons learned from a series of translocations of the archaic Hamilton's frog and Maud Island frog in central New Zealand. In P. S. Soorae (eds) *Global Re-introduction Perspectives: 2010. Additional Case Studies from Around the Globe*, IUCN/SSC Reintroduction Specialist Group, Gland, Switzerland. pp. 81–87.

(19)   Tocher M. (2011) 'State of the nation' report on New Zealand translocations including a quick overview of past translocations. *Froglog*, 99, 39–40.

(20)   Hill R. (2012) Gopher frog head-starting project reaches major milestone. *Amphibian Ark Newsletter*, 21, 9.

(21)   Wilkinson J.W. & Buckley J. (2012) Amphibian conservation in Britain. *Froglog*, 101, 12–13.

13.1.2 Wood frogs

- Two studies (including one replicated study) in the USA[1, 2] found that translocated wood frog eggs established breeding populations in 25–50% of created ponds.

- One replicated study in the USA[3] found that translocated wood frog eggs hatched and up to 57% survived as tadpoles in enclosures in restored ponds.

A before-and-after study in 1965–1986 of two created ponds in Missouri, USA (1) found that translocated wood frog *Rana sylvatica* eggs established a breeding population in one of two created ponds. At the second pond wood frogs did not establish. In 1980, wood frog eggs were translocated to two newly constructed ponds. Ponds were monitored until 1986.

A replicated, before-and-after study in 1974–1995 in Missouri, USA (2) found that one of four wood frog *Rana sylvatica* egg translocations established a breeding population. The population was stable between 1987 (311 captured) and 1995 (364). Wood frogs also colonized four other created ponds (0.9–2.4 km). In 1980, 11 wood frog egg masses were translocated 50 km into 4 created ponds. Monitoring was undertaken using drift-fencing with pitfall traps around ponds and by egg mass counts and call surveys.

A replicated study in 2005–2008 in a restored forested wetland in Lake County, Illinois, USA (3) found that translocated wood frog *Lithobates sylvaticus* eggs hatched and survived as tadpoles in enclosures in restored ponds. Tadpole survival in restored ponds was 6–57%. In

2008, two translocated wood frog egg masses were placed in separate mesh enclosures (56 x 36 x 36 cm) in each of five restored ponds. Tadpoles were monitored 2–3 times/week until metamorphosis. Tadpoles were moved if ponds dried.

(1) Sexton J. & Phillips C. (1986) A qualitative study of fish-amphibian interactions in 3 Missouri ponds. *Transactions of the Missouri Academy of Science*, 20, 25–35.
(2) Sexton O.J., Phillips C.A., Bergman T.J., Wattenberg E.W. & Preston R.E. (1998) Abandon not hope: status of repatriated populations of spotted salamanders and wood frogs at the Tyson Research Center, St.Louis County, Mo 1998. In *Status and Conservation of Midwestern Amphibians*, Universiity of Iowa Press, Iowa City, Iowa. pp. 340–344.
(3) Sacerdote A.B. (2009) Reintroduction of extirpated flatwoods amphibians into restored forested wetlands in northern Illinois: feasibility assessment, implementation, habitat restoration and conservation implications. PhD thesis. Northern Illinois University.

13.1.3 Toads

• Two of four studies (including two replicated studies) in Denmark, Germany, the UK and the USA[1, 3, 4, 6, 7] found that translocating eggs and/or adults established common toad breeding populations. One found populations of garlic toads established at two of four sites[6]. One found that breeding populations of boreal toads were not established[7].

• One before-and-after study in Denmark[9, 10] found that translocating green toad eggs to existing populations, along with aquatic and terrestrial habitat management, increased population numbers.

• Three studies (including one before-and-after study) in Germany, Italy and the USA found that 33–100% of translocated adult toads reproduced[2, 5], 19% survived up to six years[5] or some metamorphs survived over winter[8]. One replicated study in South Africa[11] found that translocated Cape platanna metamorphs survived up to 23 years at one of four sites.

Background
As there is a larger literature for natterjack toads *Bufo calamita* than other species, evidence is considered in a separate section below.

A before and after study in 1986–1992 of a created pond in wet pasture near Ahlerstedt, Germany (1, 3) found that translocated common toads *Bufo bufo* bred in the new pond every year. Population size did not differ significantly before and after resettlement (522 vs 590). In 1987, 29% of migrating toads chose the created pond rather than their original pond (across a road). By 1988 the proportion was 75% and by 1992 it was 99%. Marked individuals indicated that 83% of the population used the new pond (91% of males; 67% of females). An amphibian fence with pitfall traps was installed along 400 m of road. Toads captured were placed in the created pond (53 x 20 m). A temporary mesh fence around the pond allowed toads to reach but not leave the pond in spring 1986–1990. All animals were tagged.

A study in 1991–1992 in Wyoming, USA (2) found that a translocated pair of Wyoming toads *Bufo baxteri* bred in the first year. In 1992, tadpoles were produced from eggs laid within the breeding enclosure in the release pond. Toads did not breed in the original pond, the only remaining wild population. Female and juvenile toads were captured from the wild and over-wintered in captivity for four months. One of the females and a wild-captured male were released into a breeding enclosure within the release pond.

A replicated study in 1986–1993 of 13 created ponds in a reserve in England, UK (4) found that translocated eggs and adult common toads *Bufo bufo* established breeding populations. The first naturally laid eggs were recorded in the second year. In 1988, 64% of male and 89% of female toads captured were already marked, suggesting that most adults were introduced rather than natural colonizers. The proportion marked dropped to 15% in 1990 suggesting a 64% loss of male toads in the first year, reducing to 39% in the second and 42% in the third year. The toad population was estimated at 200–300 adults in 1993. Up to 12–13% of eggs were lost to collection and 16–39% to desiccation each year. In 1985, 13 ponds were excavated. Half a million toad eggs were introduced in 1986 and 5,911 marked adults in 1987. Adults and eggs were monitored 1–3 times/week in spring 1986–1993.

A before-and-after study in 1986–1994 of a created pond in Gifhorn, Germany (5) found that translocated common spadefoot toads *Pelobates fuscus* bred in the new pond. Mortality rate of translocated toads was high, with only 19% of toads recaptured in 1993–1994. Monitoring indicated that 33% of translocated toads reproduced in the created pond. A total of 152 juveniles were recorded in the pond in 1990. From 1989, toads were captured using drift-fencing with pitfall traps along the side of the road. Toads were marked and translocated across the road to the pond (700 m²) created for amphibians within forest in 1988. Monitoring was undertaken using drift-fencing with pitfall traps either side of the road and around the pond.

A replicated, before-and-after study in 1994–1997 in Jutland, Denmark (6) found that translocated adult and head-started tadpole garlic toads *Pelobates fuscus* established breeding populations in two restored, but not two created ponds. The authors considered that failure might have been due to predation because of the lack of vegetation and introduction of sticklebacks *Pungitius pungitius*. Forty-three toads were captured from a pond being eliminated by development. They were translocated to a restored pond. Four egg strings were laid in captivity and produced over 2,000 tadpoles. They were released at different stages before metamorphosis into the same restored and one created pond ($n = 1,000$). Two ponds had been restored and two created in 1994–1995. Toads were monitored by tadpole and call surveys.

A replicated, before-and-after study in 1995–1999 at two sites within a National Park in Colorado, USA (7) found that a breeding population of boreal toads *Bufo boreas* was not established from translocated eggs. At one site, only 12 tadpoles were recorded during the first week after release. Following that only two toads were recorded in 1997. At the other site 333 metamorphs were captured and marked in the first 2 weeks, but none were recorded in 1997–1999. Hatching success did not differ significantly between the original and release sites (69 vs 38–72%). Seventeen egg masses were collected in July 1995 and June 1996. Half of nine, and six complete egg masses were translocated to two sites, where toads had been absent for five and eight years. A small number of eggs were placed in predator-proof boxes to compare hatching success between original and release sites. Following translocation of eggs, a 0.01–0.09 km² area was searched 1–3 days/week in April–September 1995–1999. Metamorphs were toe-clipped.

A study in 1998–2000 in the Lombardy District, Italy (8) found that translocated head-started common spadefoot toad *Pelobates fuscus insubricus* tadpoles metamorphosed successfully and survived over winter. Metamorphosis occurred in both years and some juveniles were found in spring 2001. Eggs were collected from sites close to the release sites. Eggs were hatched in semi-natural conditions in captivity. In 2000, two thousand tadpoles were raised in captivity. In 2000, tadpoles with developing hind limbs were released to 6 new and restored ponds and habitat in 5 natural parks.

A before-and-after study in 1986–1997 of five ponds on coastal meadows on Avernakø island, Denmark (9, 10) found that there was a significant increase in green toad *Bufo viridis* population following translocation of eggs, along with pond creation and restoration and re-introduction of grazing. The population increased from 20 in 1988–1990 to 920 in 1995–1997. Pond occupancy increased from one to seven and the number of ponds with breeding success increased from zero to five. In 1989–1997, one pond was created and four restored by re-moving plants and dredging. Cattle grazing was reintroduced to 25 ha of coastal meadows and abandoned fields. In 1994–1995, a total of 14,500 eggs were translocated to 4 of the ponds. Populations were monitored annually in 1990–1997 during two or three call and visual surveys and dip-net surveys.

A replicated study in 1988–2011 at four water bodies in the Western Cape, South Africa (11) found that some translocated Cape platanna *Xenopus gilli* metamorphs survived for over 23 years. A year after release, seven frogs were recorded at one of the four release sites (the site that had received the most metamorphs). Nine years later in 1998, six females were cap-tured, three of which were marked. In 2008 and 2011, frogs were recorded at the same site. Two of the frogs in 2008 and one in 2011 had been marked in 1998. In 1988, 154 metamorphs were translocated 25 km to 4 water bodies (one received 69) where the species was historic-ally present. Monitoring was conducted in 1989–1990, with additional visits in 1998, 2008 and 2011. In 2008 and 2011, baited funnel traps were placed at each release point.

(1) Schlupp M., Kietz R., Podloucky R. & Stolz F.M. (1989) *Pilot Project Bracken: Preliminary Results from the Resettlement of Adult Toads to a Substitute Breeding Site*. Proceedings of the Amphibians and Roads: Toad Tunnel Conference. Rendsburg, Federal Republic of Germany. pp. 127–135.
(2) Johnson R.R. (1994) Model programs for reproduction and management: ex situ and in situ conserva-tion of toads of the family Bufonidae. In J. B. Murphy, K. Adler & J. T. Collins (eds) *Captive Management and Conservation of Amphibians and Reptiles*, Contributions to Herpetology Vol. 11, Society for the Study of Amphibians and Reptiles, Ithaca, New York. pp. 243–254.
(3) Schlupp I. & Podloucky R. (1994) Changes in breeding site fidelity: a combined study of conservation and behaviour in the common toad *Bufo bufo*. *Biological Conservation*, 69, 285–291.
(4) Cooke A.S. & Oldham R.S. (1995) Establishment of populations of the common frog *Rana temporaria* and common toad *Bufo bufo* in a newly created reserve following translocation. *Herpetological Journal*, 5, 173–180.
(5) Baumann K. (1997) The population ecology of the common spadefoot toad (*Pelobates fuscus*) near Leiferde (district Gifhorn, Germany) with special regard to the effect of its artificial relocation into a new breeding-pond. *Braunschweiger Naturkundliche Schriften*, 5, 249–267.
(6) Jensen B.H. (1997) Relocation of a garlic toad (*Pelobates fuscus*) population. *Memoranda Societatis pro Fauna et Flora Fennica*, 73, 111–113.
(7) Muths E., Johnson T.L. & Corn P.S. (2001) Experimental repatriation of boreal toad (*Bufo boreas*) eggs, metamorphs, and adults in Rocky Mountain National Park. *Southwestern Naturalist*, 46, 106–113.
(8) Gentilli A., Scali S., Barbieri F. & Bernini F. (2002) A three-year project for the management and the conservation of amphibians in Northern Italy. *Biota*, 3, 27–33.
(9) Briggs L. (2003) Recovery of the green toad *Bufo viridis* Laurenti, 1768 on coastal meadows and small islands in Funen County, Denmark. *Deutsche Gesellschaft für Herpetologie und Terrarienkunde*, 14, 274–282.
(10) Briggs L. (2004) Restoration of breeding sites for threatened toads on coastal meadows. In R. Rannap, L. Briggs, K. Lotman, I. Lepik & V. Rannap (eds) *Coastal Meadow Management – Best Practice Guidelines*, Ministry of the Environment of the Republic of Estonia, Tallinn. pp. 34–43.
(11) Measey G.J. & de Villiers A.L. (2011) Conservation introduction of the Cape platanna within the Western Cape, South Africa. In P. S. Soorae (ed) *Global Re-introduction Perspectives: 2011. More case studies from around the globe*, IUCN/SSC Re-introduction Specialist Group & Abu Dhabi Environment Agency, Gland, Switzerland. pp. 91–93.

13.1.4 Natterjack toads

- Three studies (including one review) in France and the UK found that translocated natter-jack toad eggs, tadpoles, juveniles or adults established breeding populations at one site[5] or in 30–70% of cases, some of which also released head-started or captive-bred animals or included habitat management[1, 3]. The review found that re-establishing toads on dune or saltmarsh habitat was more successful than on heathland[3]. One replicated study in the UK found that natterjack toad populations increased at sites established by translocations, particularly with replicated translocations of wild rather than captive-bred toads[4].

• Two replicated, before-and-after studies in Estonia and the UK[1, 2] found that translocating natterjack toad eggs or tadpoles resulted in breeding at 8–70% of sites, some of which had been restored.

Background

The natterjack toad *Bufo calamita* has declined, particularly in the UK where the species is now endangered. The decline is largely due to habitat destruction and the acidification of breeding sites. In Britain, attempts to reintroduce the species to sites where they had recently gone extinct started in the 1970s.

A replicated, before-and-after study in 1972–1995 at dune and heathland sites in England, UK (1) found that at least six translocations of natterjack toad *Bufo calamita* eggs resulted in expanding new populations and eight showed initial signs of success. Two of three dune sites had breeding within three years and the third (> 5 years old) established one of the largest populations. Five of 17 translocations on heathland were successful, with stable or increasing adult numbers and breeding for at least 5 years. Six less than three years old, all produced toadlets in their first year. Six failed (five pre-1980) with no successful metamorphosis. Ten successful translocations were to sites with new ponds; in heathland six were concrete and one butyl plastic lined. Nine were undertaken after 1991 and comprised translocations of eggs (2 spawn strings, i.e. 5,000 eggs) each year for 2 years. Scrub clearance was undertaken at two dune and seven heath sites. One heath site had limestone added to acidic ponds. Low-density sheep or cattle grazing (<1 animal/3 ha) was established at one dune and two heath sites. Spawn strings were counted and toadlet production estimated.

A replicated, before-and-after study in 2000–2004 of 13 coastal meadows in Estonia (2) found that translocated natterjack toad *Bufo calamita* tadpoles bred at at least one site within three years. Following translocation of tadpoles in 2000, the first calling males were heard and spawning was recorded in the spring of 2003 in Saastna. In 2001–2004, terrestrial and aquatic habitats were restored on 13 coastal meadows where natterjacks had disappeared but could be reintroduced. Approximately 30,000 tadpoles from isolated quarry populations were translocated to the restored meadows.

A review of 29 translocation programmes for the natterjack toad *Bufo calamita* in the UK (3) found that 19 of 27 translocations (70%) that could be assessed were successful in the short to medium term, with adults returning to breed successfully and self-sustaining populations established at some sites. Re-establishing toads on dune or saltmarshes was more successful than on heathland (85 vs 57% success). Translocations have resulted in populations of over 200 adults at some sites, although at other sites populations have remained small. The 10 translocations since 2000 have resulted in an increase of known natterjack sites from about 40 in 1970 to 69 in 2010. Between 1975 and 2010, toads were translocated to 29 sites. Reintroductions were mainly through translocation of eggs and tadpoles from existing populations. In some cases, head-starting of tadpoles and captive breeding was also undertaken. Habitat management has also been undertaken at some sites. Since 1985 all populations have been monitored annually.

A replicated, before-and-after study in 1985–2006 of natterjack toad *Bufo calamita* populations at 20 sites in the UK (4) found that populations increased at sites established via translocations. The average population trend for translocation sites was significantly positive (0.10) while for native sites it did not differ significantly from zero (–0.04). Numbers of years of translocations of wild (including head-started) animals, but not captive-bred animals had a significant effect on population trends. Overall, 5 of the 20 sites showed positive population trends, 5 showed negative trends and 10 trends were not significantly different from zero. Data on populations (egg string counts) and management activities over 11–21 years were obtained from the Natterjack Toad Site Register. Translocations were undertaken at seven sites using wild-sourced (including head-starting) or captive-bred toads. Habitat management for toads was also undertaken at 7 of the 20 sites.

A before-and-after study in 2001 in northern France (5) found that a breeding population of natterjack toads *Bufo calamita* was established following translocation of adults and juveniles. Annual survival was 25%, which was half the value estimated in native populations. However, there was repeated breeding over several years. In 2001, a total of 5,000 adult and juvenile toads were translocated from a development site to three receptor sites where natterjacks were still or were historically present. Toads were captured using 5 km of drift-fencing with pitfall traps and 160 plywood or carpet boards laid over a 400 ha area. Monitoring was undertaken annually by surveying potential breeding sites and radio-tracking adults.

(1) Denton J.S., Hitchings S.P., Beebee T.J.C. & Gent A. (1997) A recovery program for the natterjack toad (*Bufo calamita*) in Britain. *Conservation Biology*, 11, 1329–1338.
(2) Rannap R. (2004) Boreal Baltic coastal meadow management for Bufo calamita. In R. Rannap, L. Briggs, K. Lotman, I. Lepik & V. Rannap (eds) *Coastal Meadow Management – Best Practice Guidelines*, Ministry of the Environment of the Republic of Estonia, Tallinn. pp. 26–33.
(3) Griffiths R.A., McGrath A. & Buckley J. (2010) Reintroduction of the natterjack toad in the UK. In P.S. Soorae (ed) *Global Re-introduction Perspectives: 2010. Additional case studies from around the globe*, IUCN/SSC Re-introduction Specialist Group, Gland, Switzerland. pp. 62–65.
(4) McGrath A.L. & Lorenzen K. (2010) Management history and climate as key factors driving natterjack toad population trends in Britain. *Animal Conservation*, 13, 483–494.
(5) Beebee T., Cabido C., Eggert C., Mestre I.G., Iraola A., Garin-Barrio I., Griffiths R.A., Miaud C., Oromi N., Sanuy D., Sinsch U. & Tejedo M. (2012) 40 years of natterjack toad conservation in Europe. *Froglog*, 101, 40–43.

13.1.5 Salamanders (including newts)

• One review and three before-and-after studies in the UK and the USA found that translocated eggs or adults established breeding populations of salamanders[1–3] or smooth newts[5].

• One replicated, before-and-after study in the USA found that one of two salamander species reproduced following translocation of eggs, tadpoles and metamorphs[4]. One before-and-after study in the USA[6] found that translocated salamander eggs hatched and tadpoles had similar survival rates as in donor ponds.

Background
As there is a larger literature for great crested newts *Triturus cristatus* than other species, evidence is considered in a separate section below.

A before-and-after study in 1965–1986 of two created ponds in Missouri, USA (1) found that translocating eggs and larvae established breeding populations of spotted salamanders *Ambystoma maculatum* and ringed salamanders *Ambystoma annulatum*. A breeding population of spotted salamanders was established in one pond. At the other pond, adult ringed salamanders *Ambystoma annulatum* were recorded in 1984 and egg masses in 1986. In 1965 and 1968, eggs and larvae of spotted salamanders were translocated to a newly constructed pond. In 1977, eggs of the ringed salamander were translocated to another created pond. Ponds were monitored until 1986.

A review in 1991 of amphibian translocation programmes (2) found that three salamander translocations resulted in established breeding populations. In one study, breeding populations of two salamander species were established (1). In a second study, a breeding population of tiger salamanders *Ambystoma tigrinum* established at a created pond, with returning adults and 18–25 egg masses recorded within 4 years. In 1982–1985, 1,000 tiger salamander eggs were translocated (20 km) annually to the pond (0.2 ha) in New Jersey, USA.

A before-and-after study in 1974–1995 in Missouri, USA (3) found that translocated spotted salamander *Ambystoma maculatum* eggs established a breeding population. Numbers of salamander captures increased from 428 in 1974 to 2,301 in 1995 at the release pond. Salamanders also colonized four other created ponds (0.9–2.4 km). In 1966, spotted salamander egg masses were translocated 1 km to a newly constructed pond. Another six ponds were constructed at the site in 1965–1979. Monitoring was undertaken using drift-fencing with pitfall traps around ponds and by egg mass counts.

A replicated, before-and-after study in 1995–2000 of two created ponds in Ohio, USA (4) found that translocated spotted salamanders *Ambystoma maculatum*, but not tiger salamanders *Ambystoma tigrinum*, reproduced in created ponds. Four adult spotted salamanders and one egg mass were found in one pond in 1997 and three egg masses in the other pond in 2000. Metamorphs were produced in both ponds in 1996–1998. Tiger salamanders were not recorded following their introduction. Ponds were created in 1995–1997 and were 2–4 m deep. Vegetation, plankton and organic matter (from local wetlands) were added. Spotted salamander eggs (600–1,100), larvae (40–850) and metamorphs (4–33) and tiger salamander metamorphs (0–25) were translocated in spring 1996–1998 and 2000. Monitoring was undertaken using drift-fencing and pitfall traps surrounding ponds, dip-netting and egg counts.

A before-and-after study in 2004 of a pond in parkland in Lancashire, UK (5) found that translocated smooth newts *Triturus vulgaris* established a breeding population. Newts were translocated to the pond from a nearby building site in 2002 and monitored in spring 2004.

A replicated study in 2005–2008 in a restored forested wetland in Lake County, Illinois, USA (6) found that translocated spotted salamander *Ambystoma maculatum* eggs hatched and survived as tadpoles in enclosures in restored ponds. Overall, tadpole survival rates (without effects of pond drying) were similar in restored and donor ponds (15–65 vs 26–81%). Translocated egg masses were placed in two mesh enclosures (56 x 36 x 36 cm) in each of five restored ponds and three enclosures in three donor ponds annually in 2005–2008. Tadpoles were monitored two or three times/week until metamorphosis. Tadpoles were moved if ponds dried.

(1) Sexton J. & Phillips C. (1986) A qualitative study of fish-amphibian interactions in 3 Missouri ponds. *Transactions of the Missouri Academy of Science*, 20, 25–35.
(2) Reinert H.K. (1991) Translocation as a conservation strategy for amphibians and reptiles: some comments, concerns, and observations. *Herpetologica*, 47, 357–363.
(3) Sexton O.J., Phillips C.A., Bergman T.J., Wattenberg E.W. & Preston R.E. (1998) Abandon not hope: status of repatriated populations of spotted salamanders and wood frogs at the Tyson Research Center, St.Louis County, MO. In *Status and Conservation of Midwestern Amphibians*, Universiity of Iowa Press, Iowa City, Iowa. pp. 340–344.

(4) Weyrauch S.L. & Amon J.P. (2002) Relocation of amphibians to created seasonal ponds in southwestern Ohio. *Ecological Restoration*, 20, 31–36.
(5) Neave D.W. & Moffat C. (2007) Evidence of amphibian occupation of artificial hibernacula. *Herpetological Bulletin*, 99, 20–22.
(6) Sacerdote A.B. (2009) Reintroduction of extirpated flatwoods amphibians into restored forested wetlands in northern Illinois: feasibility assessment, implementation, habitat restoration and conservation implications. PhD thesis. Northern Illinois University.

13.1.6 Great crested newts

- Four of six studies (including one review and one replicated study) in the UK found that translocated great crested newts maintained[6] or established[1, 4, 6, 7] breeding populations. The review found that populations were present one year after release in 37% of cases[2] and one study found that although translocations maintained a population in the short term, within three years breeding failed in 48% of ponds[8]. One systematic review of 31 great crested newt studies[9] found that there was no conclusive evidence that mitigation that included translocations resulted in self-sustaining populations.

- One review in the UK[2, 3, 5] found that great crested newts reproduced following 56% of translocations, in some cases there was also release of head-started larvae and/or habitat management.

A before-and-after study in 1990–1993 of six ponds at an opencast coal site near Manchester, UK (1) found that translocated great crested newts *Triturus cristatus* established a breeding population over the first two years. The number of newts captured at the site increased from 473 in 1992 to 892 in 1993 (1,063 released). Between one and 223 metamorphs were caught leaving created ponds and 1–197 leaving existing ponds each year from 1991 to 1993. In 1990–1991, three ponds were created and three others managed for amphibians within a mitigation area for works at the mine. Artificial egg laying substrate (plastic strips) was provided in new ponds. A total of 813 newts in 1991, 250 in 1992 and 625 in 1993 were translocated from mine to conservation ponds. Newts were monitored using drift-fencing with pitfall traps around the ponds and site boundary.

A review of translocation programmes in 1990–1994 for great crested newts *Triturus cristatus* in England, UK (2), extended in later studies (3, 5), found that adults returned to ponds in most cases and bred in 61% of translocations monitored. However, longer-term monitoring over 6–18 years showed that 53% of 15 translocations before 1990 failed. In 1990–1994, adults returned in subsequent years in 92% of 92 cases monitored, although newts were already present at 10 ponds. Seventy-two translocations from development sites involved adults (average: 197; total: 13,115), juveniles (57; 914), larvae (32; 501) and many eggs. Twelve translocations involved collecting eggs and rearing and releasing larvae (average: 643) and juveniles (63) for introduction purposes. Habitat enhancement (e.g. log piles, hibernacula, tree planting) was undertaken in 79% of 28 cases where there was partial habitat destruction. Where there was complete habitat destruction, newts tended to be moved to existing sites. Licenses for all translocation projects between 1990 and 1994 were reviewed and 74 licensees contacted for information. Extra monitoring information was obtained for translocations undertaken before 1990.

In an extension of a previous review (2), a review of 178 great crested newt *Triturus cristatus* translocation programmes in 1985–1994 in the UK (3) found that populations were present one year after release in 37% of all cases (see also (5)). In 10% of cases no newts were present the following year. Over half of the projects did not have enough evidence to assess success. Success of monitored projects increased from 59% before 1990 to 78% in 1990–1994. In one project, less than 40% of 1,000 translocated newts remained within a 5 ha managed conservation area. However, those that remained produced 135 (in 3 ponds) and 567 metamorphs (6 ponds) in the first and second year respectively. Male survival over two years was estimated as 46% and translocated newts gained mass (18%). Data from translocation projects was obtained from Natural England licensing records. In the case study, newts were translocated to an adjacent conservation area in 1991–1992. Trees and shrubs had been planted and hibernacula and three ponds created. Newts were monitored using drift-fencing and pitfall traps and using bottle traps.

A before-and-after study in 1985–1993 in England, UK (4) found that a new breeding population was established from 38 translocated great crested newts *Triturus cristatus*. Although no newts were observed six years after translocation, *ad hoc* monitoring over the next few years found increasing numbers of newts. Newts were translocated 100 km from a site in Kent to Cambridgeshire because of habitat destruction during a development project. Adults, metamorphs and larvae were monitored during night spotlight counts in spring and summer each year. No newts were present at the new site prior to translocation.

In an update of previous reviews (2, 3), a review of 72 great crested newt *Triturus cristatus* translocation projects carried out to mitigate against development in 1990–2001 in England, UK (5) found that where follow-up monitoring was conducted, there was evidence of breeding at over half of sites one-year post-development (56%). However, projects did not provide data to compare numbers before and after translocation or to determine whether sustainable populations were established. Only 49% of projects monitored populations, most for two years or less. The average number of newts translocated per project declined significantly from 358 in 1990–1994 to 172 in 1995–2001. Most translocations were to areas within or adjacent to the development site (<500 m). There was a net loss in overall area of aquatic habitat. Licensing information collected by the governmental licensing authorities was analysed and a questionnaire survey sent out to a sample of 153 mitigation projects (47% provided data). Of 737 licensed projects on file, 55% contained no report of the work undertaken, although it is a condition of the licence.

A replicated, before-and-after study in 2005 of nine mitigation projects in England, UK (6) found that translocation of great crested newts *Triturus cristatus* resulted in the maintenance or establishment of populations at all sites. However, after three or more years, numbers captured at five of the nine populations were lower than that prior to translocation or less than the total number translocated. Four populations were classified as 'small' (peak count: 1–3) and 5 as 'medium' (16–86). Mitigation projects during development work had been carried out at least three years previously. Between 2 and 164 newts were translocated at each site between 1987 and 2001. Terrestrial habitat management was also undertaken at two sites and artificial refugia provided at one. Monitoring was undertaken in March–May 2005 using egg searches, torch surveys, bottle trapping and mark-recapture.

A before-and-after study in 2004 of a pond in parkland in Lancashire, UK (7) found that translocated great crested newts *Triturus cristatus* established a breeding population. Newts were translocated to the pond from a nearby building site in 2002 and monitored in spring 2004.

A before-and-after study in 2006–2009 in North Lanarkshire, Scotland, UK (8) found that translocations maintained a great crested newt *Triturus cristatus* breeding population in the short term. Breeding adult counts were higher after translocation (100–299 vs 66–140). Adult survival rate was 43% and there was some recruitment into the breeding population. However, numbers of eggs, larvae and metamorphs suggested breeding failure and low juvenile survival and recruitment. In 2008, no eggs or larvae were recorded in half of the 25 ponds. Metamorph counts decreased significantly from 39 in 2006 to 5 in 2009. The newt population was translocated from the original site to a created and restored site (29 ha; 600 m away) in 2004–2006. A total of 1,594 newts (1,012 adults) were moved. The original site had been monitored for six years before translocation. Monitoring at the release site was undertaken using torchlight sampling, egg counts and metamorph counts at the perimeter fence.

A systematic review in 2011 of the effectiveness of mitigation actions for great crested newts *Triturus cristatus* in the UK (9) found that none of the 11 studies captured nor monitoring data from licensed mitigation projects showed conclusive evidence to suggest that mitigation that included translocations resulted in self-sustaining populations or connectivity to populations in the wider countryside. Only 22 of 460 licensed projects provided post-development monitoring data and of those, 16 reported that small, 3 medium and 1 large population was sustained. Two reported a loss of the population. The review identified 11 published or unpublished studies together with 309 Natural England and 151 Welsh Assembly Government (licensing authorities) mitigation licence files. Mitigation measures were undertaken to reduce the impact of development and included habitat management such as creating or restoring ponds, as well as actions to reduce deaths including translocations.

(1)   Horton P.J. & Branscombe J. (1994) *Case Study: Lomax Brow: Great Crested Newt Project*. Proceedings of the Conservation and Management of Great Crested Newts. Kew Gardens, Richmond, Surrey. pp. 104–110.
(2)   May R. (1996) The translocation of great crested newts, a protected species. MSc thesis. University of Wales.
(3)   Oldham R.S. & Humphries R.N. (2000) Evaluating the success of great crested newt (*Triturus cristatus*) translocation. *Herpetological Journal*, 10, 183–190.
(4)   Cooke A.S. (2001) Translocation of small numbers of crested newts (*Triturus cristatus*) to a relatively large site. *Herpetological Bulletin*, 75, 25–29.
(5)   Edgar P.W., Griffiths R.A. & Foster J.P. (2005) Evaluation of translocation as a tool for mitigating development threats to great crested newts (*Triturus cristatus*) in England, 1990–2001. *Biological Conservation*, 122, 45–52.
(6)   Lewis B., Griffiths R.A. & Barrios Y. (2007) Field assessment of great crested newt *Triturus cristatus* mitigation projects in England. Natural England Report. Research Report NERR001.
(7)   Neave D.W. & Moffat C. (2007) Evidence of amphibian occupation of artificial hibernacula. *Herpetological Bulletin*, 99, 20–22.
(8)   McNeill D.C. (2010) Translocation of a population of great crested newts (*Triturus cristatus*): a Scottish case study. PhD thesis. University of Glasgow.
(9)   Lewis B. (2012) An evaluation of mitigation actions for great crested newts at development sites. PhD thesis. The Durrell Institute of Conservation and Ecology, University of Kent.

Captive breeding, rearing and releases (*ex-situ* conservation)

13.2   Breed amphibians in captivity

Background

Captive breeding involves taking wild animals into captivity and establishing and maintaining breeding populations. It tends to be undertaken when wild populations become very small or fragmented or when they are declining rapidly. Captive populations can be maintained while threats in the wild are reduced or removed and can provide an insurance policy against catastrophe in the wild. Captive breeding also potentially provides a method of increasing reproductive output beyond what would be possible in the wild. The aim is usually to release captive-bred animals back to natural habitats, either to original sites once conditions are suitable, to reintroduce species to sites that were occupied in the past or to introduce species to new sites. Some captive populations may also be used for research to benefit wild populations.

Amphibians possess a number of traits that make them potentially suitable for captive breeding programmes. They reach breeding age relatively quickly and are fertile and so captive populations can, in theory, be expanded quickly. Also their small body size and low maintenance requirements allow viable populations to be managed much more cost-effectively than many larger animals. However, captive breeding can result in problems associated with inbreeding depression, removal of natural selection and adaptation to captive conditions.

Studies in which amphibians were treated with hormones to induce sperm or egg release or in which eggs were fertilized artificially are discussed in separate sections below.

Studies that investigate the effectiveness of releasing of captive-bred amphibians are discussed in 'Release captive-bred individuals'. Those studies are not included in this section, unless specific details about captive breeding were included.

13.2.1 Frogs

- Thirty-three studies investigated the success of breeding frogs in captivity.

- Twenty-three of 33 studies, 3 of which were reviews, and 30 replicated studies across the world[1–35] found that amphibians produced eggs in captivity, in 4 cases by captive-bred females[4, 9, 16, 34]. Seven found mixed results, with some species of frogs[1, 17, 21, 24, 27] or 17–50% of captive populations[3, 8] reproducing successfully in captivity, but with other species difficult to maintain or raise to adults[17, 24]. One[5] found that frogs did not breed successfully in captivity and another[7] that all breeding frogs died. Seventeen of the studies found that captive-bred frogs were raised successfully to hatching[2], tadpoles[10, 25], froglets[13, 14, 18, 19, 22, 26, 30, 31, 34] or adults[4, 6, 9, 11, 12, 16] in captivity. One found that froglet survival was low[20] and another[29] that three species were not successfully raised to adulthood.

- Four replicated studies (including one small study) in Canada, Fiji, Hong Kong and Italy found that 30–88% of eggs hatched[12, 14, 18] or survival to metamorphosis was 75%[21], survival to froglets was 17–51%[16] or to adults was 50–90%[12] in captivity.

- One review and four replicated studies (including two small studies) in Germany, Italy and the USA found that reproductive success of frogs in captivity depended on temperature[1, 2, 25] or a simulated wet and dry season[1, 33], but not on whether frogs were housed in high or low maintenance facilities[11]. Three replicated studies (including one small study) in Germany, Australia and Canada found that egg or tadpole development in captivity was affected by parental care[21], density[33] or temperature[9, 33].

A review of captive breeding programmes (1) found that a number of amphibian species have been bred successfully in captivity. Frog species that bred successfully were: red-eyed tree frog *Agalychnis callidryas*, Asiatic treefrog *Rhacophorus leucomystax*, Malaysian leaf frog *Megophrys nasuta*, Bell's horned frog *Ceratophrys ornate* and a number of poison dart frog species (Dendrobatidae). Breeding was induced with gonadotrophin-releasing hormone in White's tree frog *Litoria caerulea* (with reduced temperatures) and African red frog *Phrynomerus bifasciatus* (with simulated wet and dry season).

A small, replicated study in 1993 of parsley frogs *Pelodytes punctatus* at Genoa University, Italy (2) found that one clutch was produced and hatched in captivity (500 eggs). One of 6 females bred following a drop in temperature from 20–24 to 17°C. All wild caught tadpoles survived. At metamorphosis the mortality rate of those animals was 19% due to dehydration, calcium deficiency and suffocation during feeding. Thirty-one tadpoles were obtained from the wild in 1993. Tadpoles were housed in a 400 L tank (20 cm water) and metamorphs in a 50 x 50 x 25 cm tank. From eight months animals were housed in a 120 x 60 x 50 cm glass breeding tank with filtered water, pebbles and moss. Eggs were moved to a separate tank.

A replicated study in 1993–1997 of captive agile frogs *Rana dalmatina* in Jersey, UK (3) found that frogs bred successfully in one of two captive populations. Breeding occurred at the Jersey Wildlife Preservation Trust in 1994, 1995 and 1997. However, breeding did not occur in the first two years within the five enclosures at the second site. In 1993, 3 males, 2 females and 17 juveniles and in 1994 an additional 8 juveniles were acquired and housed in a landscaped enclosure with a pond (20 m²). In 1995, an additional five enclosures (3 x 3–7 m) were built on private land and stocked with captive-bred tadpoles and young frogs.

A replicated study in 1994–1998 of captive green and black poison-dart frogs *Dendrobates auratus* and blue poison-dart frogs *Dendrobates azureus* in Jersey, UK (4) found that both species bred successfully in captivity. Fertile black poison-dart frogs' eggs were first recorded in December 1994 and produced five frogs. In 1995–1996, 98 mainly fertile clutches were produced. The 10 original frogs were still alive and breeding in 1998. Viable blue poison-dart frogs' eggs were first recorded in August 1996. In 1996–1998, 23 frogs were captive-bred and went on to produce eggs in 1998. Ten captive sub-adult and tadpole black poison-dart frogs and seven blue poison-dart frogs were acquired in 1994 and 1995–1996 respectively. Frogs were housed at 22–27°C in 3 x 2 x 1 m tanks with a waterfall, natural substrate and densely planted areas. Tanks had seasonal photoperiods and were misted daily. Blue poison-dart frogs were moved to smaller tanks in pairs to breed. Frogs were fed young crickets, fruit flies and wild invertebrates. Eggs were removed for rearing.

A replicated study in 1994–1996 of roseate frogs *Geocrinia rosea* at Melbourne Zoo, Australia (5) found that the frogs did not breed successfully in the first two years. Although males called from 1994, eggs were not produced until 1996. However, only one of four egg masses was fertile (25 eggs) and that was destroyed by fungus. Three of the original frogs died within 3 months; the other 2 survived 27 months in captivity. The original egg mass produced 45 froglets, 15 of which were alive at 21 months, but died within 27 months of emergence. In 1994, two male and three sub-adult frogs were housed in two outdoor tanks (120 x 60 x 60 cm) with organic substrates and water. Two egg clumps were also received and 6–7 froglets were housed in each of four indoor tanks (47 x 55 x 36 cm and 180 x 46 x 46 cm).

A replicated study in 1992–1998 in Hong Kong and Australia (6) found that Romer's frogs *Philautus romeri* reproduced in captivity. Over 180 egg clutches were produced in captivity in Hong Kong and at least 706 captive-bred frogs were produced from the captive population in Australia. A total of 1,170 frogs and 1,622 tadpoles were released in 1993–1994. In 1992, several eggs and tadpoles and 230 adults were collected from the wild. Thirty adults were sent to Melbourne Zoo and the remainder were housed at the University of Hong Kong.

A replicated study in 1993 of captive sharp snouted dayfrogs *Taudactylus acutirostris* at Melbourne and Taronga Zoos, Australia (7) found that only 1 of 109 animals taken in to captivity survived to 18 months. The one adult died within a month of the introduction of three adults from another zoo, which died at the same time. Of the others, 79 died as tadpoles, 11 during metamorphosis and 18 as metamorphs. Causes of death were largely unknown. In 1993, a total of 109 tadpoles were sent to the two zoos in five separate groups. Tadpoles at Taronga Zoo were housed at 20–23°C in three different-sized tanks with water at different depths (15–45 cm), gravel, stones, wood and pond weed. At Melbourne Zoo, tadpoles were housed at 17–24°C in groups of 8–10 in tanks (45 x 53 x 14 cm). Water was cleaned weekly. Animals were treated for dermatitis.

A replicated study in 2000–2001 of captive tarahumara frogs *Rana tarahumarae* in southern Arizona, USA (8) found that some frogs bred successfully at one of the captive breeding facilities. Wild collected eggs hatched successfully and many of the metamorphosed frogs survived to adulthood. In May 2000, part of an egg mass was collected from the wild in northern Mexico. Eggs were taken to a captive facility in Arizona and after hatching divided between at least six facilities.

A replicated study in 1993–2000 of captive great barred frogs *Mixophyes fasciolatus* at Mebourne Zoo, Australia (9) found that frogs bred successfully in captivity. In 1998, males called and three clumps of 300–500 eggs were produced. Nine egg clumps were produced in 1999–2000, some by frogs hatched in 1998. In 1998–2000, over 200 young frogs were sent to other breeders. Breeding also occurred outdoors. Tadpole growth was similar at 16–20°C and 18–22°C, but lower temperatures resulted in later metamorphosis (120–132 vs 99–108 days). A number of frogs had a metabolic bone disease that was successfully treated. In 1993–1995, 12 metamorphs were received and raised to adults. Two breeding groups of four were housed in glass aquaria (180 x 45 x 45 cm) with organic substrate, rocks, logs and water. Rain was simulated for two hours/day and night for six days in April. Tadpoles were housed in separate tanks (45 x 53 x 15 cm). In 2000, seven frogs were housed in an outdoor enclosure (300 x 300 x 220 cm).

A replicated study in 2000–2003 in Gipuzkoa province, Spain (10) found that stripeless treefrogs *Hyla meridionalis* reproduced successfully in captivity. In 2000–2003, a total of 5,767 tadpoles were bred in captivity and released (171–3,989/year).

A replicated study in 2000–2004 of captive tarahumara frogs *Rana tarahumara* in Arizona, USA (11) found that frogs bred successfully at both high and low maintenance captive facilities. Animals were collected from northern Mexico in 2000. They were reared and bred at a number of facilities ranging from high maintenance to semi-wild low maintenance. In June–October 2004, 56 adult, 229 juvenile and 328 tadpoles were released at 4 sites in south central Arizona.

A replicated study in 1997–2000 in Italy (12) found that successful captive breeding was achieved for the golden mantella *Mantella aurantiaca*, false tomato frog *Dyscophus guineti* and green burrowing frog *Scaphiophryne marmorata*. A stable breeding population of 60 golden mantellas was achieved (2 clutches/female/year; 60% eggs hatched; 80% survival to adult). Six founder false tomato frogs resulted in a breeding population of 100 frogs (1; 30%; 50%). Green burrowing frogs also bred successfully (1; 80%; 90%). The estimated cost of one captive-bred individual was: 7.50 € for golden mantillas, 3.12 € for tomato frogs and 0.54 € for green burrowing frogs. Animals were imported from Madagascar in 1997–1998 or were obtained from private breeders or other facilities. Reproduction was monitored in captivity over two years. Some data were obtained from private breeders. Costs were calculated for Italy.

A small, replicated study in 2006 of the Fijian ground frog *Platymantis vitianus* at the University of the South Pacific, Fiji (13) found that although all five froglets that hatched survived to 37 days old, they were then predated by brown house ants *Pheidole megacephala*. All froglets maintained their body weight and on average, weight and body length gradually increased. Ants were likely to have been attracted by excess ripe fruit placed in the aquarium to attract small flies as food for the froglets. Adult frogs laid eggs during the wet season in a purpose-built outdoor enclosure. The five newly hatched froglets were then transferred into a glass laboratory aquarium (0.5 x 0.3 x 0.4 m). Body weight and food supply were closely monitored.

A small, replicated study in 2006–2007 of Fijian ground frogs *Platymantis vitianus* at the University of the South Pacific, Fiji (14, 18) found that following incubation, 35 froglets hatched from one egg mass (88%; see also (15)). Two egg masses (40–42 eggs/mass) were laid by captive frogs in the wet season (December/January) of the first year. Only 11% of eggs from one mass hatched due to flooding and so the second was incubated inside a glass aquarium at 27°C. Thirty-five froglets hatched from the second batch. Five adult male and five female frogs collected from the wild were kept in outdoor wire enclosures (5 x 3 x 2 m). Newly hatched froglets were transferred to laboratory glass aquariums (0.5 x 0.3 x 0.4 m).

In a continuation of a study at the University of the South Pacific, Fiji (14, 18) a small, replicated study (15) found that adding natural structures to enclosures as potential egg laying sites resulted in the production of two clutches of eggs by Fijian ground frogs *Platymantis vitianus*. One clutch was laid underneath a moist rotting log in December and the other inside a bamboo stem lined with soil in January. Several of the five males and five females were observed in or near potential egg-laying sites throughout the breeding period. Natural structures were added as potential egg laying sites including rotting logs, hollow bamboo stems, coconut husks, rocks and decaying leaf litter. All material was sterilized prior to installation. Native plants were also added. Nocturnal activity was recorded using digital video surveillance cameras.

A replicated study in 1992–1996 of Romer's tree frog *Chirixalus romeri* at Melbourne Zoo, Australia and the University of Hong Kong (16) found that they bred successfully in captivity. Over 188 clutches of eggs were produced in Hong Kong in 1992–1996, 108 from wild and 76 from captive-bred females. Average juvenile mortality rate was 49% (range 31–71%). Five of 13 frogs survived until they were released to the wild 5 years after metamorphosis. Frogs in two terrariums died of red-leg syndrome. In Australia, 7 egg clutches and 250 froglets were produced in 1991, with a froglet mortality rate of 83% at 8 months. In 1992, 18 egg clutches (7 from wild and 11 from captive-bred females) and 530 froglets were produced. A total of 220 adults (150 males), 21 juveniles, metamorphs and tadpoles and 7 egg clutches were collected from the wild in 1991–1992. Ten to 25 adults were housed per tank (60 x 30 x 30 cm). Tadpoles were raised in small tanks (up to six/100 cm²) and froglets transferred to tanks.

A review of captive breeding programmes in 2001–2007 of priority amphibian species from Panama at 50 zoos and aquariums in the USA (17) found that maintenance and breeding in captivity had mixed success. Several of 30 species collected bred successfully in captivity including lemur leaf frog *Hylomantis lemur*, Pratt's rocket frog *Colostethus pratti*, marsupial frog *Gastrotheca cornuta*, spiny-headed treefrog *Anotheca spinosa*, Vicente's poison frog *Dendrobates vicentei*, minute poison frog *Minyobates minutus*, *Eleutherodactylus gaigae* and caretta robber frog *Pristimantis diastema*. However, some have proved difficult to raise to adulthood due to nutritional issues (e.g. marsupial frog). Species such as Palmer's treefrog *Hyloscirtus palmeri* and banded horned treefrog *Hemiphractus fasciatus* proved very difficult to maintain in captivity. Death was often related to malnutrition. Up to 40 individuals of nearly 30 species were wild caught.

A replicated study in 2007–2008 of captive amphibians at a facility in Hanoi, Vietnam (19) found that *Hylarana maosonensis*, *Rhacophorus feae*, *Rhacophorus maximus* and Chinese gliding frogs *Rhacophorus dennysi* bred successfully in captivity. Following successful breeding in 2007, 300 Chinese gliding frog froglets were released into the wild in June 2008 at the location in which the parents had been collected. *Rhacophorus feae* and *Rhacophorus maximus* were also to be released into the wild.

A replicated study in 2005–2008 of captive horned marsupial frog *Gastrotheca cornuta* in Atlanta and Panama, USA (20) found that a small number of frogs bred in captivity but froglet survival was low. In Panama, 1 female produced 14 froglets in 2007 and 4 eggs were produced in 2008. In Atlanta, 14 infertile eggs were produced in 2006, however 13 froglets were observed in 2008. Two cases of abortion were recorded and most froglet deaths occurred within 20 weeks. In total, 11 frogs survived over 1 year, but grew slowly and often had deformities. One receiving UV-B radiation for 45 minutes/day did not develop deformities. Wild frogs were collected in 2005–2006. Males and females were housed in separate enclosures (60 x 30–60 x 40–90 cm) with plants, twigs and water in Atlanta (six males, two females) and Panama. Males were introduced to females just for breeding. Frogs were misted 2–10 times/day and in Atlanta a 'dry season' was simulated. Froglets were separated for rearing.

A replicated study in 2009–2010 of captive mantella frogs in Edmonton, Canada (21) found that three of four species bred successfully. In October 2009, baron's painted mantella *Mantella baroni* and splendid mantella *Mantella pulchra* produced clutches of eggs. Eggs left in with parents developed better than those moved to dishes. Survival rate was approximately 60%. In October of 2010, golden mantella *Mantella aurantiaca* produced 2 large clutches, with 211 tadpoles hatching. Survival rate to froglets by the end of the study was 75%. Climbing mantella *Mantella laevigata* did not breed. In June 2009, all species except golden mantella were put through a three month dry cycle, with reduced temperature (<20°C), humidity and food (alternate days) and increased day length (12 hours). In September, the wet season was started (misting 4 times/day, 10 hours daylight, daily feeding). Eggs were kept in petri dishes on wet moss or left in with the parents. Tadpoles were placed into plastic containers. In 2010, one male and three female golden mantellas were put through just a wet season.

A replicated study in 2008–2012 of captive white-bellied frogs *Geocrinia alba* at Perth Zoo, Australia (22) found that the species bred successfully for the first time in captivity in 2012. Three egg clutches were laid in the breeding chambers. The eggs in one nest failed to develop but fertile eggs from the other two metamorphosed and were still alive at two months.

A review of a captive breeding programme from 2001 to 2011 for lemur leaf frogs *Agalychnis lemur* in the UK (23) found that the frogs bred successfully in captivity. Captive-bred animals were sent to breeding facilities around the world. Forty animals were sent to Europe from Bristol Zoo in 2011.

A replicated study in 2010–2012 of a captive breeding programme for *Eleutherodactylus* species of frogs at the Philadelphia Zoo, USA (24) found that six of ten species produced offspring, but only three species were raised successfully. In 2011, eight of the ten species laid eggs. Six of those species laid fertilized eggs that produced offspring. The three species that were raised successfully were Mozart's frog *Eleutherodactylus amadeus*, la hotte frog *Eleutherodactylus bakeri* and Macaya breast-spot frog *Eleutherodactylus thorectes*. By 2011, the zoo held over 650 frogs of the 10 species. In 2010, 176 frogs of 10 critically endangered species were collected from four sites in Haiti. Frogs were housed in seven biosecure enclosures with water filtration, lighting, temperature control, misting systems and heat lamps. Tanks had soil substrate and plants. Each species was kept separate, within breeding groups (3–15 frogs).

A replicated study in 2009–2011 at San Diego Zoo, California, USA (25) found that mountain yellow-legged frogs *Rana muscosa* reproduced in captivity. Breeding success improved from 2009 to 2011 as frogs matured and with winter cooling to replicate natural conditions. In 2006, 82 tadpoles were rescued from a drying stream. Breeding was attempted from 2009. In 2010, half of the frogs and in 2011 all frogs were kept at 4°C for 2–3 months over winter. In 2010–2011, 330 eggs and 336 tadpoles were released into a stream. Tadpoles were kept in cages to acclimatize for different periods of time before release.

A small, replicated study in 2010–2011 of Orlov's treefrog *Rhacophorus orlovi* in Leningrad Zoo, Russia (26) found that two pairs bred successfully in captivity. Reproduction first took place during the first two months of captivity. Fertilization rate of the first 3 egg masses were 0%, 90% and 30%; a further 8 egg masses were produced within 7 months. The following year, 50% of egg masses were infertile, but juveniles survived to at least 12 months. Two pairs of frogs were caught in the wild in Ha Tinh Province, Vietnam in 2010. They were housed at 19–24°C in glass tanks (20 x 40 x 30 cm) with branches and plants. Tadpoles were kept in plastic tanks (39 x 28 x 22 cm). Froglets were kept in groups of 15–17 in glass tanks similar to adults.

A replicated study in 2011–2012 of captive frogs in Andasibe, Madagascar (27) found that two of eight local species bred in the first year of captivity. Both *Boophis pyrrhus* and *Mantidactylus betsileanus* bred in captivity in 2012. Tadpoles and frogs were raised in captivity to determine the optimal husbandry requirements for these species so that they could be applied to maintaining threatened similar species in the future. In April 2011, eight local frog species were housed in a newly constructed biosecure captive breeding facility.

A replicated study in 2010–2011 of Darwin's frogs *Rhinoderma* species at the National Zoo of Chile, Santiago, Chile (28) found that the frogs bred successfully in captivity. Reproductive activity and the first young were produced a few months after arriving in captivity. More juveniles were produced the following year. Breeding groups of Darwin's Frogs were collected from wild populations in 2010. Frogs were housed in enclosures with automatic misting systems, climate control and pre-filtered water.

A replicated study in 2006–2012 of amphibians at two breeding facilities in Panama (29) found that two undescribed species at risk of extinction and la loma treefrog *Hyloscirtus colymba* were not successfully raised to adulthood.

A replicated study in 2005–2011 of captive cascade glass frogs *Sachatamia albomaculata* at Zoo Atlanta and Minnesota Zoo, USA (30) found that animals bred successfully in captivity. Few eggs were produced in 2005–2007. However, regular breeding occurred from 2008. Eggs were produced in most months (28–60/clutch). The first metamorphosis was recorded in March 2009 and continued throughout 2009 to 2011. Fifteen wild caught frogs were caught in 2005 and divided between two zoos. They were housed in groups of two to four in tanks (30 x 30 x 45 cm) with aquatic plants. Tanks were misted six times each day in November–April and twice a day in May–October. Eggs, larvae (30/tank), metamorphs (1/container) and froglets were kept in separate containers.

A small, replicated study in 2011–2012 of captive *Scinax alcatrazin* at São Paulo Zoo, Brazil (31) found that eggs were produced and juveniles maintained in captivity. The first breeding event occurred after 33 days in captivity. One female deposited around 140 eggs, of which 132 hatched. By July 2012, 93 froglets were still alive. Two males and a female died on the first day in captivity. Eleven animals (five males, three females, three tadpoles) were collected from the wild in October 2011 and housed in a biosecure room. Adults were kept in two glass enclosures, with plants and water. An ultra-sonic fogger was used to increase nighttime humidity to stimulate breeding. Tadpoles were housed in a plastic enclosure and froglets in plastic cups. Management and husbandry protocols had been established over two years using captive *Scinax perpusillus* (see (32)).

A small, replicated study in 2009–2010 of *Scinax perpusillus* at São Paulo Zoo, Brazil (32) found that eggs were produced in captivity. Five batches of 4–77 eggs were laid in 2010 by 1 female. Two of three adult males died during the year. The three males, one female and six larvae were wild caught in 2009. Following five months quarantine, adults were housed at 12–27°C in a glass tank (70 x 30 x 45 cm) with a water dish and plants. Tanks were misted once or twice a day and before breeding an ultra-sonic fogger was turned on for 10 hours overnight 3 times a week. Management and husbandry protocols were established using this species in preparation for attempted captive breeding of *Scinax alcatrazin* (see (31)).

A small, replicated study in 2005–2009 of captive Malayan horned frogs *Megophrys nasuta* at Cologne Zoo, Germany (33) found that frogs bred successfully. Between 50 and 300 larvae hatched/egg batch. The minimum interval between egg laying was about a month. Dry followed by phases of intense water spraying triggered reproduction. Larval development was faster at higher water temperatures and lower densities. Three males and two females obtained from the pet trade were housed in aquariums (145 x 60 x 56 cm) divided into aquatic and terrestrial sections. Larvae were moved to plastic tanks (13 L water). Two–month-old tadpoles were transferred into aquariums (54 x 65 x 30 cm) and metamorphs and juveniles, in groups of 20–30, into tanks (60 x 45 x 30 cm).

A replicated study in 1994–2004 at Taronga Zoo, Australia (34) found that captive breeding of green and golden bell frogs *Litoria aurea* was very successful. The captive breeding programme was established in 1994. Between 1996 and 2004, over 20,000 tadpoles and meta-morphs, including fourth generation captive-bred individuals, were released at 5 sites.

A replicated study in 2011–2012 of captive southern corroboree frogs *Pseudophryne corroboree* at Taronga and Melbourne Zoo, Australia (35) found that frogs reproduced successfully in captivity. Having had difficulties breeding the species in the first years of the programme, captive breeding protocols had been established that resulted in high reproductive success. In 2011 and 2012, the majority of mature females produced eggs. Eggs were separated and observed during early development. Captive breeding was undertaken as less than fifty individuals remained in the wild, mainly because of chytridiomycosis.

(1) Maruska E.J. (1986) Amphibians: review of zoo breeding programmes. *International Zoo Yearbook*, 24/25, 56–65.
(2) Emanueli L., Jesu R., Schimment G., Arillo A., Mamone A. & Lamagni L. (1997) Captive breeding programme of the parsley frog (*Pelodytes punctatus* Daudin, 1803) at Genoa aquarium (Italy). *Herpetologica Bonnensis*, 1997, 115–118.
(3) Gibson R.C. & Freeman M. (1997) Conservation at home: recovery programme for the agile frog *Rana dalmatina* in Jersey. *Dodo*, 33, 91–104.
(4) Preece D.J. (1998) The captive management and breeding of poison-dart frogs, family Dendrobatidae, at Jersey Wildlife Preservation Trust. *Dodo*, 34, 103–114.
(5) Birkett J., Vincent M. & Banks C. (1999) Captive management and rearing of the roseate frog, *Geocrinia rosea*, at Melbourne Zoo. *Herpetofauna*, 29, 49–56.
(6) Dudgeon D. & Lau M.W.N. (1999) Romer's frog reintroduction into a degraded tropical landscape, Hong Kong, P.R. China. *Re-introduction News*, 17, 10–11.
(7) Banks C.B. & McCracken H.E. (2002) Captive management and pathology of sharp-snouted torrent frogs, Taudactylus acutirostris, at Melbourne and Taronga Zoos. In A.E.O. Nattras (ed) *Frogs in the Community*, Queensland Frog Society Inc., Brisbane. pp. 94–102.
(8) Rorabaugh J. & Humphrey J. (2002) The Tarahumara frog: return of a native. *Endangered Species Bulletin*, 27, 24–26.
(9) Banks C., Birkett J., Young S., Vincent M. & Hawkes T. (2003) Breeding and management of the great barred frog, *Mixophyes fasciolatus*, at Melbourne Zoo. *Herpetofauna*, 33, 2–12.
(10) Rubio X. & Etxezarreta J. (2003) Plan de reintroducción y seguimiento de la ranita meridional (*Hyla meridionalis*) en Mendizorrotz (Gipuzkoa, País Vasco) (1998-2003). *Munibe*, 16, 160–177.
(11) Rorabaugh J. (2005) Re-establishment of the Tarahumara frog into Arizona, USA. *Re-introduction News*, 24, 43–44.
(12) Mattioli F., Gili C. & Andreone F. (2006) Economics of captive breeding applied to the conservation of selected amphibian and reptile species from Madagascar. *Natura Societa Italiana di Scienze Naturale e Museo Civico di Storia Naturale Milan*, 95, 67–80.
(13) Narayan E., Christi K. & Morley C. (2007) Captive management of newly hatched Fijian ground frog *Platymantis vitianus* froglets: lessons learnt from an unanticipated invertebrate predator invasion, Suva, Fiji. *Conservation Evidence*, 4, 58–60.
(14) Narayan E., Christi K. & Morley C. (2007) Improvement in ex-situ egg hatchability of Fijian ground frog *Platymantis vitianus* by laboratory incubation of egg masses, University of the South Pacific, Suva, Fiji. *Conservation Evidence*, 4, 25–27.

(15)   Narayan E., Christi K. & Morley C. (2007) Provision of egg-laying sites for captive breeding of the endangered Fijian ground frog *Platymantis vitianus*, University of the South Pacific, Suva, Fiji. *Conservation Evidence*, 4, 61–65.

(16)   Banks C.B., Lau M.W.N. & Dudgeon D. (2008) Captive management and breeding of Romer's tree frog *Chirixalus romeri*. *International Zoo Yearbook*, 42, 99–108.

(17)   Gagliardo R., Crump P., Griffith E., Mendelson J., Ross H. & Zippel K. (2008) The principles of rapid response for amphibian conservation, using the programmes in Panama as an example. *International Zoo Yearbook*, 42, 125–135.

(18)   Narayan E., Christi K. & Morley C. (2009) Captive propagation of the endangered native Fijian frog *Platymantis vitiana*: implications for ex-situ conservation and management. *Pacific Conservation Biology*, 15, 47–55.

(19)   Truong N.Q., The D.T., Cuong P.T., Tao N.T. & Ziegler T. (2009) Amphibian breeding station in Hanoi: a trial model for linking conservation and research with sustainable use. *Froglog*, 91, 12–15.

(20)   Gagliardo R., Griffith E., Hill R., Ross H., Mendelson J., Timpe E. & Wilson B. (2010) Observations on the captive reproduction of the horned marsupial frog *Gastrotheca cornuta* (Boulenger 1898). *Herpetological Review*, 41, 52–58.

(21)   Woods W. (2010) Mantella breeding success. *Amphibian Ark Newsletter*, 13, 21–22.

(22)   Bradfield K. (2011) Geocrinia captive breeding and rear for release programs at Perth Zoo. *Amphibian Ark Newsletter*, 17, 9.

(23)   Gray A.R. (2011) Lemur leaf frog update. *Amphibian Ark Newsletter*, 17, 11.

(24)   Martínez Rivera C.C., Bell J. & Parker J. (2011) Haiti's disappearing frogs. *Amphibian Ark Newsletter*, 16, 15–16.

(25)   Medlin D.D. (2011) San Diego Zoo release more Southern California Mountain yellow-legged frogs. *Amphibian Ark Newsletter*, 17, 12.

(26)   Wildenhues M.J., Bagaturov M.F., Schmitz A., Dao T.T.A., Hendrix R. & Ziegler T. (2011) Captive management and reproductive biology of Orlov's treefrog, *Rhacophorus orlovi* Ziegler & Köhler, 2001 (Amphibia: Anura: Rhacophoridae), including larval description, colour pattern variation and advertisement call. *Der Zoologische Garten*, 80, 287–303.

(27)   Edmonds D. & Claude J. (2012) An update from Andasibe, Madagascar. *Amphibian Ark Newsletter*, 19, 8.

(28)   Fenolio D. (2012) The Darwin's frog conservation initiative. *Amphibian Ark Newsletter*, 18, 22–23.

(29)   Gratwicke B. (2012) Amphibian rescue and conservation project – Panama. *Froglog*, 102, 17–20.

(30)   Hill R.L., Kaylock J.B., Cuthbert E., Griffith E.J. & Ross H.L. (2012) Observations on the captive maintenance and reproduction of the cascade glass frog, *Sachatamia albomaculata* (Taylor, 1949). *Herpetological Review*, 43, 601–604.

(31)   Lisboa C.S. (2012) Conservation of *Scinax alcatraz* (Anura: Hylidae): captive breeding and in situ monitoring of a critically endangered treefrog species. *Amphibian Ark Newsletter*, 20, 6–8.

(32)   Lisboa C.S. & Vaz R.I. (2012) Captive breeding and husbandry of *Scinax perpusillus* at São Paulo Zoo: preliminary action for ex situ conservation of *Scinax alcatraz* (Anura: Hylidae). *Herpetological Review*, 43, 435–437.

(33)   Wildenhues M., Rauhaus A., Bach R., Karbe D., van der Straeten K., Hertwig S.T. & Ziegler T. (2012) Husbandry, captive breeding, larval development and stages of the Malayan horned frog *Megophrys nasuta* (Schlegel, 1858) (Amphibia: Anura: Megophryidae). *Amphibian and Reptile Conservation*, 5, 15–28.

(34)   McFadden M. (2012a) Release of green and golden bell frog tadpoles from Taronga Zoo. *Amphibian Ark Newsletter*, 18, 20.

(35)   McFadden M. (2012b) Captive-bred southern corroboree frog eggs released. *Amphibian Ark Newsletter*, 19, 10.

13.2.2 Toads

- Ten replicated studies (including three small studies) in Germany, Italy, Spain, the UK and the USA[1, 2, 4, 5, 7–12] found that toads produced eggs in captivity, in one case by second generation captive females[10]. Eight found that captive-bred toads were raised successfully to tadpoles[2, 11], toadlets[4, 5, 7, 12] or adults[4, 9, 10] in captivity. Two found that most toads died after hatching[8] or after metamorphosis[1]. Two reviews found mixed results with four species of toad[3] or 21% of captive populations of Puerto Rican crested toad[6] breeding successfully in captivity.

- Four replicated studies in Germany, Spain and the USA found that reproductive success of captive toads was affected by tank humidity[10] and was higher in outdoor enclosures than indoor tanks[8, 9, 11]. One replicated study in Germany[9] found that survival of European red-bellied toad eggs, tadpoles and juveniles was higher in captivity than the wild.

Background
As there is a larger literature for Mallorcan midwife toads *Alytes muletensis* and for harlequin toads *Atelopus sp.* than other species, evidence is considered in separate sections below for these species.

A small, replicated study in 1973–1974 of captive Colombian giant toads *Bufo blombergi* at Brownsville Zoo, USA (1) found that toads reproduced successfully in captivity. Three months after being housed together, one male started calling and one of three females produced eggs. Of the few eggs that were fertile (separated prematurely by keepers), 5 hatched and 2 tadpoles survived metamorphosis, dying at day 68 and 173. The following year one female produced eggs that hatched into approximately 600 tadpoles. Five male and three female toads were housed together. Eggs were placed in spring water at 19°C and tadpoles in spring water with sphagnum moss.

A small, replicated study in 1981–1984 of captive natterjack toads *Bufo calamita* in Norfolk, UK (2) found that two egg strings were produced by nine females. These hatched successfully. The majority of the tadpoles were released back to the wild. In the first year only one of nine wild toadlets survived overwintering in captivity and died shortly after. The following year all 25 wild tadpoles survived to adulthood. In 1981, 12 and in 1982, 25 tadpoles were collected from the wild. Toadlets were housed in a tank with water, a sloping sandy substrate and moss. In the first year toadlets were housed outdoors, after that all tadpoles were kept indoors for at least their first year. Adults were housed in an outdoor enclosure with sandy substrate, a pool, wood and heathland plants.

A review of captive breeding programmes (3) found that a number of amphibian species have been bred successfully in captivity. Toad species that were bred successfully were: Surinam toad *Pipa pipa,* Colombian giant toad *Bufo blombergi,* Houston toad *Bufo houstonensis* and Puerto Rican crested toad *Peltophryne lemur.*

A replicated study in 1984–1989 of captive Puerto Rican crested toads *Peltophryne lemur* in Toronto Zoo, Canada and Buffalo Zoo, USA (4) found that they bred successfully in captivity. Over 3,000 captive-bred toadlets and 12 2-year-old toads were released and 400 toadlets sent to other zoos. A small land area or 'beach' was created at one end of each tank by slowly reducing the water level, to simulate pond drying. Shelter habitat, such as halved coconuts, were provided for emerging toadlets to prevent dessication.

A small, replicated study in 1988–1990 of captive common midwife toads *Alytes obstetricans* in Norfolk, UK (5) found that nine egg strings were produced from two captive females. Strings contained 14–32 eggs; less than 3 per batch were infertile. Toadlets were observed leaving the water from at least the first five batches. One female was introduced to a number of males in 1988 and 1990. Toads were housed in an outdoor enclosure (120 x 75 cm) with sandy soil, wood, plants and a small pool. Tadpoles were moved to separate tanks.

A review in 1994 of captive breeding programmes for the Puerto Rican crested toad *Peltophryne lemur* (6) reported that the species had bred at 3 of the 14 zoos and institutions with captive populations.

A study in 1993 of a captive European fire-bellied toad *Bombina bombina* in England, UK (7) found that 94 tadpoles hatched from eggs produced by the one captive female. A total of 52 toadlets survived at least one month after metamorphosis. Deaths were largely due to cannibalism. Two male and one female toad were obtained from the wild in 1993. They were housed in a 60 x 30 cm tank with aquatic plants and shelters. The water depth was increased from 8 to 25 cm for breeding. Adults were removed following hatching.

A replicated study in 1991–1994 of Wyoming toads *Bufo hemiophrys baxteri* in a zoo in Colorado, USA (8) found that breeding was moderately successful in field enclosures but in captivity although females produced eggs, tadpoles did not survive. Protected breeding pens in the field were considered by the authors to be moderately successful. In captivity, five of seven hormonally induced females produced thousands of fertile eggs in 1994. However, the majority of tadpoles that hatched died within 72 hours. Deaths were considered by the authors to have been due to water quality. In captivity, two to four wild-caught and captive-bred toads were housed per tank (40 x 61 x 23 cm) at 20°C. Cork bark, sheet moss, sand, water, artificial plants and a basking lamp were provided. In 1991–1993, toads were transported to breeding enclosures at the edge of the lake. In 1994, five toads were overwintered for six weeks at 4.5°C. Seven females were hormonally induced and paired in captivity.

A replicated study in 1999–2003 of captive European red-bellied toads *Bombina bombina* in northern Germany (9) found that toads bred successfully, particularly outdoors where they were more active, started calling earlier and had higher reproductive success than those indoors. In 2001, 20 indoor females produced an average of 31 eggs/batch (range: 15–40), compared to a total of 1,100 eggs from 3 outdoor females. Mortality of eggs (8–20%; n = 380), tadpoles (4–7%; n = 1,680) and juveniles (8%; n = 250) was lower in captivity than the field. However, disease could kill all juveniles within 3–5 weeks. Few adults died in captivity, with some living 12 years. All toads successfully over-wintered (at 4–7°C). Breeding enclosures were glass tanks (150 x 60 x 60 cm) with aquatic and terrestrial areas, each housing 2 adult males and 3–5 females. Eggs were moved to plastic dishes and then outdoor aquaria for hatching. Metamorphs were moved to tanks (60 x 30 x 30 cm). Day temperature was increased to 21°C and daylight periods lengthened to induce breeding.

A replicated study in 2000–2006 of Kihansi spray toads *Nectophrynoides asperginis* at zoos in the USA (10) found that toads bred successfully in captivity. Within the first 6 months, 82% of 269 founders and 43% of toadlets died. However, by 2006 this captive population was 159 toads. A total of 401 toadlets were born in the first year, with second generation toads born the following year. A second captive population of 230 founders initially doubled, declined to 32 toads and then increased to 130 by 2006. Ceasing or reducing misting inhibited reproductive activity. Primary diseases were lungworm infection and Gram-negative septicaemia; other health issues were also recorded. In November 2000, 499 adults were collected from the one remaining wild population. Following quarantine, the 269 toads maintained at Bronx Zoo were separated into 13 groups of 20–31 toads. Aquaria (38–76 L) were misted 4–9 times/day. Toadlets were transferred into smaller aquariums. The other 230 toads were transported to the National Amphibian Conservation Center (3 zoos).

A replicated study in 2006–2011 of captive common midwife toads *Alytes obstetricans* near Madrid, Spain (11) found that toads bred successfully. Housing adults outdoors, under semi-captive conditions, was most effective for achieving mating. Over 180 tadpoles were produced in captivity. Tadpoles were collected in the wild and treated against the chyrtid fungus using elevated temperature (> 21°C) and baths in antifungal drugs (itraconazole). Tadpoles were reared in indoor aquariums in similar environmental conditions to the wild.

A replicated study in 2008–2011 of captive Apennine yellow-bellied toads *Bombina varie-gata pachypus* at the University of Genoa, Italy (12) found that by 2011 the captive breeding programme had succeeded in raising several tadpoles to metamorphosis. The plan was to go on to release animals to purpose-built breeding sites.

(1)   Burchfield P.M. (1975) Breeding the Colombian giant toad *Bufo blombergi* at Brownsville Zoo. *International Zoo Yearbook*, 15, 89–90.

(2)   Jones M. (1984) Captive rearing and breeding of Norfolk natterjacks, *Bufo calamita*. *British Herpetological Society Bulletin*, 10, 43–45.

(3)   Maruska E.J. (1986) Amphibians: review of zoo breeding programmes. *International Zoo Yearbook*, 24/25, 56–65.

(4)   Johnson B. & Paine F. (1989) *The Release of Puerto Rican Crested Toads: Captive Management Implications and the Cactus Connection*. Proceedings of the Regional Meetings of the American Association of Zoological Parks and Aquariums. pp. 962–967.

(5)   Billings D. (1991) Keeping and breeding the midwife toad (*Alytes obstetricans*) in captivity. *British Herpetological Society Bulletin*, 35, 12–16.

(6)   Johnson R.R. (1994) Model programs for reproduction and management: ex situ and in situ conservation of toads of the family Bufonidae. In J.B. Murphy, K. Adler & J.T. Collins (eds) *Captive Management and Conservation of Amphibians and Reptiles*, Contributions to Herpetology Vol. 11, Society for the Study of Amphibians and Reptiles, Ithaca, New York. pp. 243–254.

(7)   Wilkinson J. W. (1994) An account of successful captive reproduction of *Bombina bombina*, the European fire-bellied toad. *British Herpetological Society Bulletin*, 35, 12–16.

(8)   Burton M.S., Thorne E.T., Anderson A. & Kwiatkowski D.R. (1995) Captive management of the endangered Wyoming toad at the Cheyenne Mountain Zoo. *Bulletin of the Association of Reptilian and Amphibian Veterinarians*, 5, 6–8.

(9)   Kinne O., Kunert J. & Zimmermann W. (2004) Breeding, rearing and raising the red-bellied toad *Bombina bombina* in the laboratory. *Endangered Species Research*, 1, 11–23.

(10)   Lee S., Zippel K.C., Ramos L. & Searle J. (2006) Captive breeding program for the Kihansi spray toad (*Nectophrynoides asperginis*) at the Wildlife Conservation Society, Bronx, New York. *International Zoo Yearbook*, 40, 241–253.

(11)   Martín-Beyer B., Fernández-Beaskoetxea S., García G. & Bosch J. (2011) Re-introduction program for the common midwife toad and Iberian frog in the Natural Park of Peñalara in Madrid, Spain: can we defeat chytridiomycosis and trout introductions? In P.S. Soorae (ed) *Global Re-introduction Perspectives: 2011. More case studies from around the globe*, IUCN/SSC Re-introduction Specialist Group & Abu Dhabi Environment Agency, Gland, Switzerland. pp. 81–84.

(12)   Canessa S. (2012) Trying to reverse the decline of the Apennine yellow-bellied toad in northern Italy. *Froglog*, 101, 24–25.

13.2.3 Mallorcan midwife toad

• Two replicated studies in the UK[1, 2] found that Mallorcan midwife toads produced eggs that were raised to metamorphs or toadlets successfully in captivity. One[1] found that clutches dropped by males were not successfully maintained artificially.

• One replicated study in the UK[1] found that survival to metamorphosis was 85%. One randomized, replicated, controlled study in the UK[3] found that toads bred in captivity for nine or more generations had slower tadpole development, reduction in one predator defence trait and decreased genetic diversity.

A replicated study in 1985–1988 of Mallorcan midwife toads *Alytes muletensis* at Jersey Zoo, UK (1) found that toads bred successfully in captivity. No breeding occurred in the first two years. However, in the third year, 17 egg clutches were produced, with an average of 12 eggs (range: 9–15). Three clutches dropped by the males were not successfully maintained artificially. Most eggs hatched and the first six clutches had metamorphosed by October 1988. The average survival to metamorphosis of those clutches was 85% (range: 22–100%). From 1985, 6 to 14 toads were housed in 2 glass tanks (1 x 0.6 x 1 m) in an unheated room (7–30°C). Tanks contained rocks, branches, tiles and a small pond and were misted to prevent drying. In April 1988, six toads were moved to an outdoor tank (5–12°C), but were returned indoors in May. Tadpoles and toadlets were reared in separate tanks.

A replicated study in 1997–1999 of Mallorcan midwife toads *Alytes muletensis* at Jersey Zoo, UK (2) found that three captive populations bred successfully. In 1999, 40 clutches (433 eggs) were produced by the three breeding groups. Tadpoles hatched and reached meta-morphosis. In 1997, 25 tadpoles were collected from each of 3 wild populations and were housed separately. Housing was in plastic tanks with gravel substrate, hides, water and simulated rainfall.

A randomized, replicated, controlled study of captive Mallorcan midwife toads *Alytes muletensis* in the UK (3) found that there was a significant reduction in one predator defence trait (lower tail fin depth) in animals maintained in captivity for 9–12 compared to 1–2 generations. Long-term stock tadpoles also developed more slowly and had a significant loss of genetic variation. Tail length did not differ between populations. Forty tadpoles from a population captive-bred for 1–2 or for 9–12 generations (different ancestry) were divided between 2 treatments: chemical cues from viperine snakes or a control. Tadpoles were meas-ured every 15 days. DNA was analysed.

(1)   Tonge S.J. & Bloxam Q.M.C. (1989) Breeding the Mallorcan midwife toad *Alytes muletensis* in captivity. *International Zoo Yearbook*, 28, 45–53.
(2)   Buley K.R. & Gonzalez-Villavicencio C. (2000) The Durrell Wildlife Conservation Trust and the Mallorcan midwife toad, *Alytes mulentensis* – into the 21st century. *Herpetological Bulletin*, 72, 17–20.
(3)   Kraaijeveld-Smit F.J.L., Griffiths R.A., Moore R.D. & Beebee T.J.C. (2006) Captive breeding and the fitness of reintroduced species: a test of the responses to predators in a threatened amphibian. *Journal of Applied Ecology*, 43, 360–365.

13.2.4 Harlequin toads (*Atelopus* species)

• One review and three of five replicated studies (including one small study) in Colombia, Ecuador, Germany and the USA[1-6] found that harlequin toads reproduced in captivity. One[4] found that eggs were only produced in captivity by simulating a dry and wet season and one[3] found that successful breeding was difficult. One[6] found that captive-bred har-lequin toads were raised successfully to metamorphosis in captivity. Two found that most toads died before[3] or after hatching[4].

A replicated study of captive Panamanian golden frogs *Atelopus zeteki* in the USA (1) found that the frogs bred successfully in captivity. Hundreds of offspring were bred in captivity and sent to other zoos in the breeding programme. Small assurance populations of the species were maintained and bred in case of extinction in the wild from threats that included chytri-diomycosis. Captive-bred frogs were to be released if wild populations became extinct.

A review of captive breeding programmes in 2001–2007 of priority amphibian species from Panama at 50 zoos and aquariums in the USA (2) found that Panamanian golden frogs *Atelopus zeteki* bred successfully in captivity. By 2007, 41 of 111 wild-caught Panamanian golden frogs *Atelopus zeteki* were surviving and there were over 1,500 frogs in the captive-breeding population. In 2001–2005, 111 Panamanian golden frogs were collected, including 26 pairs and 59 newly metamorphosed froglets. Strict quarantine and hygiene protocols were enforced. Animals were tested for chytrid and treated with Itraconazole.

A small, replicated study in 2011–2012 of captive harlequin toads *Atelopus* in Ecuador (3) found that although one clutch of eggs was produced by each species, maintaining healthy adults, successfully breeding and rearing juveniles was difficult. One of three breed-ing attempts for elegant stubfoot toad *Atelopus elegans* and the one attempt for Pebas stubfoot toad *Atelopus spumarius* and *Atelopus spumarius-pulcher* complex resulted in a clutch of eggs. However, most *Atelopus spumarius* embryos were dead within eight days. This was considered by the authors to be due to a drop in water temperature one night. Three tadpoles survived the first month and just one over four months. Nineteen *Atelopus spumarius-pulcher* complex toadlets survived over eight months (from 500 eggs). Causes of death were unknown. Twenty adult elegant stubfoot toads, 8 Pebas stubfoot toads and 30 *Atelopus spumarius-pulcher* complex toads were wild caught. Breeding tanks were 60 x 35 x 30 cm with stones, plants and an open system of filtered water. One of three female elegant stubfoot toads and one Pebas stubfoot toad were stimulated with human chorionic gonadotrophin (0.05 ml).

A replicated study in 2008–2011 of captive harlequin toads *Atelopus flavescens* at Cologne Zoo, Germany (4) found that egg deposition was stimulated by maintaining toads in a drier environment followed by a period of intensive irrigation. However, no toadlets survived past day 142. Three breeding trials resulted in no egg production. Following the simulation of a dry then wet season, 2 clutches of eggs were produced with 400–500 eggs (5–10% unfertilized). On day 43 after egg deposition, only 2 larvae survived. One tadpole survived to day 112, the other died as a froglet at day 142. Males were housed in three groups of 12–15 in tanks (100 x 60 x 60 cm) with artificial streams. Four females were transferred to the tanks for breeding. Tanks were misted several times each day. A dry season with reduced water and misting was then simulated for three months followed by a wet season with increased water and misting.

A replicated study in 2006–2012 of amphibians at two breeding facilities in Panama (5) found that the majority of the priority conservation species bred successfully in captivity. At one breeding facility, 10 of the 15 priority amphibian species collected from chytrid-infected areas in 2006 reproduced in captivity, with varying success rates. This included the Panamanian golden frog *Atelopus zeteki*. At a second facility, captive Limosa harlequin frog *Atelopus limosus*, Toad Mountain harlequin frog *Atelopus certus* and Pirre Mountain frog *Atelopus glyphus* reproduced successfully. The Panama Amphibian Rescue and Conservation project, launched in 2009, aimed to establish assurance colonies of species in extreme danger of extinction and to reduce impacts of the chytrid fungus.

A replicated study in 2009–2012 of harlequin frogs *Atelopus* at Cali Zoo, Colombia (6) found that toads bred and survived to metamorphosis in captivity. Eggs were laid in the first year, with each female producing 200–300 eggs. Tadpoles hatched within ten days. In 2009–2012 there were nine successful hatchings, eight of which fully metamorphosed. Seventeen adults, 10 males and 7 females were collected from the wild in June 2009.

(1)   Zippel K.C. (2002) Conserving the Panamanian golden frog: Proyecto Rana Dorada. *Herpetological Review*, 33, 11–12.
(2)   Gagliardo R., Crump P., Griffith E., Mendelson J., Ross H. & Zippel K. (2008) The principles of rapid response for amphibian conservation, using the programmes in Panama as an example. *International Zoo Yearbook*, 42, 125–135.
(3)   Coloma L.A. & Almeida-Reinoso D. (2012) Ex situ management of five extant species of *Atelopus* in Ecuador – progress report. *Amphibian Ark Newsletter*, 20, 9–12.
(4)   Gawor A., Rauhaus A., Karbe D., Van Der Straeten K., Lötters S. & Ziegler T. (2012) Is there a chance for conservation breeding? Ex situ management, reproduction, and early life stages of the harlequin toad *Atelopus flavescens* Duméril & Bibron, 1841 (Amphibia: Anura: Bufonidae). *Amphibian & Reptile Conservation*, 5, 29–44.
(5)   Gratwicke B. (2012) Amphibian rescue and conservation project – Panama. *Froglog*, 102, 17–20.
(6)   Silva C. (2012) A conservation program for *Atelopus* species at the Cali Zoo, Colombia. *Amphibian Ark Newsletter*, 19, 7.

13.2.5 Salamanders (including newts)

* Four of six replicated studies (including four small studies) in Japan, Germany, the UK and the USA[1, 2, 4, 5, 7, 8] found that eggs were produced successfully in captivity, in one case by

one captive-bred female[5]. Two found that production of eggs depended on tank habitat[5] or was more successful in semi-natural compared to laboratory conditions[7]. Captive-bred salamanders were raised to yearlings[4] or a small number of larvae[1, 2, 8] or adults[5] in captivity. One review[3] found that four salamander species bred successfully in captivity, but slimy salamanders produced eggs that did not hatch.

- One replicated study in Japan[4] found that 60% of Japanese giant salamander eggs survived to hatching in captivity. Two replicated studies (including one small study) in Mexico and the USA found that larval development[1, 6], body condition and survival[6] of captive-bred amphibians were affected by water temperature, density and whether they were raised under laboratory or semi-natural conditions.

A small, replicated study in 1979–1980 of captive Texas blind salamander *Typhlomolge rathbuni* in Cincinnati Zoo, USA (1) found that the species bred successfully in captivity. Three clutches of 8–21 eggs were produced. Three larvae from the first clutch survived and all 14 eggs from the third clutch hatched. Embryonic development required a constant water temperature of 20–21°C as lower temperatures resulted in deformities. Adults had been in captivity since 1975 and were moved to separate tanks once eggs were produced.

A small, replicated study in 1981 of captive Pyrenean mountain salamanders *Euproctus asper asper* in the UK (2) found that eggs were produced by one of two pairs in captivity. Eight eggs were produced by the pair in June 1982. Three of the eggs hatched one month later. One tadpole survived to at least six months having eaten the other two. Mating behaviour had been observed all year. Two pairs were obtained in 1981 and housed in two aerated 30 x 20 x 20 cm tanks. Gravel substrate and cover were provided.

A review of captive breeding programmes (3) found that a number of amphibian species have been bred successfully in captivity. Salamander species that were bred successfully in captivity were: Texas blind salamander *Typhlomolge rathbuni*, Tennessee cave salamander *Gyrinophilus palleucus*, Japanese giant salamander *Andrais japonicas* and Anderson's salamander *Ambystoma andersoni*. Slimy salamander *Plethodon glutinosus* produced eggs but they did not hatch.

A replicated study in 1978–1988 of captive Japanese giant salamanders *Andrias japonicas* in a zoo in Hiroshima, Japan (4) found that the salamanders bred successfully in captivity. Between one and three females produced eggs in each of the three breeding groups each year. A total of 36 egg masses were produced between 1979 and 1988 (336–2434 eggs/mass). Survival to hatching was approximately 60% (range: 0–97%). By 1988, there were 1,035 captive larvae and young. Males were observed eating eggs in the smaller enclosures. Between 1978 and 1983, three breeding groups were established with two to four males and three females. Groups were housed in two to four connected outdoor tanks (90 x 70 x 45 cm) with sand and water. Larvae were reared in separate outdoor tanks (65 x 38 x 15 cm).

A small, replicated study in 1990–1994 of Texas salamanders *Eurycea neotenes* at the Dallas Aquarium, USA (5) found that captive breeding was successful under certain conditions. In 1991, the female in a planted tank deposited 19 eggs. Eggs were transferred to a dark tank and four hatched. After one year, two of the three surviving captive-bred salamanders laid fully developed eggs. No further reproductive behaviour was seen for 1.5 years. The one original and one captive-bred female placed in an artificial aquifer laid eggs in 1–2 years. Larvae left in the aquifer were not predated by the parents over two months. Three pairs of wild-caught salamanders were housed in separate 4 L aquaria with water flow (22°C). One had gravel substrate, one contained plants and the other partially buried rocks. Fourteen additional animals were housed in a 189 L aquarium with water flow and pipe sections. One original and one captive-bred pair were placed in an artificial 1.2 m long aquifer.

A replicated study in 2003 of Mexican axolotls *Ambystoma mexicanum* in Xochimilco, Mexico (6) found that survival was lower but growth and body condition greater in captive-bred animals reared in semi-natural compared to laboratory conditions. Survival was significantly lower from day 10 to 30 under low-maintenance, semi-natural conditions in canals than under high-maintenance laboratory conditions (8–42% vs 88–98%). Survival was independent of density. By day 30, axolotls reared in canals were significantly larger than those in the laboratory. Those raised at low densities were significantly larger than those at high densities (canal: 0.44 vs 0.39; lab: 0.17 vs 0.11 g). Body condition was significantly better under semi-natural conditions. Average maintenance costs/axolotl/year for seven captive-breeding facilities were US$14–340. One hundred and fifty larvae were divided between 6 aquaria (45 L; 17–19°C) and 6 cages in a canal (45 L; 19–24°C) at densities of either 5 or 20 larvae. Aquaria had artificial plants and were cleaned every 10 days. The canal had filter systems to prevent aquatic predators from entering. Larvae were measured every 10 days.

A replicated study in 1994–2004 in the Luhe valley, Germany (7) found that captive breeding in an outdoor enclosure was significantly more effective than in indoor tanks for great crested newts *Triturus cristatus* and smooth newts *Triturus vulgaris*. Initially, 24 of each species were housed in 100 L indoor tanks (one male and three females/tank). In following years, eggs produced in an outdoor enclosure were collected and transferred indoors for rearing. Sixty great crested newt and 90 smooth newt larvae and juveniles were released into 2 created ponds annually.

A small, replicated study in 2009–2011 of southern dwarf sirens *Pseudobranchus axanthus* at the Central Florida Zoo and Botanic Gardens, USA (8) found that breeding occurred in captivity. Single eggs were recorded in December 2010 and larvae in February 2011. The 12 larvae observed were moved to a separate tank and all survived to at least 9 months. Nine wild caught animals were housed in two aquaria (38 L) with sand and leaf litter substrate. In May 2010, two males and two females were moved to outdoor cattle troughs filled with rainwater, with sand and leaf litter substrate. Local aquatic plants were added and invertebrates colonized naturally.

(1) Maruska E.J. (1982) *The Reproduction and Husbandry of Salamanders in Captivity with Special Emphasis on the Texas Blind Salamander,* Typhlomolge rathbuni. Proceedings of the 5th Annual Reptile Symposium on Captive Propagation and Husbandry. Oklahoma City Zoo. pp. 151–161.
(2) Wisniewski P.J. & Paull L.M. (1982) A note on the captive maintenance and breeding of the Pyrenean mountain salamander (*Euproctus asper asper* Dugès). *British Herpetological Society Bulletin,* 6, 46–47.
(3) Maruska E.J. (1986) Amphibians: review of zoo breeding programmes. *International Zoo Yearbook,* 24/25, 56–65.
(4) Kuwabara K., Suzuki N., Wakabayashi F., Ashikaga H., Inoue T. & Kobara J. (1989) Breeding the Japanese giant salamander at Asa Zoological Park. *International Zoo Yearbook,* 28, 22–31.
(5) Roberts D.T., Schleser D.M. & Jordan T.L. (1995) Notes on the captive husbandry and reproduction of the Texas salamander *Eurycea neotenes* at the Dallas Aquarium. *Herpetological Review,* 26, 23–25.
(6) McKay J.E. (2003) An evaluation of captive breeding and sustainable use of the Mexican axolotl (*Ambystoma mexicanum*). MSc thesis. University of Kent.
(7) Kinne O. (2004) Successful re-introduction of the newts *Triturus cristatus* and *T. vulgaris. Endangered Species Research,* 1, 25–40.
(8) Stabile J.L. (2012) Captive propagation of the southern dwarf siren (*Pseudobranchus axanthus*). *Herpetological Review,* 43, 600–601.

13.3   Use hormone treatment to induce sperm and egg release

• One review and nine of ten replicated studies, including two randomized, controlled studies, in Austria, Australia, China, Latvia, Russia and the USA[2, 3, 5, 6, 8–12, 14, 17] found that hormone treatment of male amphibians stimulated[6, 8, 10, 12, 17] or increased sperm production[11, 14] or resulted in successful breeding[3, 5, 9] in captivity. One[2] found that hormone treatment of males and females did not result in breeding. Four found that the amount[10, 12, 14, 17] and viability[12] of sperm produced was affected by the type, amount or number of doses of hormone.

- One review and 9 of 14 replicated studies (including six randomized and/or controlled studies) in Australia, Canada, China, Ecuador, Latvia and the USA[1–10, 13–16] found that hormone treatment of female amphibians had mixed results, with 30–71% of females producing viable eggs following treatment[1, 4, 6, 10], or with egg production depending on the combination[7, 13], amount[13] or number of doses[5, 6, 8, 14] of hormones. Three found that hormone treatment stimulated egg production[16] or successful breeding[3, 9] in captivity. Two found that hormone treatment did not stimulate[2] or increase[15] egg production.

- Five replicated studies (including one controlled study) in Canada, Latvia and the USA found that eggs induced by hormone treatment were raised successfully to tadpoles[9], toadlets[1] or froglets[5, 13, 15] in captivity. Two replicated studies, one of which was small, in Ecuador and the USA found that most toads died before[16] or soon after hatching[4].

Background
Captive animals do not always breed successfully under artificial conditions. Reproductive technologies such as hormone treatment to induce ovulation or sperm production are techniques that can be used in an attempt to achieve or increase breeding success by amphibians in captive facilities. Hormone stimulation protocols are often species specific.

A replicated study in 1983 of Puerto Rican toads *Peltophryne lemur* at Buffalo Zoological Gardens, USA (1) found that following hormone treatment, one of three females produced viable eggs. Another female produced hundreds of infertile eggs and the third a few unfertilized eggs. Over 150 tadpoles hatched from the viable clutch and tadpole survival was 100%. In 1983, 75 of the toadlets were released at an artificial pond in Puerto Rico. Three male and female captive bred-toads were housed in an enclosure. Breeding was induced by lutenizing hormone-releasing hormone (0.01–0.05 ml/10 g body weight). Tadpoles were transferred to aquaria for rearing and metamorphs to containers.

A small, replicated study in 1983 of Puerto Rican crested toads *Bufo lemur* in the USA (2) found that hormone treatment of males and females did not induce successful breeding in captivity. Two pairs displayed breeding behaviour for three weeks but no eggs were produced. Adults were housed in wooden cages (85 x 90 x 75 cm) with bark chips, wood, plants and a pool. To induce breeding, humid conditions were created by placing toads in glass tanks (76 x 70 x 200 cm) with an overnight misting system. Lutenizing hormone-releasing hormone was given under the skin to females (0.1 ml/100 g body weight) and males (0.01 ml/100 g).

A review of captive breeding programmes (3) found that breeding was induced with gonadotrophin-releasing hormone in White's tree frog *Litoria caerulea* and African red frog *Phrynomerus bifasciatus*.

A replicated study in 1991–1994 of Wyoming toads *Bufo hemiophrys baxteri* in a zoo in Colorado, USA (4) found that five of seven hormonally induced females produced thousands of fertile eggs in 1994. However, the majority of tadpoles that hatched died within 72 hours. Deaths were considered by the authors to have been due to water quality. In captivity, two to four wild-caught and captive-bred toads were housed per tank (40 x 61 x 23 cm) at 20°C. Cork bark, sheet moss, sand, water, artificial plants and a basking lamp were provided. In 1991–1993, toads were transported to breeding enclosures at the edge of the lake. In 1994, five toads were overwintered for six weeks at 4.5°C. Seven females were hormonally induced and paired in captivity.

A replicated study in 1988–1992 of captive European tree frogs *Hyla arborea* in Latvia (5) found that following hormone treatment of males and females, frogs bred successfully in captivity. In several cases three or four hormone injections were required to induce spawning. Each females produced 200–800 eggs. An average of 60–90% of larvae metamorphosed in captivity. The period of metamorphosis was shorter in captivity than the wild (30–60 vs 90 days). Over 4,000 froglets were produced. Wild-caught frogs were housed in outdoor tanks and were overwintered in a refrigerator. From February, daylight period, UV light and feeding was increased. Two males and one female were placed in separate 35 L aquaria with water and plants. Breeding was stimulated with hormone injections (100 mg luliberin-surphagon/ml of solution) in March or May. Females received 15–20 mg and males 10 mg. The injection was repeated after 24 hours if spawning did not start. Larvae, metamorphs and toadlets were raised in separate tanks.

A small, replicated, controlled study of captive Chinese giant salamanders *Andrias davidianus* in China (6) found that injection with reproductive hormones induced egg and sperm production. Eggs were produced by 60% of females given injections of lutenizing hormone-releasing hormone-a or human chorionic gonadotrophin (400–500 eggs). However, mating was not observed. Eggs were laid earlier following injection with human chorionic gonadotrophin compared to lutenizing hormone-releasing hormone and earlier with higher water temperatures. A single injection was more effective than repeated injections. Females produced eggs between 96–120 hours and sperm was produced after 80 hours. A 1°C drop in water temperature resulted in a 10 hour delay. Animals not injected with hormones did not produce eggs or sperm and reproductive organs degenerated and were absorbed. Wild-caught salamanders were housed in 16 m² tanks.

A randomized, replicated study in 2005 of captive Fowler toads *Bufo fowleri* in the USA (7) found that treatments of progesterone along with other hormones were effective at inducing egg production in a high proportion of toads and resulted in high egg numbers. Successful progesterone (5 mg) treatments were: progesterone and lutenizing hormone-releasing hormone-a (LHRHa; 60 µg) alone (71% produced eggs; 2,004 eggs/toad), or with dopamine-2 receptor antagonist pimozide (0.25 mg) and human chorionic gonadotrophin (500 IU; 85%; 1,078), or progesterone, LHRHa (20 µg) and pimozide (0.25 mg; 58%; 2,486). Two repeated doses of 5 mg progesterone or a single dose of 20 µg LHRHa did not result in egg production. Egg production was low with 4 µg LHRHa and 500 IU human chorionic gonadotrophin (29%; 2,283) or 20 µg LHRHa and 0.25 mg pimozide (14%; 627). Second doses of 60 µg LHRHa or 500 IU human chorionic gonadotrophin given 24 or 48 hours after initial doses resulted in low egg numbers. Wild caught toads were housed in 50 x 40 x 10 cm tanks. Females were randomly assigned to the seven treatments with seven females/treatment. Treatments were given in 100 µl of saline.

A replicated study in 2005 of captive Wyoming toad *Bufo baxteri* in the USA (8) found that one or two priming doses of hormones were required to induce egg production, but not sperm production. Eight of 10 males receiving a single dose of 300 IU human chorionic gonadotrophin (hCG) produced spermic urine within 5 hours. Females given a single dose of hCG plus lutenizing hormone-releasing hormone-a (LHRHa) produced no eggs. Compared to one priming dose, two priming doses resulted in a greater proportion of females spawning (70 vs 88%) and significantly higher average number of eggs produced (1,647 vs 3,280) and numbers produced/female at a given time (4 vs 7). The total number of eggs/female did not differ with treatment. Toads were housed in 45 L tanks. Ten females were primed with 500 IU hCG and 4 μg LHRHa. After 72 hours, the 10 females and an additional 10 females were given 100 IU hCG and 0.8 μg LHRHa, followed 96 hours later by 500 IU hCG and 4 μg LHRHa.

A replicated study in 1999–2006 of Wyoming toads *Bufo hemiophrys baxteri* in Saratoga, Wyoming, USA (9) found that hormone treatment of males and females induced successful breeding in captivity. Between 1999 and 2006, an average of 6,863 toads were bred and released each year. In 2006, an 18% increase in hatch rate was achieved. This was thought to be due to over-wintering at cooler temperatures, to simulate the harsh weather faced in the wild. Breeding pairs were carefully selected from a studbook of the 150 captive toads. Pairs were housed in separate water tanks. Toads were injected with hormones to induce production of eggs and sperm. Over 20 breeding events were undertaken each year. Most toads are released as tadpoles in autumn.

A randomized, replicated, controlled study in 2009 of southern corroboree frogs *Pseudophryne corroboree* at Monash University, Australia (10) found that hormone treatment successfully induced sperm release and to a lesser extent egg production. Human chorionic gonadotropin (hCG) and luteinizing hormone-releasing hormone (LHRHa) both induced significantly higher proportions of males to release sperm than controls (82 vs 0%). LHRHa treated males released significantly higher numbers of sperm (670 vs 50) and concentration of sperm (4,500 vs 800 x 103/ml) over a longer period than those treated with human chorionic gonadotropin. There was no significant difference in numbers of females releasing eggs following LHRHa and controls (30 vs 0%). Eggs were released 24–48 hours post-treatment (peak 36 hours). Average clutch size was 15. Six randomly selected males were given a dose of either 20 μg/g bodyweight of human chorionic gonadotropin or 5 μg/g of LHRHa in simplified amphibian Ringer solution (SAR) or a control of 0.1 ml of SAR. Sperm response was tested in urine 7 times up to 72 hours post-treatment. Seventeen females received a priming (1 μg/g) and ovulatory dose (5 μg/g) of LHRHa in SAR. Eight received a control of 0.1 ml of SAR. Ovulation was tested every 12 hours for 5 days.

A replicated study in 2009 of captive European common frogs *Rana temporaria* in Austria (11) found that injecting males with human chorionic gonadotrophin increased sperm production. Males stimulated with hormones had greater sperm production than untreated males (0.004 vs 0.002 testis/body weight). The same was true for the sperm cell concentration (80 vs 11 x 10^6/ml in 1.5 ml motility-inhibiting saline/testes). Males received injections of 150 IU of human chorionic gonadotrophin and were killed after 15 hours. Testes were removed, weighed and macerated in motility-inhibiting saline.

A replicated, controlled study in 2009 of captive Günther's toadlets *Pseudophryne guentheri* in Western Australia (12) found that hormone treatment successfully induced sperm release. Luteinizing hormone-releasing hormone (LHRHa) in doses of 1, 2, 4 or 8 μg/g induced 100% of males to produce sperm, compared to 10–30% of controls. Numbers of sperm released was significantly higher following 2 μg/g of LHRHa (25 x 10^3) than 8 μg/g (5 x 10^3) or controls (0); other doses did not differ significantly (8–12 x 10^3). Sperm viability was significantly higher following the 1 μg/g compared to 8 μg/g treatment. Arg^8-vasotocin acetate salt (4 μg/g) alone or with 2 μg/g LHRHa resulted in similar numbers of males releasing sperm as a single 2 µg/g dose of LHRHa (71; 71; 100% respectively). However, sperm numbers were significantly lower (0 vs 25 x 10³). Male toadlets were given a single dose of 1, 2, 4 or 8 µg/g bodyweight of LHRHa in simplified amphibian Ringer solution, or a control of 100 µL of simplified amphibian Ringer solution (n = 7–10/treatment). Sperm release was tested at 3, 7 and 12 hours post-treatment.

A replicated, controlled study in 2008 of captive frogs in Ottawa, USA (13) found that injection with a gonadotropin-releasing hormone (GnRH) agonist and a dopamine antagonist was effective at inducing egg production. After one week in captivity GnRH-A (0.4 µg/g body weight) and metoclopramide (10 µg/g) was more effective at inducing egg production in northern leopard frogs Lithobates pipiens (100%) than GnRH-A and pimozide (10 µg/g; 50%), GnRH-B (0.4 µg/g) and pimozide (42%) or no treatment (0%). After one month in captivity, GnRH-A with 10 µg/g of metoclopramide was significantly more effective than with 5 µg/g (60 vs 44%). Out-of-season breeding was induced with GnRH-A and metoclopramide in five pairs, with 25% of females producing eggs (and metamorphs). Egg production (and metamorphs) was also induced in Argentine horned frog Ceratophrys ornate (1 pair), Cranwell's horned frog Ceratophrys cranwelli (1 pair) and escuercitos Odontophrynus americanus (10 males, 5 females). A week after collection in April, 12 female and 18 male leopard frogs were given one of 4 initial treatments. Controls were given saline and dimethyl sulfoxide. A month after collection, 9 females and 15 males were given 0.4 µg/g GnRH-A and either 5 or 10 µg/g of metoclopramide, or were controls. Following collection in September, artificial overwintering was induced in 8 females and 15 males. In October, males were primed with two injections of GnRH-A (0.025 then 0.05 µg/g a week later). Frogs were then injected with GnRH-A (0.4 µg/g) and metoclopramide (10 µg/g).

A randomized, replicated, controlled study in 2009 of captive Günther's toadlets Pseudophryne guentheri in Western Australia (14) found that hormone treatment successfully induced sperm and egg release. Proportions of males producing sperm with no, one or two priming injections of luteinizing hormone-releasing hormone (LHRHa) did not differ (100%), but were significantly higher than controls (25%). Amount of sperm produced decreased with priming treatments (none: 1.8 x 10⁴; one: 0.6 x 10⁴; two: 0.3 x 10⁴). Sperm viability did not differ between hormone treatments (0.6–0.7 sperm/total) and was highest at 3 hours. Significantly higher numbers released eggs with one or two priming treatments (priming: 100%; none: 25%; control: 0%). The same was true for the number of eggs (priming: 217–220; none: 19; control: 0). Mass of eggs from two priming treatments was significantly greater than from no priming (0.007 vs 0.001 g; one priming: 0.006 g). Thirty-two wild collected males and females were randomly assigned to four treatments: a single dose of 2 µ/g LHRHa in simplified amphibian Ringer solution, or a dose preceded by one or two priming injections of 0.4 µ/g LHRHa (an hour apart), or a control of 100 µ/g of simplified amphibian Ringer solution. Sperm release was tested at 3, 7 and 12 hours after treatment. Ovulation was tested at 10–11 hours.

A replicated study in 2010–2011 of captive Oregon spotted frogs Rana pretiosa in Vancouver, Canada (15) found that frogs bred successfully in captivity and that treatment with hormones did not increase the proportion of females producing eggs or numbers of eggs. The two hormonal substances tested did synchronize timing of egg production. The small number of mature frogs produced 291 tadpoles in the first year. In 2011, a larger number of frogs bred and over 9,000 eggs were produced, of which 3,000 hatched. Providing a seasonal daylight and temperature regime was considered by the authors to be crucial to breeding success. Metamorphs and tadpoles were released in spring 2011. Eggs were collected each year from the wild to increase genetic diversity of the captive population.

A small, replicated, controlled study in 2011–2012 of captive harlequin toads *Atelopus* in Ecuador (16) found that following treatment with human chorionic gonadotrophin, females produced eggs. The one female elegant stubfoot toad *Atelopus elegans* and one Pebas stubfoot toad *Atelopus spumarius* treated with hormones produced a clutch of eggs. However, most Pebas stubfoot toad embryos were dead within eight days. Two untreated elegant stubfoot toads did not produce eggs. Twenty adult elegant stubfoot toads and eight Pebas stubfoot toads were wild caught. Breeding tanks were 60 x 35 x 30 cm with stones, plants and an open system of filtered water. One of three female elegant stubfoot toads and one Pebas stubfoot toad were stimulated with human chorionic gonadotrophin (0.05 ml).

A replicated, controlled study in 2011 of captive amphibians in Russia (17) found that the greatest sperm production was induced with high dose lutenizing hormone-releasing hormone-a (LHRHa) for common frogs *Rana temporaria* and priming with LHRHa prior to human chorionic gonadotrophin (hCG) for common toads *Bufo bufo*. In common frogs, 1.2 µg/g bodyweight LHRHa induced significantly higher sperm numbers (650×10^6/ml) than pituitary extract (485×10^6) or 0.12 µg/g LHRHa (444×10^6), which produced significantly higher numbers than 23 IU/g hCG (170×10^6) and 12 IU/g hCG (39×10^6). High dose LHRHa had the highest percentage of samples with sperm concentrations above 200×10^6/ml (high LHRH: 40%; pituitaries: 36%; low LHRH: 15%; hCG: 0%). Sperm motility was similar with all treatments (76–90%). Priming common toads resulted in significantly higher numbers (11.6×10^6 vs 8.0×10^6/ml) and quality of sperm (motility: 85 vs 73%), but not higher sperm concentration (1.5×10^6 vs 1.8×10^6/ml). Four wild-caught frogs received each of the five hormone injection treatments. There were also ten controls. Four wild-caught toads were primed with 0.13 µg/g LHRHa 24 hours before receiving 13 IU/g hCG; controls received only the second dose. Spermic urine was monitored.

(1)   Miller T.J. (1985) Husbandry and breeding of the Puerto-Rican toad (*Peltophryne lemur*) with comments on its natural history. *Zoo Biology*, 4, 281–286.
(2)   Paine F. (1985) *Husbandry, Management, and Reproduction of the Puerto Rican Crested Toad*. Proceedings of the Eighth International Herpetological Symposium. Thurmont, Maryland. pp. 59–71.
(3)   Maruska E.J. (1986) Amphibians: review of zoo breeding programmes. *International Zoo Yearbook*, 24/25, 56–65.
(4)   Burton M.S., Thorne E.T., Anderson A. & Kwiatkowski D.R. (1995) Captive management of the endangered Wyoming toad at the Cheyenne Mountain Zoo. *Bulletin of the Association of Reptilian and Amphibian Veterinarians*, 5, 6–8.
(5)   Zvirgzds J., Stašuls M. & Vilnitis V. (1995) Reintroductions of the European tree frog (*Hyla arborea*) in Latvia. *Memoranda Societatis pro Fauna et Flora Fennica*, 71, 139–142.
(6)   Xiao H.-B., Liu J.-Y., Yang Y.-Q. & Lin X.-Z. (2006) Artificial propagation of tank-cultured Chinese giant salamander (*Andrias davidianus*). *Acta Hydrobiologica Sinica*, 30, 533–539
(7)   Browne R.K., Seratt J., Li H. & Kouba A. (2006a) Progesterone improves the number and quality of hormonally induced fowler toad (*Bufo fowleri*) oocytes. *Reproductive Biology and Endocrinology*, 4, 1–7.
(8)   Browne R.K., Seratt J., Vance C. & Kouba A. (2006b) Hormonal priming, induction of ovulation and in-vitro fertilization of the endangered Wyoming toad (*Bufo baxteri*). *Reproductive Biology Endocrinology*, 4, 34.
(9)   Springer C. (2007) Hatchery breeds Wyoming's rarest toad. *Endangered Species Bulletin*, 32, 26–27.
(10)   Byrne P.G. & Silla A.J. (2010) Hormonal induction of gamete release and in-vitro fertilisation in the critically endangered Southern Corroboree Frog, *Pseudophryne corroboree*. *Reproductive Biology and Endocrinology*, 8, 144.
(11)   Mansour N., Lahnsteiner F. & Patzner R.A. (2010) Motility and cryopreservation of spermatozoa of European common frog, *Rana temporaria*. *Theriogenology*, 74, 724–732.
(12)   Silla A.J. (2010) Effects of luteinizing hormone-releasing hormone and arginine-vasotocin on the sperm-release response of Günther's toadlet, *Pseudophryne guentheri*. *Reproductive Biology and Endocrinology*, 8, 139–147.
(13)   Trudeau V.L., Somoza G.M., Natale G.S., Pauli B., Wignall J., Jackman P., Doe K. & Schueler F.W. (2010) Hormonal induction of spawning in 4 species of frogs by coinjection with a gonadotropin-releasing hormone agonist and a dopamine antagonist. *Reproductive Biology and Endocrinology*, 8, 1–9.

(14) Silla A.J. (2011) Effect of priming injections of luteinizing hormone-releasing hormone on sper-
      miation and ovulation in Günther's toadlet, *Pseudophryne guentheri*. *Reproductive Biology and
      Endocrinology*, 9, 68–76.
(15) Thoney D.A. (2011) Oregon Spotted Frog – Endangered in British Columbia. *Amphibian Ark Newsletter*, 17, 13.
(16) Coloma L.A. & Almeida-Reinoso D. (2012) Ex situ management of five extant species of Atelopus
      in Ecuador – progress report. *Amphibian Ark Newsletter*, 20, 9–12.
(17) Uteshev V.K., Shishova N.V., Kaurova S.A., Browne R.K. & Gakhova E.N. (2012) Hormonal
      induction of spermatozoa from amphibians with *Rana temporaria* and *Bufo bufo* as anuran models.
      Reproduction, Fertility and Development, 24, 599–607.

13.4   Use artificial fertilization in captive breeding

- Three replicated studies (including two randomized studies) in Australia and the USA
  found that the success of artificial fertilization depended on the type[1] and number of
  doses[1, 2, 4] of hormones used to stimulate egg production. One replicated study in
  Australia[3] found that 55% of eggs were fertilized artificially, but soon died.

Background

Reproductive technologies such as artificial fertilization are techniques that can be used
in an attempt to achieve or increase breeding success by captive amphibians. Many amphibians have external fertilization, making this technique relatively simple. However,
consideration must be given to the storage and ratio of sperm and eggs, effects of
temperature, solution strength and egg jelly on the potential for fertilization.

A randomized, replicated study in 2005 of captive Fowler toads *Bufo fowleri* in the USA (1)
found that the proportion of eggs fertilized artificially was affected by hormone treatment
used to stimulate egg production. Only treatments with lutenizing hormone-releasing
hormone-a (LHRHa; 20 μg or more) plus another hormone resulted in fertilized eggs. The
proportion of fertilized eggs was significantly higher following treatment with progesterone
(5 mg) and 60 μg LHRHa (73%) than progesterone with 20 μg LHRHa and dopamine-2 receptor antagonist pimozide (35%) or progesterone with 60 μg LHRHa, pimozide and human
chorionic gonadotrophin (500 IU; 20%). Following treatment with LHRHa but no progesterone only one toad produced eggs, of which 34% became fertilized. Second doses of 60 μg
LHRHa or 500 IU human chorionic gonadotrophin given 24 or 48 hours after initial doses
resulted in low egg numbers and fertilization. Wild-caught toads were housed in 50 x 40 x
10 cm tanks. Females were randomly assigned to treatments with seven females/treatment.
Treatments were given in 100 μl of saline. Eggs were fertilized in a dish with spermic urine.
   A replicated study in 2005 of captive Wyoming toad *Bufo baxteri* in the USA (2) found
that the proportion of eggs that became fertilized artificially was similar following one or
two priming dose of hormones, but two priming doses resulted in higher numbers of viable eggs. Females given two priming doses produced significantly more tadpoles than those
given one priming dose (2,300 vs 84). Toads were housed in 45 L tanks. Ten females were
primed with 500 IU human chorionic gonadotrophin and 4 μg lutenizing hormone-releasing
hormone (LHRHa). After 72 hours, the 10 females and an additional 10 females were given
100 IU human chorionic gonadotrophin and 0.8 μg LHRHa, followed 96 hours later by 500 IU
human chorionic gonadotrophin and 4 μg LHRHa. Eggs produced during the fertile period
(12–18 hours after hormone treatment) were fertilized in a dish with spermic urine.
   A replicated study in 2009 of southern corroboree frogs *Pseudophryne corroboree* at Monash
University, Australia (3) found that artificial fertilization resulted in 55% of eggs being fer-

tilized, but embryos failed prior to gastrulation. Fertilization and the stage that the embryo failed varied between and within females. Hormone treatment was used to induce sperm and egg release. Artificial fertilization was attempted by combining spermic urine (1.1–2.9 x 10^2) with eggs from five females in a dilute solution of simplified amphibian Ringer solution at 10°C. Embryonic development was checked every 6–12 hours for 7 days.

A randomized, replicated study in 2009 of captive Günther's toadlets *Pseudophryne guentheri* in Western Australia (4) found that hormone treatment with one priming injection resulted in high artificial fertilization rates (91–100%), whereas eggs with zero or two priming treatments failed to fertilize. Twenty-four females were randomly assigned to three treatments: a single dose of 2 µ/g lutenizing hormone-releasing hormone-a in simplified amphibian Ringer solution, or a dose preceded by one or two priming injections of 0.4 µ/g lutenizing hormone-releasing hormone (one hour apart). Twenty eggs/female were fertilized with sperm from macerated testis of wild caught males in simplified amphibian Ringer solution.

(1) Browne R.K., Seratt J., Li H. & Kouba A. (2006a) Progesterone improves the number and quality of hormonally induced fowler toad (*Bufo fowleri*) oocytes. *Reproductive Biology and Endocrinology*, 4, 1–7.
(2) Browne R.K., Seratt J., Vance C. & Kouba A. (2006b) Hormonal priming, induction of ovulation and in-vitro fertilization of the endangered Wyoming toad (*Bufo baxteri*). *Reproductive Biology Endocrinology*, 4, 34.
(3) Byrne P.G. & Silla A.J. (2010) Hormonal induction of gamete release and in-vitro fertilisation in the critically endangered Southern Corroboree Frog, *Pseudophryne corroboree*. *Reproductive Biology and Endocrinology*, 8, 144.
(4) Silla A.J. (2011) Effect of priming injections of luteinizing hormone-releasing hormone on spermiation and ovulation in Günther's toadlet, *Pseudophryne guentheri*. *Reproductive Biology and Endocrinology*, 9, 68–76.

13.5   Freeze sperm or eggs for future use

- Nine replicated studies (including three controlled studies) in Austria, Australia, Russia, the UK and USA found that following freezing frog and toad sperm viability depended on species[1, 5, 7] and/or cryoprotectant used[1–3, 5–10]. One[3] found that although sperm viability was low following freezing, it could be frozen for up to 58 weeks. Five of the studies and one additional replicated study in Australia found that following freezing viability of sperm, and in one case eggs[4], also depended on storage temperature[3, 4], storage method[4, 5], freezing[7–9] or thawing rate[8, 9].

- Seven replicated studies (including three controlled studies) in Austria, Australia, the UK and USA found that frog and toad sperm viability was greatest following freezing with the cryoprotectant dimethyl sulfoxide[1–3, 8], glycerol[2, 6], sucrose[6, 7] or dimethyl formamide[10].

Background

Conservation breeding programmes are being used more frequently for threatened amphibian species. However, captive breeding often results in loss of genetic variation. This can mean that animals that were bred for release back in to the wild have reduced fitness. Freezing, or 'cryopreservation', of sperm and eggs, allows them to be stored until they are needed. Gene banks can therefore be created for amphibians ensuring that species' genetic variation is preserved. It also means that the number of a particular species needed in captivity can be reduced and genes can be swapped between captive facilities. Fewer animals in captivity means that fewer amphibians need to be taken from the wild. Freezing can damage cells and so a cryoprotectant, such as dimethyl sulphoxide or glycerol is usually required to protect the cells.

A replicated study in 1997–1998 of captive amphibians in the USA (1) found that recovery of viable sperm following freezing was significantly lower for leopard frogs *Rana pipiens* and American toads *Bufo americanus* compared to freeze-tolerant wood frogs *Rana sylvatica*. Sperm recovery was 59%, 48% and 81% respectively. Survival and viability of wood frog sperm was significantly greater using the cryoprotectant dimethyl sulfoxide and supplement of fetal bovine serum (survival: 96%; viability: 45%) than the other three protectants with glutathione (survival: 34–54%; viability with methanol: 10%) or without protectants (survival: 44–54%; viability with methanol: 16%). Testes from wild or commercially obtained males were macerated in a buffer solution. Sperm solutions from wood frogs were mixed with 0.5 M cryoprotectant (dimethyl sulfoxide, methanol, glycerol or ethylene glycol), a supplement (fetal bovine serum or glutathione) or a combination of these. Using the most successful cryopreservation treatment, sperm from each species was incubated on ice for 15 minutes, then frozen to –80°C for 1 hour (rate: 130°C/minute). Thawing was in warm water.

A replicated study of cane toads *Bufo marinus* in Australia (2) found that sperm retained motility and fertilizing capacity following cryopreservation, provided that cryoprotectants were used. Sperm frozen in sucrose alone retained no motility. The highest rates of recovery of sperm motility and fertilizing capacity were observed following storage with 15% dimethyl sulfoxide (motility: 69%; fertilization: 61%) and 20% glycerol (motility: 58%; fertilization: 81%). However, storage with different concentrations of dimethyl sulfoxide or glycerol all showed some motility (dimethyl sulfoxide: 35–69%; glycerol: 15–58%) and fertilizing capacity (dimethyl sulfoxide: 3361%; glycerol: 15–81%). Sperm from macerated testes of four toads were cryopreserved in suspensions of 10% sucrose alone or with 10, 15 or 20% dimethyl sulfoxide or glycerol. Suspensions were cooled slowly to –196 °C. Sperm was thawed in air and tested within five minutes for motility and fertilization capacity (eggs from two females).

A replicated, controlled study in 1997 of captive wood frogs *Rana sylvatica* in the USA (3) found that some sperm recovered following freezing for up to 58 weeks provided that the cryoprotectant dimethyl sulfoxide or glycerol was used. Sperm viability was significantly reduced after freezing compared to chilling for 1–30 hours with glycerol (13–17 vs 50–55%) or dimethyl sulfoxide (10–13 vs 60%). However, viability was zero without a cryoprotectant. Viability was not significantly affected by cryoprotectant concentration. There was no significant difference in viability following freezing for 1–30 hours compared to 58 weeks. Whole testes frozen in dimethyl sulfoxide had significantly higher sperm viability than those in glycerol (14 vs 5%). When chilled, sperm in had lower survival than controls and so glucose was excluded. Testes from five wild-caught frogs were macerated in a buffer. Sperm solutions from each were mixed with glucose (2 M), glycerol or dimethyl sulfoxide (1.5 or 3 M), chilled for 20 minutes and then half were frozen to –80°C in ethanol/dry ice (rate 130°C/minute) for 1–30 hours or 58 weeks. Thawing was in a 30°C water bath. Four intact testes were frozen at –80°C in glucose or dimethyl sulfoxide for five days.

A replicated study of captive cane toads *Bufo marinus* in Australia (4) found that storage method and temperature affected sperm and egg viability. Sperm stored in testes showed greater than 50% motility for seven days at 0°C and five days at 4°C. By day 15 only sperm stored at 0°C showed any motility (3%). In suspension, the longest retention of motility and fertilizing capacity was following storage in concentrated (1:1 dilution) anaerobic suspensions (up to 25–30 days). However, fertilization rates were significantly higher following storage in 1:5 dilution (day 5: 85% vs 55% for other concentrations). Egg viability was significantly higher following storage at 15°C compared to other temperatures (8 hours: 90% vs 0–60%). Storage at 5°C resulted in a decline to 0% viability after two hours. Sperm from wild toads were stored in intact testes at 0 or 4°C for 15 days (n = 6/treatment) or in suspension (macerated testes; $n = 24$) with Simplified amphibian Ringer solution at 0°C for 30 days. Dilutions were 1:1, 1:5 or 1:10 (testes:solution) and storage tubes were either opened or sealed. Immediately after ovulation, eggs from three females were stored in simplified amphibian Ringer solution at 5, 10, 15, 20 and 25°C (1,500 eggs/female). Fertilization rate was monitored up until 12 hours.

A replicated study of captive frogs in Australia (5) found that following storage at −80°C, sperm from tree frog species (Hylidae) showed greater motility than myobatrachid species (0–100 vs 1–20%). For tree frogs, sperm storage at −80°C in 15% dimethyl sulfoxide resulted in the highest motility (15%: 45–100%; 20%: 80%; glycerol 15%: 0–100%; glycerol 20%: 10–87%). Striped marsh frog *Limnodynastes peronii* sperm maintained higher motility when stored at 0°C in suspension compared to testes (3 days: 41 vs 6%). Motility of whistling treefrog *Litoria verreauxi* sperm did not differ with storage method (3 days: 83 vs 86%; 6 days: 41 vs 40%). Recovery of tree frog sperm did not differ with testes weight. Sperm from six frogs of two species were stored in intact testes and sperm from four frogs of three species were stored in suspension (macerated testes) for three or six days at 0°C. Sperm from nine tree frog and four myobatrachid species were cryopreserved in suspensions of 10% sucrose with dimethyl sulfoxide or glycerol (15 or 20%). Sperm were frozen slowly to −80°C, thawed in air and observed for three minutes.

A replicated study of captive Puerto Rican frogs *Eleutherodactylus coqui* in the USA (6) found that cryopreservation of sperm was successful with a cryoprotectant and fetal bovine serum (FSB). FBS alone resulted in only 8% viability. However, sperm viability was significantly higher with the addition of sucrose or glycerol to FBS (sucrose: 28%; glycerol: 30%; dimethyl sulfoxide: 20%). Viability did not differ significantly with dimethyl sulfoxide. Prior to freezing sperm had a viability of 56% and so normalized viabilities were: 14% for FBS alone and 35%, 50% and 54% with added dimethyl sulfoxide, sucrose and glycerol respectively. Testes of wild caught frogs were macerated in solution. Sperm was then mixed with a cryoprotectant solution (six replicates/treatment): heat inactivated FBS alone, FBS with 2M sucrose, FBS with 2M glycerol or FBS with 2M dimethyl sulfoxide. Mixtures were frozen at −80°C for 24 hours and then thawed rapidly in a 20°C water bath. Fluorescent dye was used to examine sperm.

A replicated, controlled study in 2004 of captive African clawed frog *Xenopus laevis* and western clawed frog *Xenopus tropicalis* in the UK (7) found that although sperm lost viability following freezing to −80°C, sufficient survived to fertilize eggs. Relative sperm motility after freezing, compared to a control was 30–40% (141–178 days) for African clawed frog and 39–70% (22–182 days) for western clawed frog. Optimum motility was obtained with a cooling rate of 10°C/minute in 0.2 m sucrose. Sodium bicarbonate was less effective and pentoxyfylline not effective at protecting sperm during a freeze-thaw cycle. Frozen sperm half-life was approximately one year for both species. Fertilization efficiency was greater in sodium chloride solution concentrations of 0.4 compared to 0.1 for western clawed frogs. Fertilization was similar with varying concentrations (4–40 mM) for African clawed frogs. Testes were macerated in sodium chloride solutions. Cryoprotectants (with egg yolk) were: 0.2, 0.4 or 0.6 M sucrose, sodium bicarbonate or pentoxyfylline. Sperm was frozen to −80°C at rates of 0.5–50°C/minute. Samples were defrosted rapidly in a water bath at 30°C. Fresh eggs (40–100/test) were fertilized and success recorded after 5 hours.

A replicated study in 2008 of captive African clawed frog *Xenopus laevis* in Austria (8) found that the most effective cryopreservation protocol was sperm in motility-inhibiting saline (MIS) with 5% dimethyl sulfoxide and sucrose, frozen 10 cm above liquid nitrogen and thawed at room temperature for 40 seconds. Sperm motility and viability was significantly higher following incubation (>10 mins) at 4°C in 10% dimethyl sulfoxide (motility: 40–50%; viability: 65–75%) than in 5% glycerol (10–30%; 15–55%) or 10% methanol (0–15%; 0–35%). Sperm in 10% dimethyl sulfoxide frozen 10 cm above liquid nitrogen (motility: 20%; viability: 50%) and thawed at room temperature for 40 seconds (20%; 48%) had significantly higher motility and viability than sperm frozen 5 cm (1%; 8%) or 8 cm (8%; 16%) above liquid nitrogen and thawed at 5, 25, or 30°C for 10, 15 or 60 seconds respectively (1–8%; 6–20%). Sperm frozen in MIS with 5% dimethyl sulfoxide resulted in higher hatching rate (29%) than sperm frozen in sucrose or glucose (300 mmol/L) containing 5% or 10% dimethyl sulfoxide (6–19%) or in MIS containing 10% dimethyl sulfoxide (9%). Viability did not differ (24–38%). Addition of 73 mmol/L sucrose to MIS with 5% dimethyl sulfoxide increased sperm motility (18 to 46%) and hatching rate (29 to 48%). Testes from three males were macerated and tested/treatment. Fertilization was tested using 25–30 eggs at 18°C.

A replicated study in 2009 of captive European common frogs *Rana temporaria* in Austria (9) found that the most effective cryopreservation protocol was sperm in motility-inhibiting saline (MIS) with 5% glycerol, 2.5% sucrose and 5% hen egg yolk, frozen 10 cm above liquid nitrogen and thawed at 22°C for 40 seconds. Sperm motility was maintained following incubation for 40 minutes at 4°C in MIS with 10% dimethyl sulfoxide (71%), 5% glycerol (69%) or 10% methanol (59%), but not 10% propandiol (0%). When frozen, in combination with sucrose, dimethyl sulfoxide resulted in significantly greater sperm motility and viability (10%; 42% respectively) than glycerol (8%; 25%). With MIS, motility and viability was similar with either dimethyl sulfoxide (13%; 27%) or glycerol (10%; 29%). Sperm frozen in MIS with sucrose and methanol had no motility. Sperm frozen 5 cm above liquid nitrogen had no motility, whereas at 10 cm motility was 30–35%. Addition of 5% (vs 10%) egg yolk and 2.5% sucrose to MIS with glycerol significantly increased hatching rate compared to all other treatments (23 vs 2–12%). Motility and viability did not differ. Testes from wild males were macerated (3/treatment). Sperm was frozen in liquid nitrogen. Fertilization was tested using 25–30 eggs.

A replicated, controlled study in 2009 of European common frogs *Rana temporaria* in the Moscow Region, Russia (10) found that recovery of sperm after cryopreservation was high with certain cryoprotectants. Sperm motility was significantly greater with the cryoprotectant dimethyl formamide (motility: 65%; fertilization: 90%) compared to dimethyl sulphoxide (36–44%; 82–90%). High concentrations of dimethyl sulphoxide (6 vs 2–4%) significantly reduced hatching (54 vs 80%) and larval survival (49 vs 70–76%), but not fertilization (80 vs 86–90%). Motility-inhibiting saline and glycerol cryoprotectant resulted in low motility (28%) and zero fertility. Tris buffer in cryoprotectants did not significantly increase motility (43–48 vs 45%) or fertilization (70–81 vs 84%). Maximum fertilization was achieved with spermic urine from hormonally induced males (luteinizing hormone-releasing hormone) at concentrations of 15 x 10^6/ml (93%). Spermic urine or macerated testes from wild frogs were mixed with simplified amphibian Ringer solution or saline and cryodiluents: 2–12% dimethyl sulphoxide or 12% dimethyl formamide or motility-inhibiting saline and 5% glycerol, with 2.5, 6.5 or 10% sucrose with or without Tris buffer or 5–10% egg yolk. Spermic urine (1.0 x 10^8 cell/ml) and cryodiluents were frozen at 5–7°C/minute and then stored in liquid nitrogen. Thawing was in a 40°C water bath. Spermic urine, sperm from macerated testes (different concentrations) or thawed sperm in cryodiluents were added to eggs from hormonally induced wild females. Fertilization was assessed after 4–6 hours.

(1) Beesley S.G., Costanzo J.P. & Lee R.E. (1998) Cryopreservation of spermatozoa from freeze-tolerant and intolerant anurans. *Cryobiology*, 37, 155–162.

(2) Browne R.K., Clulow J., Mahony M. & Clark A. (1998) Successful recovery of motility and fertility of cryopreserved cane toad (*Bufo marinus*) sperm. *Cryobiology*, 37, 339–345.

(3) Mugnano J.A., Costanzo J.P., Beesley S.G. & Lee R.E. (1998) Evaluation of glycerol and dimethyl sulfoxide for the cryopreservation of spermatozoa from the wood frog (*Rana sylvatica*). *Cryo-Letters*, 19, 249–254.

(4) Browne R.K., Clulow J. & Mahony M. (2001) Short-term storage of cane toad (*Bufo marinus*) gametes. *Reproduction*, 121, 167–173.

(5) Browne R.K., Clulow J. & Manony M. (2002) The short-term storage and cryopreservation of spermatozoa from hylid and myobatrachid frogs. *Cryo Letters*, 23, 129–136.

(6) Michael S.F. & Jones C. (2004) Cryopreservation of spermatozoa of the terrestrial Puerto Rican frog, *Eleutherodactylus coqui*. *Cryobiology*, 48, 90–94.

(7) Sargent M.G. & Mohun T.J. (2005) Cryopreservation of sperm of *Xenopus laevis* and *Xenopus tropicalis*. *Genesis*, 41, 41–46.

(8) Mansour N., Lahnsteiner F. & Patzner R.A. (2009) Optimization of the cryopreservation of African clawed frog (*Xenopus laevis*) sperm. *Theriogenology*, 72, 1221–1228.

(9) Mansour N., Lahnsteiner F. & Patzner R.A. (2010) Motility and cryopreservation of spermatozoa of European common frog, *Rana temporaria*. *Theriogenology*, 74, 724–732.

(10) Shishova N.R., Uteshev V.K., Kaurova S.A., Browne R.K. & Gakhova E.N. (2010) Cryopreservation of hormonally induced sperm for the conservation of threatened amphibians with *Rana temporaria* as a model research species. *Theriogenology*, 75, 220–232.

13.6   Release captive-bred individuals

- One review[1] found that 41% of release programmes of captive-bred or head-started amphibians showed evidence of breeding in the wild for multiple generations, 29% showed some evidence of breeding and 12% evidence of survival following release.

Background

Captive breeding is usually undertaken to provide individuals for release into the wild, either to reintroduce the species to part of their former range, or to increase the size of an existing population.

Amphibians possess a number of traits that make them potentially suitable for captive breeding and reintroduction programmes. They reproduce relatively quickly and their small size and low maintenance requirements allow viable populations to be managed much more cost-effectively than many larger animals. Also unlike higher vertebrates that possess many learned behaviours that may reduce survival in the wild, the hard-wired physiology and behaviour of amphibians means that pre- and post-release training are not required. However, before release consideration must be given to genetic management, health screening, acclimation of animals, long-term monitoring and involvement of local stakeholders.

Studies investigating captive breeding are discussed in 'Breed amphibians in captivity'.

A review in 2008 of the effectiveness of 39 release programmes of captive-bred or head-started amphibians (1) found that 14 of 17 programmes that could be assessed were considered successful. Seven species (2 toad; 3 frog; 2 newt) showed evidence of breeding in the wild for multiple generations (high success), 5 species (3 toad; 2 frog) showed some evidence of breeding (partial success) and 2 species (1 toad; 1 frog) only showed evidence of survival following release (low success). Three programmes were considered unsuccessful and the outcome was not known for the other 19. Species from 16 countries were involved in these release programmes, with a bias towards temperate countries. Half of the species were classified in the top four highest IUCN threat categories (i.e. vulnerable to extinct in the wild).

(1) Griffiths R.A. & Pavajeau L. (2008) Captive breeding, reintroduction, and the conservation of amphibians. *Conservation Biology*, 22, 852–861.

13.6.1 Frogs

- Four of five studies (including one replicated study and one review) in Europe, Hong Kong and the USA found that captive-bred frogs released as tadpoles, juveniles or adults established populations[2] or stable breeding populations at 100%[1, 4, 6] or 88%[5] of sites, and in some cases colonized new sites[1, 4, 5]. One study[11] found that stable breeding populations were not established. One before-and-after study in Spain[7] found that released captive-bred, captive-reared and translocated frogs established breeding populations at 79% of sites.

- Three replicated studies in Australia and the USA found that a high proportion of captive-bred frogs released as eggs survived to metamorphosis[12], some released as tadpoles survived at least the first few months[10] and few released as froglets survived[9]. Three studies (including two replicated studies) in Australia, Italy and the UK and a review in the USA found that captive-bred frogs reproduced at all[9] or 31–33% of release sites[8, 11], or that there was very limited breeding by released frogs[3].

Background
As there is a larger literature for green and golden bell frogs *Litoria aurea* than other species, evidence is considered in a separate section below.

A before-and-after study in 1988–1997 in ponds on abandoned farmland in Liepâja, Latvia (1, 4) found that released captive-bred European tree frog *Hyla arborea* froglets established stable breeding populations at release sites and frogs colonized new breeding sites. Males were recorded calling from 1990 and tadpoles were observed from 1991. From 1993, calling males were heard outside the release site. By 1994 there were 7 ponds with calling males up to 2 km away and by 1997 this had increased to 48 ponds within a 20 km radius of the release site. Breeding was recorded in at least 10 of those ponds. At least four generations had been produced in the wild by 1997. A total of 4,110 froglets were released into ponds in a Nature Conservation Area (300 ha) in June–August 1988–1992. Ponds were monitored using call surveys in spring and by counting tadpoles and froglets in autumn.

A before-and-after study of projects in 1986–1997 that released captive-bred amphibians into restored and created ponds in Denmark (2) found that released European tree frogs *Hyla arborea* established populations. European tree frogs established populations in 10 restored and 13 created ponds. A questionnaire was sent to all those responsible for pond projects across Denmark to obtain data. Animals were reared in captivity and then released into ponds as tadpoles or juveniles. For a pond to be defined as 'colonized' a species had to be present but not breeding.

A replicated, before-and-after study in 1994–1997 in Jersey, UK (3) found that there was limited breeding by released captive-bred agile frogs *Rana dalmatina*. The first egg mass was recorded two years after the first release and eggs were head-started due to the risk of predation by palmate newts *Triturus helveticus*. However, there was no breeding at the site the following year, although adults were recorded. In 1994–1996, 100–200 well-developed tadpoles each year and in 1996 twenty young frogs were released into 2 ponds.

A replicated, before-and-after study in 1992–1998 in Hong Kong (5) found that released captive-bred Romer's frog *Philautus romeri* tadpoles and adults established and maintained populations at seven of eight sites for four to five years after release. However, populations remained small and only one expanded its range significantly. In 1992, a total of 230 adults, several eggs and tadpoles were collected from the wild. Thirty adults were sent to Melbourne Zoo and the remainder were housed at the University of Hong Kong. A total of 1,170 frogs and 1,622 tadpoles were released in 1993 at 3 sites and in 1994 at 8 sites. Additional small ponds were constructed at some sites to provide fish-free habitat. Frogs were monitored annually by call and visual surveys.

A before-and-after study in 1992–2000 in Jersey, UK (6) found that released captive-bred agile frog *Rana dalmatina* tadpoles established a breeding population. The first egg mass was found in 1996, two years after the first release. Breeding also occurred in the release pond in 1998–2000. Mortality during the embryonic stage was 50% in captivity compared to 40% in the wild. One or two egg clumps were taken from the wild for captive breeding in 1992–1993 and 1997–2000. Captive-bred tadpoles were released into a pond in 1994–2000.

A before-and-after study in 1998–2003 in Gipuzkoa province, Spain (7) found that released captive-bred and captive-reared stripeless tree frog *Hyla meridionalis* juveniles and translocated adults established breeding populations in 11 of 14 created ponds. Metamorphosis, mating, eggs and well-developed larvae were observed in 11 of the ponds, froglets were also recorded in some ponds. Translocated adults survived in good numbers and returned to 12 of 14 ponds. Introduced predators, dense vegetation, eutrophication and drying resulted in reduced survival and reproduction in some ponds. A small number of additional ponds were colonized by the species. Thirteen ponds were created and one restored, with vegetation planted in 1999–2000. In 2000–2003, a total of 5,767 tadpoles were bred in captivity and released (171–3,989/year). Eggs were also collected, reared in captivity (in outdoor ponds) and then released as 871 metamorphs and 19,478 tadpoles into 8 of the ponds. In 1998–2003, a total of 1,405 adults were translocated to the ponds.

A replicated, before-and-after study in 1999–2006 in 18 ponds in Lombardy, Northern Italy (8) found that captive-bred Italian agile frogs *Rana latastei* released as tadpoles reproduced in 6 of the ponds. At least one egg mass (1–14) and/or calling males (4–8 in two ponds) were recorded in 6 of the 18 ponds. Four the ponds with breeding were new ponds and two were unmanaged. Up to four adults were found in three of the ponds. Breeding success was negatively affected by human disturbance and predator presence and positively affected by woodland, shore incline and pond permanence. Human disturbance was noted at 89% of the sites and potential predators, mainly fish, were found in 39% of ponds. New ponds were excavated in 6 Natural Parks in 1999–2001. In 2000 and 2001, tadpoles were released in 13 new ponds and 5 existing unmanaged ponds that had not recently been used for breeding. In February–April 2006, ponds were monitored during 45 visual and call surveys (average 2.5/pond).

A before-and-after study in 2008–2010 in New South Wales, Australia (9) found that only 4 of 610 released captive-bred booroolong frogs *Litoria booroolongensis* were found a year after release. A total of 105 frogs were captured after release, 29 of which survived to sexual maturity and engaged in breeding activity. At sexual maturity, released frogs were similar in size and condition to wild frogs at the site. A high infection rate of chytridiomycosis was recorded in the population. A total of 610 2- to 4-month-old frogs were marked and released along a 1.5 km section of a creek in February 2008. The creek was surveyed four times during the two months following the release and six times in October and February 2008–2010.

A replicated study in 2009–2011 at San Diego Zoo, California, USA (10) found that mountain yellow-legged frog *Rana muscosa* tadpoles survived for at least the first few months after release. All tadpoles survived in acclimation cages prior to release. In 2011, a number of tadpoles released that year survived at least until the autumn. In 2006, 82 tadpoles were rescued from a drying stream and breeding was attempted from 2009. In 2010, 30 eggs and 36 tadpoles and in 2011, 300 eggs and 300 tadpoles were released into screen cages in a stream within a reserve. Tadpoles were kept in cages to acclimatize for different periods of time before release. Regular monitoring was undertaken.

A review of two release programmes of captive-bred chiricahua leopard frogs *Lithobates chiricahuensis* in Arizona, USA (11) found that one programme resulted in breeding at 4 of 13 release sites and at 4 new localities, whereas the other programme failed. In one programme, breeding was first observed 10 months after release and a total of 32 egg masses were recorded. In the second programme, multiple releases at four sites over a number of years did not result in the establishment of populations as no frogs were detected from 2009. In the first programme, 3,542 metamorphs and late-stage tadpoles were released at 13 sites throughout a watershed in 2009–2010. In the second programme, frogs were released at three sites from 1996 and four from 2000 to 2011. Most releases comprised fewer than 100 frogs. Surveys were undertaken shortly after release and then two to three times annually.

A replicated study in 2012 of southern corroboree frogs *Pseudophryne corroboree* at Taronga and Melbourne Zoo, Australia (12) found that a high proportion of captive-bred frogs that were released as eggs reached metamorphosis and exited the ponds. Over 750 eggs were released into ponds at 3 remote sites. Captive breeding was undertaken as fewer than 50 individuals remained in the wild, mainly because of chytridiomycosis.

(1) Zvirgzds J., Stašuls M. & Vilnìtis V. (1995) Reintroductions of the European tree frog (*Hyla arborea*) in Latvia. *Memoranda Societatis pro Fauna et Flora Fennica*, 71, 139–142.
(2) Fog K. (1997) A survey of the results of pond projects for rare amphibians in Denmark. *Memoranda Societatis pro Fauna et Flora Fennica*, 73, 91–100.
(3) Gibson R.C. & Freeman M. (1997) Conservation at home: recovery programme for the agile frog *Rana dalmatina* in Jersey. *Dodo*, 33, 91–104.
(4) Zvirgzds J. (1998) Treefrog reintroduction project in Latvia. *Froglog* 27, 2–3.
(5) Dudgeon D. & Lau M.W.N. (1999) Romer's frog reintroduction into a degraded tropical landscape, Hong Kong, P.R. China. *Re-introduction News*, 17, 10–11.
(6) Racca L. (2002) The conservation of the agile frog *Rana dalmatina* in Jersey (Channel Islands). *Biota*, 3, 141–147.
(7) Rubio X. & Etxezarreta J. (2003) Plan de reintroducción y seguimiento de la ranita meridional (*Hyla meridionalis*) en Mendizorrotz (Gipuzkoa, País Vasco) (1998–2003). *Munibe*, 16, 160–177.
(8) Pellitteri-Rosa D., Gentilli A., Sacchi R., Scali S., Pupin F., Razzetti E., Bernini F. & Fasola M. (2008) Factors affecting repatriation success of the endangered Italian agile frog (*Rana latastei*). *Amphibia-Reptilia*, 29, 235–244.
(9) McFadden M., Hunter D., Harlow P., Pietsch R. & Scheele B. (2010) Captive management and experimental re-introduction of the booroolong frog on the South Western Slopes region, New South Wales, Australia. In P. S. Soorae (ed) *Global Re-introduction Perspectives: 2010. Additional case studies from around the globe*, IUCN/SSC Re-introduction Specialist Group, Gland, Switzerland. pp. 77–80.
(10) Medlin D.D. (2011) San Diego Zoo release more Southern California Mountain yellow-legged frogs. *Amphibian Ark Newsletter*, 17, 12.
(11) Sredl M.J., Akins C.M., King A.D., Sprankle T., Jones T.R., Rorabaugh J.C., Jennings R.D., Painter C.W., Christman M.R., Christman B.L., Crawford C., Servoss J.M., Kruse C.G., Barnitz J. & Telles A. (2011) Re-introductions of Chiricahua leopard frogs in southwestern USA show promise, but highlight problematic threats and knowledge gaps. In P. S. Soorae (ed) *Global Re-introduction Perspectives: 2011. More case studies from around the globe*, IUCN/SSC Re-introduction Specialist Group & Abu Dhabi Environment Agency, Gland, Switzerland. pp. 85–90.
(12) McFadden M. (2012b) Captive-bred southern corroboree frog eggs released. *Amphibian Ark Newsletter*, 19, 10.

13.6.2 Green and golden bell frog

• One review and two before-and-after studies in Australia found that captive-bred green and golden bell frogs released mainly as tadpoles did not establish breeding populations[2, 3], or only established stable breeding populations following one of four release programmes[4].

- One study in Australia found that a small proportion of captive-bred green and golden bell frogs released as tadpoles survived at least 13 months after release[1].

A study in 2000–2007 in New South Wales, Australia (1) found that 2 captive-bred green and golden bell frogs *Litoria aurea* released as tadpoles survived at least 13 months after release. Twelve tadpoles were recorded soon after release, followed by two metamorphs. The area was in drought following release. The release site was a pond within a wetland system. Potential predators were removed from the site (eels: 50 kg; red foxes: 17 removed). Approximately 3,500 tadpoles were released in December 2005, 1,500 in February 2006 and 1,000 in April 2007. Nocturnal visual count surveys were undertaken 5 times in the first 2 weeks, 12 times within the first 2 months and then monthly August–May (32 visits). Monitoring was to continue in 2008–2012.

A before-and-after study in 1998–2004 at a created wetland on a golf course in Long Reef, Sydney, Australia (2) found that captive-bred green and golden bell frogs *Litoria aurea* released as tadpoles did not establish a self-sustaining population. Once tadpole releases had stopped the number of frogs declined to zero. Only 45 adult frogs were recorded. A few males were heard calling, but breeding was not recorded. Releases did not result in any metamorph or immature frogs if they occurred during autumn, involved low numbers of tadpoles, if ponds dried out soon after release or if fish were present. Successive releases into fish-free ponds were decreasingly successful in terms of numbers of metamorphs and immatures. Sixteen ponds, 12 interconnected (20–200 cm), were created in 1996–1997 with planting of aquatic emergent vegetation and shrubs. A total of 9,000 captive-bred 3–4-week-old tadpoles were released into the ponds over 11 occasions in 1998–2003. Amphibian monitoring was undertaken at 1–4 week intervals using artificial shelters around ponds, dip-netting and visual count surveys.

A before-and-after study in 2004–2006 of three created ponds in a restored wetland in New South Wales, Australia (3) found that captive-bred green and golden bell frogs *Litoria aurea* released as tadpoles did not result in the establishment of a stable population due to deaths from chytridiomycosis. Tadpole survival was high following release and some metamorphs survived for up to a year. However, numbers declined over the first 13 months and no frogs were recorded from March 2006. Four of six dead frogs found in 2005 and 53% of a sample of 60 juveniles captured tested positive for chytridiomycosis. In summer 2005, 850 tadpoles were released into 3 ponds created in 2002. A fence was installed surrounding the ponds and adjacent grassland (2,700 m²) to contain the frogs and in an attempt to exclude competing species, predators and the chytrid fungus. Visual encounter surveys were carried out 2–4 times each month. A sample of frogs were captured and tested for chytrid.

A review of four release programmes near Sydney, Australia (4) found that only one resulted in the establishment of a stable population of captive-bred green and golden bell frogs *Litoria aurea*. That population, which had been supplemented with 3 translocated and at least 5 colonizing adults was estimated at over 50 adults within 4 years. At Botany, frogs were detected the following spring, but none survived the summer. Non-native fish killed all individuals from the second release (fish were then eradicated). At Long Reef, 45 adults were recorded, but without continued releases the population declined to zero (for more details see (2)). At Marrickville, breeding took place after the second release, but only two survived to adults and the population became infected with chytridiomycosis. All individuals were predated after the third release. At Arncliffe, 200 captive-bred tadpoles were released in 2 created ponds in 2000–2001. At Botany, there were 4 releases each of 500–1,500 tadpoles and 0–50 juveniles into 2 ponds in 1996–2000. At Long Reef Golf Course, 9,300 tadpoles, 70 juveniles and 5 adults were released into 16 ponds over 11 occasions in 1998–2004. At Marrickville, a total of 162 tadpoles were released into a created pond over 3 occasions in 1998–2000.

(1) Daly G., Johnson P., Malolakis G., Hyatt A. & Pietsch R. (2008) Reintroduction of the green and golden bell frog *Litoria aurea* to Pambula on the south coast of New South Wales. *Australian Zoologist*, 34, 261–270.
(2) Pyke G.H., Rowley J., Shoulder J. & White A.W. (2008) Attempted introduction of the endangered green and golden bell frog to Long Reef Golf Course: a step towards recovery? *Australian Zoologist*, 34, 361–372.
(3) Stockwell M.P., Clulow S., Clulow J. & Mahony M. (2008) The impact of the amphibian chytrid fungus *Batrachochytrium dendrobatidis* on a green and golden bell frog *Litoria aurea* reintroduction program at the Hunter Wetlands Centre Australia in the Hunter region of NSW. *Australian Zoologist*, 34, 379–386.
(4) White A.W. & Pyke G.H. (2008) Frogs on the hop: translocations of green and golden bell frogs *Litoria aurea* in Greater Sydney. *Australian Zoologist*, 34, 249–260.

13.6.3 Toads

• Two of three studies (including two replicated studies) in Denmark, Sweden and the USA found that captive-bred toads released as tadpoles, juveniles or metamorphs established populations[6], in one case at 70% of sites[5]. One of the studies[2,3] found that populations were not established from captive-bred and head-started toads.

• Two studies in Puerto Rico found that survival of released captive-bred Puerto Rican crested toads was low[4] and that 25% were predated within two days of release[1].

Background
As there is a larger literature for Mallorcan midwife toads *Alytes muletensis* than other species, evidence is considered in a separate section below.

A study in 1988 in Guanica, Puerto Rico (1) found that 4 of 12 captive-bred Puerto Rican crested toads *Peltophryne lemur* were predated by non-native Indian mongoose *Herpestes palustris* within 2 days of release. Twelve two-year-old captive-bred toads were fitted with radio-transmitters and were released into the breeding ponds that their parents had been collected from.

A replicated study in 1982–1986 in Attwater Prairie Chicken National Wildlife Refuge in Texas, USA (2, 3) found that released captive-bred and head-started Houston toads *Bufo houstonensis* did not establish populations. Eight released males but no females were recorded during five years of monitoring. Two egg strings were found in 1985. Survival was low as many tadpoles were predated. Over 5 years, 62 adult, 6,985 metamorphs and 401,384 eggs were released at 1–10 sites/year. Animals were either captive-bred or eggs were collected in the wild and raised in captivity (indoors and outdoors) before release. Monitoring was undertaken nightly in February–June 1982–1986.

A study in western Guánica, Puerto Rico (4) reported that a small number of captive-bred Puerto Rican crested toads *Peltophryne lemur* survived after release. Two of a group of 640 released were observed in 1989 and others sighted in 1992 and 1993. Predation by mongooses had a significant effect on the survival of radio-tracked released adults. Three thousand newly metamorphosed toads were released in 1988. A further 12 captive-bred adults were released with radio-transmitters.

A replicated, before-and-after study in 1982–1993 in Sweden (5) found that released captive-bred fire-bellied toad *Bombina bombina* metamorphs established populations at 7 of 10 sites, with some populations increasing over 3 years. By 1993, toads were found in 75 ponds. Numbers of ponds with calling males decreased from 26 in 1990 to 15 in 1993, although numbers calling increased to 80. Metamorphs were recorded at 0–10 ponds. The total minimum population size varied from 150 to 300. In 1982–1984, eggs from Denmark were raised to adults for breeding. Over 6 years, captive-bred larvae were released into net cages in ponds at 10 sites, each with 2–15 suitable breeding ponds. Some eggs were raised to metamorphs before release, all metamorphs were set free in ponds. Toads were monitored in 1990–1993.

A before-and-after study of projects in 1986–1997 that released captive-bred amphibians into restored and created ponds in Denmark (6) found that European fire-bellied toads *Bombina bombina* and green toads *Bufo viridis* established populations. Released fire-bellied toads established populations in 18 restored and 22 created ponds and green toads in 3 created ponds. A questionnaire was sent to all those responsible for pond projects across Denmark to obtain data. Animals were reared in captivity and then released into ponds as tadpoles or juveniles. For a pond to be defined as 'colonized' a species had to be present but not breeding.

(1) Johnson B. & Paine F. (1989) *The Release of Puerto Rican Crested Toads: Captive Management Implications and the Cactus Connection.* Proceedings of the Regional Meetings of the American Association of Zoological Parks and Aquariums. pp. 962–967.
(2) Quinn H., Peterson K., Mays S., Freed P. & Neitman K. (1989) *Captive Propagation/Release and Relocation Program of the Endangered Houston Toad, Bufo houstonensis.* Proceedings of the 1989 American Association of Zoological Parks and Aquariums National Conference, Wheeling, WV. pp. 457–459.
(3) Dodd C.K.J. & Seigel R.A. (1991) Relocation, repatriation, and translocation of amphibians and reptiles: are they conservation strategies that work? *Herpetologica*, 47, 336–350.
(4) Johnson R.R. (1994) Model programs for reproduction and management: ex situ and in situ conservation of toads of the family Bufonidae. In J. B. Murphy, K. Adler & J. T. Collins (eds) *Captive Management and Conservation of Amphibians and Reptiles*, Contributions to Herpetology Vol. 11, Society for the Study of Amphibians and Reptiles, Ithaca, New York. pp. 243–254.
(5) Andren C. & Nilson G. (1995) Re-introduction of the fire-bellied toad *Bombina bombina* in Southern Sweden. *Memoranda Societatis pro Fauna et Flora Fennica*, 71, 82–83.
(6) Fog K. (1997) A survey of the results of pond projects for rare amphibians in Denmark. *Memoranda Societatis pro Fauna et Flora Fennica*, 73, 91–100.

13.6.4 Mallorcan midwife toad

- Three studies (including one replicated study and one review) in Mallorca found that captive-bred midwife toads released as tadpoles, toadlets or adults established breeding populations at 38%[1], 80%[2] or 100% of sites[4].

- One randomized, replicated, controlled study in the UK[3] found that predator defences were maintained, but genetic diversity reduced in a captive-bred reintroduced population.

A review of release programmes for captive-bred Mallorcan midwife toads *Alytes muletensis* in Mallorca (1) found that breeding populations were established at three of the eight release sites. Four of the other sites had calling males that were expected to breed by the end of 1994. Captive-bred toads, from three institutions, were released on eight occasions starting in 1989.

A before-and-after study in 1985–2002 in Mallorca (2) found that captive-bred Mallorcan midwife toads *Alytes muletensis* released as larvae and adults established breeding populations at 12 of 15 sites. Between 4 and 721 larvae were counted per site in 2001 and 4,000–5,000 were counted at 1 site in 2002. Two of the three populations failed because of predation by viperine snakes *Natrix maura*. From 1985, captive breeding was undertaken by a number of national and international centres. Adults (0–387/site) and larvae (0–227) were released at 15 sites over 1–9 years in 1985–1997. Populations were surveyed in 2001.

A randomized, replicated, controlled study in captivity in the UK (3) found that predator defences were maintained in a captive-bred reintroduced population of Mallorcan midwife toads *Alytes muletensis*, but genetic diversity was reduced. There was no significant difference in morphological responses to predators in a population that had been captive-bred for 3–8 generations and released in a predator-free pond and the ancestral natural population. Tail length, lower tail fin shape and development did not differ. In terms of genetic diversity, although heterozygosity was similar between populations, the reintroduced population had lower allelic richness. Forty-eight tadpoles from the natural and reintroduced population (with the same ancestry) were captured. Treatments were: chemical cues from viperine snakes *Natrix maura* or green frogs *Rana perezi* or a control. Tadpoles were measured every 15 days. DNA was analysed.

A replicated study in 1989–2001 in Mallorca (4) found that released captive-bred Mallorcan midwife toads *Alytes muletensis* established stable and in some cases increasing, breeding populations at all 18 release sites. Seventy-six captive-bred tadpoles were released at 2 sites in 1989. Toadlets and tadpoles were then released on an annual basis up to 1997 and less regularly until 2001. Toads were screened for disease before release. Tadpoles were counted annually at the 18 release sites.

(1) Bloxam Q.M.C. & Tonge S.J. (1995) Amphibians: suitable candidates for breeding-release programmes. *Biodiversity and Conservation*, 4, 636–644.
(2) Román A. (2003) El ferreret, la gestión de una especie en estado crítico. *Munibe*, 16, 90–99.
(3) Kraaijeveld-Smit F.J.L., Griffiths R.A., Moore R.D. & Beebee T.J.C. (2006) Captive breeding and the fitness of reintroduced species: a test of the responses to predators in a threatened amphibian. *Journal of Applied Ecology*, 43, 360–365.
(4) Griffiths R.A., García G. & Oliver J. (2008) Re-introduction of the Mallorcan midwife toad, Mallorca, Spain. In P. S. Soorae (ed) *Global Re-introduction Perspectives: 2008. Re-introduction case-studies from around the globe*, IUCN/SSC Re-introduction Specialist Group, Abu Dhabi. pp. 54–57.

13.6.5 Salamanders (including newts)

- One before-and-after study in Germany[1] found that captive-bred great crested newts and smooth newts released as larvae, juveniles and adults established stable breeding populations.

A before-and-after study in 1994–2004 of created ponds in wet meadows in the Luhe valley, Germany (1) found that released captive-bred great crested newts *Triturus cristatus* and smooth newts *Triturus vulgaris* established stable breeding populations. Newts colonized new ponds within four years. By 2004, they bred in 9 of 14 ponds, inhabited all terrestrial habitats (at low densities) and had moved up to 4 km away. Fourteen ponds and many small pools were created and planted with aquatic species, as well as fish removal from existing ponds and terrestrial habitat management. Initially, 24 of each species were housed in indoor tanks. In following years, eggs were produced in an outdoor enclosure and then collected and transferred indoors for rearing. Sixty captive-bred great crested newt and 90 smooth newt larvae and juveniles were released into 2 created ponds annually. In 2000–2004, 5–10 adults were also released into the 2 ponds.

(1) Kinne O. (2004) Successful re-introduction of the newts *Triturus cristatus* and *T. vulgaris*. *Endangered Species Research*, 1, 25–40.

13.7   Head-start amphibians for release

- Twenty-two studies head-started amphibians from eggs and monitored them after release.

- Six of ten studies (including five replicated studies) in Denmark, Spain, the UK and the USA and a global review found that released head-started tadpoles, metamorphs or juveniles established breeding frog populations[12] or increased populations of frogs[16, 26] or toads[3, 5, 13]. Two found mixed results with breeding populations established in 12 of 17 studies reviewed[15] or at 2 of 4 sites[7]. Two found that head-started metamorphs or adults did not prevent a frog population decline[6] or establish a breeding toad population[9]. For five of the studies, release of captive-bred individuals, translocation or habitat management were also carried out[3, 7, 12, 13, 15].

- Nine of ten studies (including nine replicated studies) in Australia, Canada, Europe and the USA found that head-started amphibians released as tadpoles, metamorphs or adults metamorphosed successfully[10, 18], tended to survive the first season[25], winter[21] or year[19, 22] or bred successfully[2, 4, 14, 21]. One found adult survival was 1–17% over four years[18] and one found limited breeding following the release of adults[11].

- Four replicated studies in Australia, the UK and the USA found that frog survival to metamorphosis[8, 16, 17] and size at metamorphosis[17, 18] was greater and time to metamorphosis shorter[8] in head-started compared to wild animals. One replicated study in Canada[14] found that young head-started leopard frogs were smaller than those in the wild. One replicated study in Australia[8] found that corroboree frog tadpoles released earlier had higher survival, but metamorphosed two weeks later than those released a month later.

- Three studies (including one replicated study) in the USA[1, 20, 23, 24] only provided results for head-starting in captivity. Two found that Houston toad eggs could be captive-reared to tadpoles, but only one found that they could be successfully reared to adults[1, 20, 23]. Three studies (including two replicated studies) in Canada and the USA found that during head-starting, amphibian growth rate, size, stress levels and survival were affected by the amount of protein provided[14], housing density[24] or enclosure location[1]. One found that mass, stress levels and survival were not affected by the amount of food or habitat complexity[24].

Background

Head-starting is a management technique that raises early stage amphibians (eggs, larvae, juveniles) to later life stages (sub-adults, adults) in captivity before releasing them into native habitats. The early life stages may either have been collected in the wild or have been bred in captivity. Here we only include those that were collected from the wild. For those that were bred in captivity see 'Breed amphibians in captivity' and 'Release captive-bred individuals'.

A replicated study in 1978–1979 of Houston toads *Bufo houstonensis* at Houston Zoo, USA (1) found that tadpoles were raised successfully from eggs, but problems were encountered raising adults. Mortality rates of tadpoles were 5–9%. Several experimental groups under different conditions demonstrated that toads raised in naturally planted outdoor enclosures grew faster and had significantly higher survival rates than those raised indoors.

A replicated, before-and-after study in 1985–1987 of 20 restored and created ponds near Aarhus, Denmark (2) found evidence of breeding by European tree frogs *Hyla arborea* a year after head-started metamorphs were released. In 1986, 17–21 males were heard calling in 4 ponds, but no females, eggs or tadpoles were recorded. In 1987, up to 50 males were heard calling in 13 ponds. Four egg masses were found in one pond and tadpoles in six ponds. One hundred and fifty egg masses were collected from a local wild population. Animals were captive-reared in hot houses. Over 6,000 metamorphs were released into 9 created and 11 restored ponds over 10 km² in 1985–1986.

A before-and-after study in 1972–1991 of natterjack toads *Bufo calamita* on heathland in Hampshire, UK (3) found that captive-rearing and releasing toadlets, along with aquatic and terrestrial habitat management, tripled the population (see also (13)). Egg string counts, i.e. the female population increased from 15 to 43, with a maximum 48 in 1989. Captive-reared toadlets raised from eggs were released in 1975 (8,800), 1979, 1980 and 1981 (1,000/year). Nine small ponds were created (< 1,000 m²) and four restored by excavation. In addition, scrub, bracken and swamp stonecrop *Crassula helmsii* were removed and limestone was added to one acidic pond annually in 1983–1989. Toads were monitored every 10 days in March and August each year.

A replicated, before-and-after study in 1991–1993 of created ponds on restored opencast mining land in England, UK (4) found that released head-started great crested newt *Triturus cristatus* tadpoles returned as adults and bred in the second year. Adults returned to at least five of eight ponds and larvae were caught in three of five ponds netted in 1993 (2–5 tadpoles/pond). Newt eggs were collected and reared to tadpoles in aquaria. In 1991, 630 tadpoles were released into 4 ponds and in 1992, 1,366 tadpoles into 8 ponds (66–243/pond). Ponds were surveyed using a dip-net in July 1993. Sixteen ponds (30 x 20 m) with shelved edges and terrestrial habitat had been created on restored land. Ponds were planted with submerged and edge plants. Terrestrial habitat created included scrub/woodland, rough grassland, ditches and hedgerows.

A before-and-after study in 1986–1997 of restored and created ponds at six sites in Funen County, Denmark (5) found that releasing head-started toadlets increased the population of European fire-bellied toads *Bombina bombina* over ten years. The total adult population increased from 82 in 1986–1988 to 542 in 1995–1997 (from 1–30 to 8–170 toads/site). Numbers of ponds occupied by adults increased from 8 to 62 and by tadpoles from 1 to 18. The population only declined at one site that was flooded with salt water. Wild-caught toads were paired in separate nest cages in ponds and eggs collected and reared in aquaria. Metamorphs and 1-year-olds were released into 69 restored and created ponds. Each year, ponds were monitored for calling males and breeding success (capture-recapture estimate) in 1987–1997.

A before-and-after study in 1987–1994 of ponds on Jersey, UK (6) found that releasing head-started agile frogs *Rana dalmatina* did not prevent a decline in breeding within the population. Over 300 head-started toadlets were released. However, frog activity at the release site decreased over the years and there was no breeding in 1991 or 1994. In 1987–1989 and 1992, eggs were collected in the wild, reared to froglets and released back into the wild.

A replicated, before-and-after study in 1994–1997 of garlic toads *Pelobates fuscus* in Jutland, Denmark (7) found that released head-started tadpoles established breeding populations in the two restored, but not two created ponds. Forty-three adults were also translocated to one of the restored ponds. The authors considered that the failure of created ponds may have been due to predation, because of the lack of vegetation and introduction of sticklebacks *Pungitius pungitius*. Four egg strings were laid in captivity and produced over 2,000 tadpoles. One thousand tadpoles were released at different stages before metamorphosis into one restored and one created pond. Two ponds had been restored and two created in 1994–1995. Tadpole and call surveys were undertaken.

A replicated study in 1997 at three sites in the Snowy Mountains, Australia (8) found that southern corroboree frog *Pseudophryne corroboree* survival from eggs to metamorphosis was significantly higher for captive-reared compared to wild tadpoles (53–70% vs 0–13%). That was the case at two of the three sites, at the third, the same trend was seen for average clutch survivorship (captive: 33%; field: 15%). Tadpoles released earlier had higher survival than those released later. However, late-release tadpoles metamorphosed two weeks before early-release tadpoles. Field-reared tadpoles metamorphosed two weeks later than both. A total of 374 eggs were collected from the wild. Late stage tadpoles were returned to field enclosures within their original pools in two batches one month apart. Survival of field and captive-reared tadpoles was monitored by dip-netting once a fortnight until metamorphosis. Water levels were maintained to avoid pool drying.

A before-and-after study in 1995–1999 of boreal toads *Bufo boreas* in a National Park in Colorado, USA (9) found that captive-reared and released toads did not establish a stable breeding population. Eighteen of the 800 released metamorphs were recorded one week after release, but none in 1997–1999. Unmarked metamorphs were found in 1996–1997. Fifty-six of the adult toads released were recaptured during the first three months, but none were seen in following years. Nine eggs masses were collected from the wild in July 1995. Half of each egg mass was captive reared. In September 1995, 800 captive-reared metamorphs were toe-clipped and released. The site was monitored for the following week, twice monthly in May–June 1996 and then weekly. One hundred toads were reared and released in July 1996. These were monitored on alternate days in July–September and then weekly until November. Toads had been absent from the release site for five years.

A replicated study in 1998–2000 of Italian agile frog *Rana latastei* in the Lombardy District, Italy (10) found that tadpoles were raised successfully in captivity and metamorphosed once they were released. In 2000, 1,200 agile frog tadpoles were raised successfully. In 2001, the number raised was 28,000. Animals metamorphosed once released in both years. Eggs were collected and hatched in semi-natural conditions in captivity. Tadpoles with developing hind limbs were released back to their original ponds. In 2000, frogs were released back to two sites and in 2001 to six sites. Half of the tadpoles were translocated to new and restored ponds.

A replicated study in 1999–2002 of northern leopard frogs *Rana pipiens* in Alberta, Canada (11) found limited evidence of breeding following captive-rearing and release of frogs. At one site, seven released frogs were recaptured, a further three were heard calling and one egg mass was observed. Survival to metamorphosis in captivity was 17–33% each year. Three to six egg masses were collected from the wild each year and reared to froglets in two man-made outdoor ponds. Predation was prevented where possible by exclusion or removal of predators. Between 1999 and 2002, a total of 6,500 captive-reared frogs were tagged and released at 3 new sites. Surveys were undertaken at one release site in May–July 2002.

A replicated, before-and-after study in 1998–2003 of stripeless tree frogs *Hyla meridionalis* in Gipuzkoa province, Spain (12) found that released, captive-reared juveniles, with captive-bred juveniles and translocated adults, established breeding populations in 11 of 14 created ponds. Metamorphs, breeding behaviour, eggs and well-developed larvae were observed in 11 of the ponds. Froglets were also recorded in some ponds. Translocated adults survived in good numbers and returned to 12 of 14 ponds. Introduced predators, dense vegetation, eutrophication and drying resulted in reduced survival and reproduction in some ponds. A small number of additional ponds were colonized by the species. Thirteen ponds were created and one restored with vegetation planted in 1999–2000. Eggs were collected and reared in captivity in outdoor pools. A total of 871 metamorphs and 19,478 tadpoles were released into 8 of the ponds. An additional 5,767 tadpoles were bred in captivity and released and 1,405 adults translocated to the ponds.

A replicated, before-and-after study in 1972–1999 of natterjack toads *Bufo calamita* at two sites in England, UK (13) found that captive-rearing toadlets, along with pond creation and restoration and vegetation clearance, increased populations over 20 years. At one site, the continuation of a study in 1972–1991 (3) until 1999 indicated that there was a doubling of the population. Egg string counts (i.e. female population) increased from 15 in 1972 to 32 in 1999, with a maximum number of 48 in 1989. At a second site, where head-starting had been undertaken most years since 1980, egg string counts increased from 1 in 1973 to 8 in 1999, with a maximum number of 29 in 1997. Ponds were created and restored by excavation, scrub and bracken was cleared and captive-reared toadlets raised from eggs and released. Toads were monitored annually.

A replicated study in 2000–2005 at two wetlands in British Columbia, Canada (14) found that captive-reared and released northern leopard frog *Rana pipiens* tadpoles and metamorphs survived over winter and bred successfully. At one site, 7 juveniles, 3 adults and 13 unmarked young of the year were recorded the year after release. At the other site three egg masses and numerous young-of-year were recorded in one area, but no frogs were caught in the second area. In 2005, population estimates for young of the year/site were 1,361 and 3,874 respectively. Wild young were significantly larger than captive-reared young in all but two years (13 vs 8 g). Average survival in captivity was 82%. An increased protein diet resulted in increased size at metamorphosis and decreased time to metamorphosis (reduced 75 days). In 2001–2005, 30,065 hatchlings from 27 egg masses were collected and reared in captivity. In total, 10,147 tadpoles and 14,487 metamorphs were marked and released back to the source population and at 2 restoration sites. Monitoring was undertaken using visual encounter and call surveys.

A review of the effectiveness of 39 release programmes for head-started or captive-bred amphibians (15) found that 14 of 17 programmes that could be assessed were considered successful. Seven species (2 toad; 3 frog; 2 newt) showed evidence of breeding in the wild for multiple generations (high success), 5 species (3 toad; 2 frog) showed some evidence of breeding (partial success) and 2 species (1 toad; 1 frog) only showed evidence of survival following release (low success). Three programmes were considered unsuccessful and the outcome was not known for the other 19. Species from 16 countries were involved in these release programmes, with a bias towards temperate countries. Half of the species were classified in the top four highest IUCN threat categories (i.e. vulnerable to extinction in the wild).

A replicated, before-and-after study in 1995–2007 of chiricahua leopard frogs *Lithobates chiricahuensis* at the Phoenix Zoo, USA (16) found that head-starting and releasing tadpoles and froglets increased populations. With releases, some populations had recovered enough to produce hundreds of egg masses by 2001. By 2007, the number of ponds where frogs had become or were becoming established had increased four-fold. In captivity, over 90% of egg masses survived to froglets or late stage tadpoles, compared to only about 5% reaching metamorphosis in the wild. Egg masses were collected from the wild from the late 1990s. Between 1995 and 2007, over 7,000 tadpoles and frogs were head-started. Froglets and late stage tadpoles were released back to the wild.

A replicated study in 2008 of Jersey agile frogs *Rana dalmatina* on Jersey, UK (17) found that survival to metamorphosis was higher for head-started animals than those in the wild (15–22 vs 9–17%). However, those with initial protection in the wild (17%) had similar survival to those head-started. There was no significant difference in survival from release to dispersal of head-started tadpoles released earlier (32–46%) or later (40–46%). However, those released later were larger (0.6–0.7 vs 0.5–0.6 g). Head-started metamorphs were larger than those in the wild (0.5–0.7 vs 0.3–0.5 g). Eleven egg masses were collected and raised in aquaria. Tadpoles (n = 4,468) were marked and released back to two ponds in two groups, ten days apart. Egg masses left in the wild were either protected in mesh bags for two weeks, or for four weeks in bags followed by protection pens. In June–July, frogs were monitored daily using pitfall traps 3 m apart along drift-fences surrounding ponds.

A replicated study in 2006–2010 in Kosciuszko National Park, Australia (18) found that 1–66% of released captive-reared southern corroboree frogs *Pseudophryne corroboree* survived. Survival was 1–17% over four years for released adults. Breeding males were recorded at one site in 2008 and 2010 and both sites in 2009. Survivorship from eggs to metamorphosis in artificial pools was 35–66% over two years. Tadpoles and metamorphs tended to be larger in artificial compared to natural pools. Chytrid fungus was detected in 1 of 11 artificial pools in 2008 and 1 frog in 2009. In January 2006, 196 4-year-old and 15 5-year-old frogs, largely reared from wild-collected eggs, were marked and released across two sites. Six call surveys were undertaken per site in January 2007–2010. In April–May 2008–2010, 50 wild-collected eggs were placed in 20 artificial pools (400 L tubs) across 4 natural bog sites. Tubs had a constant water flow, a layer of pond silt and *Sphagnum* moss. Survival, size and chytrid infection was assessed just before metamorphosis.

A replicated study in 2010–2012 of white-bellied frogs *Geocrinia alba* at Perth Zoo, Australia (19) found that about 70% of head-started frogs released survived for at least a year. Eggs were collected from the wild and reared in captivity for 12 months. A total of 70 frogs were released in 2010 and 31 in 2011 at the same site.

A replicated study in 2007–2012 of Houston toads *Anaxyrus houstonensis* in Texas, USA (20, 23) found that eggs were successfully reared to toads in captivity. In 2007, 35% of juveniles survived in captivity and by 2010 survival had increased to 50–55%. By 2012, approximately 700 toads were held in 3 captive breeding facilities. In 2007, 500 toads were released, in 2009 it was 4,194 and in 2010, 14,728 were released back into the wild at 10 sites. Thirty-one egg strands were collected from the wild and tadpoles raised in biosecure rooms and four outdoor exclosure tubs.

A replicated study in 2006–2011 of common midwife toads *Alytes obstetricans* and Iberian frogs *Rana iberica* in a National Park near Madrid, Spain (21) found that released head-started midwife toads bred successfully and Iberian frog metamorphs survived their first winter. Fifteen radio-tagged adult midwife toads, two males carrying eggs, one pregnant female and a number of tadpoles were recorded. Mortality of metamorphs during the winter was high. A number of Iberian frogs released the previous year and earlier in the year were located. From 2006, all toad tadpoles found in 250 ponds were collected. Larvae were treated against the chytrid fungus using elevated temperatures (> 21°C) and baths in antifungal drugs (itraconazole). Tadpoles were reared in indoor aquariums, in similar environmental conditions to in the wild. Juveniles and adults were released where they were captured. Frog egg masses and tadpoles were also collected from a stream and head-started in aquariums. Tadpoles, juveniles and adults were released in several streams where fish had been removed by electro-fishing. Animals were monitored twice a week during the summer.

A replicated study in 2008–2009 of Ozark hellbenders *Cryptobranchus alleganiensis bishop* in Missouri, USA (22) found that 64% of captive-reared animals survived up to a year following release. Five deaths occurred within 30 days of release, 3 within 50–92 days and 5 within 126–369 days. Most hellbenders stayed within a small area (90–94%), with only 7% moving over 20 m/day. Home ranges varied widely in the first 4–7 months after release (51–987 m²), but were significantly smaller during the following six months (11–31 m^2). Of those that were known to have established a home range, 69% dispersed less than 50 m from the release point. Overall, 77% had entered their core home range within 21 days of release. Thirty-six hellbenders were captive-reared from eggs and were released back to the two original sites. Animals were radio-tracked from May 2008 to August 2009.

A replicated study in 2010 of spotted salamanders *Ambystoma maculatum* in the USA (24) found that housing larvae at low densities resulted in bigger salamanders, higher survival and lower stress levels, similar to larvae in the wild. At different larval densities there were significant differences in body mass (6/tank: 1.8 g; 12/tank: 1.6 g; 30/tank: 0.9 g), survival (94%; 67%; 33% respectively) and stress levels (white blood cell ratios: 0.4; 1.5; 2.2 respectively). At medium larval densities, increased food or habitat complexity had no significant effect on body mass (food: 1.4 g; environment: 1.7 g), survival (89%; 50% respectively), or stress levels (1.3; 0.7 respectively). Egg masses were collected from the wild. Larvae were reared in 3 replicates of 5 treatments: starting densities of 6, 12 or 30 larvae/1,000 L tank, increased food (12 larvae/tank with triple the zooplankton) or increased habitat complexity (tank filled with sticks and refugia). All tanks had leaf litter on the bottom. Metamorphs were weighed and blood sampled for stress hormone levels.

A replicated study in 2010–2012 of gopher frogs *Lithobates capito* in southwest Georgia, USA (25) found that some head-started froglets survived once released. In 2012, some froglets released earlier that year were observed and a large adult female that had been released in 2010 was re-captured. Portions of egg masses were collected from one of the remaining breeding sites and transferred to institutions for rearing to metamorphosis. Tadpoles were reared outdoors in large tanks with plant matter from the egg collection site. Tadpoles were offered some supplemental feeding, but largely ate the plants provided. Over 4,300 froglets were marked and released onto restored Nature Conservancy land, which lacked a natural population. In 2012, froglets were released directly into burrows as protection from drought. Monitoring began in summer 2012.

A replicated, before-and-after study of agile frogs *Rana dalmatina* at two sites on Jersey, UK (26) found that following the release of head-started metamorphs, breeding increased at both sites. The number of egg clumps increased by approximately 500% and the number of breeding ponds occupied increased compared to five years previously. Tadpoles were held in captivity until metamorphosis and then released at existing, re-profiled and newly created ponds at two sites.

(1) Quinn H. (1980) Captive propagation of endangered Houston toads. *Herpetological Review*, 11, 109.
(2) Skriver P. (1988) A pond restoration project and a tree-frog *Hyla arborea* project in the municipality of Aarhus Denmark. *Memoranda Societatis pro Fauna et Flora Fennica*, 64, 146–147.
(3) Banks B., Beebee T.J.C. & Denton J.S. (1993) Long-term management of a natterjack toad (*Bufo calamita*) population in southern Britain. *Amphibia-Reptilia*, 14, 155–168.
(4) Bray R. (1994) *Case Study: A Programme of Habitat Creation and Great Crested Newt Introduction to Restored Opencast Land for British Coal Opencast*. Proceedings of the Conservation and Management of Great Crested Newts. Kew Gardens, Richmond, Surrey. pp. 113–125.
(5) Briggs L. (1997) Recovery of *Bombina bombina* in Funen County, Denmark. *Memoranda Societatis pro Fauna et Flora Fennica*, 73, 101–104.
(6) Gibson R.C. & Freeman M. (1997) Conservation at home: recovery programme for the agile frog *Rana dalmatina* in Jersey. *Dodo*, 33, 91–104.
(7) Jensen B.H. (1997) Relocation of a garlic toad (*Pelobates fuscus*) population. *Memoranda Societatis pro Fauna et Flora Fennica*, 73, 111–113.
(8) Hunter D., Osborne W., Marantelli G. & Green K. (1999) Implementation of a population augmentation project for remnant populations of the southern corroboree frog (*Pseudophryne corroboree*). In E. A. Campbell (ed) *Declines and Disappearances of Australian Frogs*, Environment Australia, Canberra. pp. 158–167.
(9) Muths E., Johnson T.L. & Corn P.S. (2001) Experimental repatriation of boreal toad (*Bufo boreas*) eggs, metamorphs, and adults in Rocky Mountain National Park. *Southwestern Naturalist*, 46, 106–113.

(10)   Gentilli A., Scali S., Barbieri F. & Bernini F. (2002) A three-year project for the management and the conservation of amphibians in Northern Italy. *Biota*, 3, 27–33.

(11)   Kendell K. (2003) Northern leopard frog reintroduction: year 4 (2002). Alberta Sustainable Resource Development & Fish and Wildlife Service Report. Alberta Species at Risk Report.

(12)   Rubio X. & Etxezarreta J. (2003) Plan de reintroducción y seguimiento de la ranita meridional (*Hyla meridionalis*) en Mendizorrotz (Gipuzkoa, País Vasco) (1998–2003). *Munibe*, 16, 160–177.

(13)   Buckley J. & Beebee T.J.C. (2004) Monitoring the conservation status of an endangered amphibian: the natterjack toad *Bufo calamita* in Britain. *Animal Conservation*, 7, 221–228.

(14)   Adama D.B. & Beaucher M.A. (2006) Population monitoring and recovery of the northern leopard frog (*Rana pipiens*) in southeast British Columbia. Report to the Columbia Basin Fish and Wildlife Compensation Program Report.

(15)   Griffiths R.A. & Pavajeau L. (2008) Captive breeding, reintroduction, and the conservation of amphibians. *Conservation Biology*, 22, 852–861.

(16)   Sprankle T. (2008) Giving leopard frogs a head start. *Endangered Species Bulletin*, 33, 15–17.

(17)   Jameson A. (2009) An assessment of the relative success of different conservation strategies for the Jersey agile frog (*Rana dalmatina*). MSc thesis. University of Kent.

(18)   Hunter H., Marantelli G., McFadden M., Harlow P., Scheele B. & Pietsch R. (2010) Assessment of re-introduction methods for the southern corroboree frog in the Snowy Mountains region of Australia. In P. S. Soorae (ed) *Global Re-introduction Perspectives: 2010. Additional case studies from around the globe*, IUCN/SSC Re-introduction Specialist Group, Gland, Switzerland. pp. 72–76.

(19)   Bradfield K. (2011) Geocrinia captive breeding and rear for release programs at Perth Zoo. *Amphibian Ark Newsletter*, 17, 9.

(20)   Forstner M.R.J. & Crump P. (2011) Houston toad population supplementation in Texas, USA. In P. S. Soorae (ed) *Global Re-introduction Perspectives: 2011. More case studies from around the globe*, IUCN/SSC Re-introduction Specialist Group & Abu Dhabi Environment Agency, Gland, Switzerland. pp. 71–76.

(21)   Martín-Beyer B., Fernández-Beaskoetxea S., García G. & Bosch J. (2011) Re-introduction program for the common midwife toad and Iberian frog in the Natural Park of Peñalara in Madrid, Spain: can we defeat chytridiomycosis and trout introductions? In P. S. Soorae (ed) *Global Re-introduction Perspectives: 2011. More case studies from around the globe*, IUCN/SSC Re-introduction Specialist Group & Abu Dhabi Environment Agency, Gland, Switzerland. pp. 81–84.

(22)   Bodinof C.M., Briggler J.T., Junge R.E., Beringer J., Wanner M.D., Schuette C.D., Ettling J., Gitzen R.A. & Millspaugh J.J. (2012) Postrelease movements of captive-reared Ozark hellbenders (*Cryptobranchus alleganiensis bishopi*). *Herpetologica*, 68, 160–173.

(23)   Crump P. (2012) The recovery program for the Houston Toad. *Amphibian Ark Newsletter*, 21, 13–14.

(24)   Davis A.K. (2012) Investigating the optimal rearing strategy for Ambystoma salamanders using a hematological stress index. *Herpetological Conservation and Biology*, 7, 95–100.

(25)   Hill R. (2012) Gopher frog head-starting project reaches major milestone. *Amphibian Ark Newsletter*, 21, 9.

(26)   Wilkinson J.W. & Buckley J. (2012) Amphibian conservation in Britain. *Froglog*, 101, 12–13.

14 Education and awareness raising

Background
Compared to other groups of animals such as mammals and birds, amphibians tend to be small, cryptic animals, of little economic value, which are rarely in the public eye and therefore often not valued by the general public. This means that there are significant challenges in terms of gaining understanding or involvement from the public for conservation purposes.

A number of amphibian conservation programmes have attempted to raise awareness in the public. Ideally a quantitative change in awareness, perception or behaviour would be measured within the project. However, such data are often not collected. We therefore include measures of the number of people who were engaged in the project, in other words the 'uptake', or measures of the spatial distribution or extent of engagement.

For other interventions that involve engaging volunteers to help manage amphibian populations or habitats see 'Threat: Transportation and service corridors – Use humans to assist migrating amphibians across roads' and 'Threat: Agriculture – Engage landowners and other volunteers to manage land for amphibians'.

Key messages

Raise awareness amongst the general public through campaigns and public information
Two studies, including one replicated, before-and-after study, in Estonia and the UK found that raising public awareness, along with other interventions, increased amphibian breeding habitat and numbers of toads. One before-and-after study in Mexico found that raising awareness in tourists increased their knowledge of axolotls. However, one study in Taiwan found that holding press conferences had no effect on a frog conservation project.

Provide education programmes about amphibians
One study in Taiwan found that education programmes about wetlands and amphibians, along with other interventions, doubled a population of Taipei frogs. Four studies, including one replicated study, in Germany, Mexico, Slovenia, Zimbabwe and the USA found that education programmes increased the amphibian knowledge of students.

Engage volunteers to collect amphibian data (citizen science)
Five studies in Canada, the UK and the USA found that amphibian data collection projects engaged up to 10,506 volunteers and were active in 16–17 states in the USA. Five studies in the UK and the USA found that volunteers surveyed up to 7,872 sites, swabbed almost 6,000 amphibians and submitted thousands of amphibian records.

14.1   Raise awareness amongst the general public through campaigns and public information

- Two studies (including one replicated, before-and-after study) in Estonia and the UK found that raising public awareness, along with other interventions, increased numbers of natterjack toads[1] and created 1,023 ponds for amphibians[5].

- One before-and-after study in Mexico[2] found that raising awareness in tourists increased their knowledge of axolotls. One study in Taiwan[3] found that holding press conferences to publicize frog conservation had no effect on a green tree frog project.

- Two studies in Panama and the UK found that awareness campaigns reached over 50,000 members of the public each year[4] or trained 1,016 people at 57 events over 4» years[5].

Background

Raising awareness about amphibians, the threats that they face and about their conservation can help to change public perceptions and therefore have an indirect effect on the conservation of amphibians.

This intervention involves general information and awareness campaigns in response to a range of threats. Studies describing educational and data collection programmes are described in 'Provide education programmes about amphibians' and 'Engage volunteers to collect amphibian data'.

There are a number of studies that describe projects that involve raising awareness in which measures of success such as uptake or impacts on amphibian populations have not been monitored (e.g. Zippel 2002; Griffiths *et al.* 2004). Examples of global campaigns are the 'Year of the Frog' in 2008 and the 'Search for Lost Frogs'. The 'Year of the Frog' was led by Amphibian Ark and raised awareness among governments, media, educators and the general public, and raised funds for amphibian conservation programmes (Pavajeau *et al.* 2008). The 'Search for Lost Frogs' campaign was launched in 2010 and supported expeditions to 21 countries across 5 continents to search for species not seen for a decade or more.

Griffiths R.A., Graue V., Bride I.G. & McKay J.E. (2004) Conservation of the axolotl (*Ambystoma mexicanum*) at Lake Xochimilco, Mexico. *Herpetological Bulletin*, 89, 4–11.

Pavajeau L., Zippel K.C., Gibson R. & Johnson K. (2008) Amphibian Ark and the 2008 Year of the Frog campaign. *International Zoo Yearbook*, 42, 24–29.

Zippel K.C. (2002) Conserving the Panamanian Golden Frog: *Proyecto Rana Dorada. Herpetological Review*, 33, 11–12.

A replicated, before-and-after study in 2001–2004 of 16 coastal meadows in Estonia (1) found that raising awareness, along with habitat restoration and translocation, increased numbers of natterjack toads *Bufo calamita*. Toad numbers increased on 1 island, declines were halted on 2 islands and 1 of 13 translocated populations was recorded breeding. Information on the natterjack population and conservation management was published, information boards put up and a documentary film on coastal meadows produced. In 2001–2004, habitats were restored on 3 coastal meadows where the species still occurred and on 13 where natterjacks could be reintroduced. Two hundred volunteers helped during 14 work camps. Restoration included reed and scrub removal, mowing (cuttings removed) and implementation of grazing. Sixty-six breeding ponds and natural depressions were cleaned, deepened and restored.

Approximately 30,000 tadpoles from isolated quarry populations were translocated to the 13 restored meadows.

A before-and-after study in 2002–2007 of a project to develop nature tourism at Lake Xochimilco, Mexico (2) found that activities such as training local boatmen in environmental interpretation increased visitor awareness of axolotls *Ambystoma mexicanum*. The proportion of visitors that knew what an axolotl was increased (from 35% to 57%), as did knowledge about the species (1–11% to 8–35%). Once boatmen had attended workshops, visitors regarded boatmen, rather than videos as the best source of information about the lake and its wildlife (pre-workshop: 10 vs 55%; after: 37 vs 18% respectively). Fifty-five boatmen completed workshops and 64 other locals attended conservation or souvenir production workshops. Content was informed by baseline data collected on visitation, souvenir markets and from boatmen. Press releases, brochures, souvenirs and an art calendar competition (1,300 entries) were also used to raise awareness. The profile of the project was also raised within the Mexican and UK government. A survey of 11 boatmen was undertaken for 1 month following training.

A study in 2001–2008 of raising awareness about amphibian conservation in Taiwan (3) found that holding press conferences to publicize frog conservation did not appear to help a green tree frog *Rhacophorus arvalis* project. Authors felt that the public was only interested in 'good news' stories about animals and that the media only focused on political issues. Press conferences were held and news released about the progress of the conservation project to stimulate concern and interest amongst local people. Radio, television, the Internet, websites, emails, blogs and magazines were used to increase awareness. Documentaries and special programmes for educating the general public were also broadcast.

A study in 1999–2012 of the Amphibian Rescue and Conservation Project in Panama (4) found that a large audience was reached via a multi-media award-winning awareness campaign. The website, in Spanish and English, received approximately 50,000 new visitors annually. In addition, the project engaged approximately 5,000 Facebook fans and 1,500 Twitter followers. The network of supporters provided a resource for fundraising and recruitment of volunteers. The campaign also resulted in about 50 news stories about the project each year, a weekly blog and a documentary film. In addition, legislation was passed in 2010 declaring August 14th National Golden Frog Day in Panama. In 2009, there was a legal resolution to draft and implement a national action plan for the conservation of the amphibians of Panama.

A study in 2012 of the Million Ponds Project in England and Wales, UK (5) found that 1,023 ponds were created, over 60 organizations were involved and more than 1,016 people were trained in pond creation at 57 events. In 2008–2012, the project team worked with landowners and managers to create the 1,023 ponds for rare and declining pond species. There was good coverage of the project in national and regional media and articles in more specialist publications. The aim of the 50-year initiative was to change attitudes so that pond creation becomes a routine activity in land management practices. Policy makers and the media were targeted to help raise awareness. Pond creation and management training courses were provided to partner and non-partner organizations. Over 50 factsheets were produced for an online toolkit and funding for pond creation was also provided.

(1) Rannap R. (2004) Boreal Baltic coastal meadow management for *Bufo calamita*. In R. Rannap, L. Briggs, K. Lotman, I. Lepik & V. Rannap (eds) *Coastal Meadow Management – Best Practice Guidelines*, Ministry of the Environment of the Republic of Estonia, Tallinn. pp. 26–33.
(2) Bride I.G., Griffiths R.A., Melendez-Herrada A. & McKay J.E. (2008) Flying an amphibian flagship: conservation of the Axolotl *Ambystoma mexicanum* through nature tourism at Lake Xochimilco, Mexico. *International Zoo Yearbook*, 42, 116–124.

(3) Chang J.C.-W., Tang H.-C., Chen S.-L. & Chen P.-C. (2008) How to lose a habitat in 5 years: trial and error in the conservation of the farmland green tree frog *Rhacophorus arvalis* in Taiwan. *International Zoo Yearbook*, 42, 109–115.
(4) Gratwicke B. (2012) Amphibian rescue and conservation project – Panama. *Froglog*, 102, 17–20.
(5) Million Ponds Project (2012) Million Ponds Project pond conservation – year 4 report. Pond Conservation Report.

14.2 Provide education programmes about amphibians

- One study in Taiwan[4] found that education programmes about wetlands and amphibians, along with other interventions, doubled a population of Taipei frogs.

- Three studies, including one replicated study, in Germany, Mexico, Zimbabwe and the USA found that education programmes increased the amphibian knowledge of students (1, 6), boatmen and their tourists (2). Two studies, including one replicated study, in Germany and Slovenia found that students who were taught using live amphibians and had previous direct experience, or who participated in outdoor amphibian conservation work, gained greater knowledge (1, 5), had improved attitudes towards species and retained knowledge better (5) than those than those taught indoors with pictures.

- Four studies in Mexico, Taiwan, Zimbabwe and the USA found that courses on amphibians and the environment were attended by 119–6,000 participants (2, 4, 6) and amphibian camps by 700 school children (3).

Background
Providing education programmes about amphibians can be a valuable way to raise awareness about the threats to species and the habitats that they live in and about what can be done to help. They can also help to change perceptions and may therefore indirectly help towards conserving species.

A replicated, before-and-after study in 2005 of amphibian education at a school in Baden-Württemberg, Germany (1) found that although knowledge improved significantly for all students, those who participated in outdoor conservation work performed significantly better. Achievement scores increased from two to four for indoor students and to five for students who had also captured and identified animals outdoors. Emotions did not vary between groups. Students expressed high interest and well-being and low anger, anxiety and boredom. Forty-six 9–11-year-olds were taught about amphibians indoors. A small booklet guided children through learning activities covering identification, development, habitat requirements, predation, migration and conservation. Half of the students also helped to preserve migrating amphibians, handling them outdoors. Species identification and emotional variables were tested before, one week after and four to five weeks after the programme.

A before-and-after study in 2002–2007 of a project to develop nature tourism at Lake Xochimilco, Mexico (2) found that training local boatmen in environmental interpretation resulted in increased relevant knowledge, job satisfaction, incomes and visitor awareness of axolotls *Ambystoma mexicanum*. Fifty-five boatmen completed workshops and 64 other locals attended conservation or souvenir production workshops. Trained boatmen and other students became facilitators and project assistants. Following workshops, visitors regarded boatmen rather than videos as the best source of information about the lake and its wildlife (pre-workshop: 10 vs 55%; after: 37 vs 18% respectively). Boatmen incomes increased if they provided environmental interpretation to tourists (without: 100; with: 165 pesos/trip). In

2002–2007, eight workshops were held. Workshops covered amphibian biology and conservation, conservation education, souvenir production and five were on environmental interpretation for boatmen. Content was informed by baseline data collected on visitation, souvenir markets and from boatmen. Other activities were also used to raise awareness in tourists. A survey of 11 boatmen was undertaken for 1 month following training.

A study in 2001–2008 of an educational programme for children in Taiwan (3) found that 700 school children attended 'Froggy Camps'. In 2001–2002, summer camps were two days and one night and from 2003 three days and two nights. Since 2005, camps were held twice a year. Children were given lessons on amphibians and insects in Taiwan and were taken into the field to observe frogs and other wildlife. They were taught to identify all 32 species of frogs and about how to protect natural resources.

A study in 1999–2006 of paddy fields in Taipei County, Taiwan (4) found that educating and raising awareness in a local community, along with other interventions, doubled a population of Taipei frogs *Rana taipehensis*. In 2002, over 80 locals, largely teachers and social workers attended a 5 day wetland conservation course. A further 5 courses were held in 2003–2007 with over 6,000 students attending. Three participants from the first course said they would provide farmland for wetland restoration and Taipei frog relocation. By August 2003, the Taipei frog population in the field had more than doubled (from 28 to 85) and the farmer adopted organic-farming practices. Pollution from river construction work resulted in a drastic decline in the population in 2004–2005 (20 to 4), but by 2006 the population appeared to be recovering (19). With the help of the local community, by selling a proportion of a farmer's crop and paying for any additional expenses, he was persuaded to stop using herbicides and pesticides on his field, which formed the centre of the breeding habitat. Habitat-improvement work was also undertaken with participation from a local school and agricultural foundation.

A replicated study in 2004–2005 of amphibian education in schools in Ljubljana, Slovenia (5) found that students who were taught using live animals and had previous direct experience of amphibians had the greatest knowledge and knowledge retention. Four months after the lesson, there were no significant differences between pupils taught with pictures and those with no previous experience taught with live animals. Knowledge decreased more rapidly over time in those taught with pictures. Using live animals significantly improved students' attitudes to species, with or without previous experience. Teaching with pictures significantly improved attitudes only for those that had no previous direct experience of amphibians. Twenty-one classes of 11–12-year-olds from 10 schools were given a 45-minute lesson about amphibians by the same teacher. For 127 pupils, pictures were used. The other 265 pupils handled seven live species of amphibians. Attitude towards and knowledge about amphibians was tested before and one week, two months and four months after the lesson.

A study in 2010–2012 of the Toad Trackers education programme in Houston, USA and Zimbabwe (6) found that over 190 participants completed the course and as a result were more aware of the threats to and conservation of amphibians. Since 2010, 172 participants in the USA and approximately 20 in Zimbabwe completed the course. At the end of the course, 95% of participants could list human threats to amphibians. All participants could identify simple ways they could help amphibians, such as organic gardening, helping with habitat restoration and protection, volunteering for citizen science programmes and educating others. The Houston Zoo Conservation Department developed the programme with professional herpetologists. It was aimed mainly at 8–18-year-olds, but also used for teachers, zoo workers and college students. Classroom workshops and field-based experiences covered topics such as amphibian ecological roles, conservation issues, native frog diversity and data collection. At the end of fieldwork, students completed an evaluation.

(1)   Randler C., Ilg A. & Kern J. (2005) Cognitive and emotional evaluation of an amphibian conservation program for elementary school students. *Journal of Environmental Education*, 37, 43–52.
(2)   Bride I.G., Griffiths R.A., Melendez-Herrada A. & McKay J.E. (2008) Flying an amphibian flagship: conservation of the axolotl *Ambystoma mexicanum* through nature tourism at Lake Xochimilco, Mexico. *International Zoo Yearbook*, 42, 116–124.
(3)   Chang J.C.-W., Tang H.-C., Chen S.-L. & Chen P.-C. (2008) How to lose a habitat in 5 years: trial and error in the conservation of the farmland green tree frog *Rhacophorus arvalis* in Taiwan. *International Zoo Yearbook*, 42, 109–115.
(4)   Lin H.-C., Cheng L.-Y., Chen P.-C. & Chang M.-H. (2008) Involving local communities in amphibian conservation: Taipei frog *Rana taipehensis* as an example. *International Zoo Yearbook*, 42, 90–98.
(5)   Tomažič I. (2008) The influence of direct experience on students' attitudes to, and knowledge about amphibians. *Acta Biologica Slovenica*, 51, 39–49.
(6)   Rommel R.E. (2012) Toad trackers: amphibians as gateway species to biodiversity stewardship. *Herpetological Review*, 43, 417–421.

14.3   Engage volunteers to collect amphibian data (citizen science)

- Five studies in Canada, the UK and the USA found that amphibian data collection projects engaged 100–10,506 volunteers (1, 5–8) and were active in 16–17 states in the USA (2, 5).

- Five studies in the UK and the USA found that volunteers undertook 412 surveys (7, 8), surveyed 121–7,872 sites (3–5), swabbed almost 6,000 amphibians (3) and submitted thousands of amphibian records (6).

Background

Projects in which volunteers are engaged to collect data on amphibians help to raise awareness and change public perceptions and may therefore have an indirect effect on the conservation of species.

A study in 2002 of a northern leopard frog *Rana pipiens* reintroduction programme in Alberta, Canada (1) found that over 100 volunteers became involved in the project. Volunteers included members of the general public and individuals from wildlife and commercial organizations. The project was also publicized on the radio, television and in three newspapers. Volunteers helped with frog surveys and the collection, marking and release of captive-reared frogs. Many of the volunteers, naturalist groups and school groups were given formal and informal presentations about the captive rearing programme and the natural history of Alberta's reptiles and amphibians.

A study in 2005 of the volunteer-based North American Amphibian Monitoring Program (2) found that within 8 years of establishment the programme was active in 17 states. The programme was established in 1997. Regional coordinators recruited and trained volunteer observers. Monitoring followed a specific protocol and involved stratified randomly placed roadside call surveys. These were undertaken along 25 km transects at night during 3 seasonal sampling periods.

A study in 2008 of a project investigating the distribution of the amphibian chytrid fungus *Batrachochytrium dendrobatidis* in the UK (3) found that 96 breeding ponds in England, 7 in Scotland, 16 in Wales and 2 in Jersey were sampled by volunteers engaged in the project. Almost 6,000 amphibians were swabbed for the disease. The secondary aim of the project had been to raise awareness of amphibians, the threat of the disease and the biosecurity measures that should be taken when visiting breeding ponds. Field work was largely carried out by volunteers, who were recruited and trained through voluntary county groups known as Amphibian and Reptile Groups. Ponds were visited to sample amphibians by skin swabbing (30/pond) in spring and summer.

A study in 1996–2003 of the Michigan Frog and Toad Survey in the USA (4) found that volunteers collected annual amphibian data for over 3,000 wetlands across Michigan. Between 255 and 350 routes were monitored by volunteers each year, each of which included 10 wetland sites. The programme was started in 1996 to collect data and to educate and raise awareness of amphibians and their habitats. Volunteers throughout the state monitored routes that included 10 wetlands separated by 400 m. These were surveyed using call surveys at night, three times in spring–summer.

A study in 1998–2010 of the FrogWatch USA citizen science programme (5) found that there were 10,506 registered volunteers and 20 organizations hosting FrogWatch groups in 16 states. Roughly a quarter of registered volunteers submitted data in 1998–2010 (17–43% per state). Monitoring of amphibians was undertaken at 7,872 sites by volunteers, with up to 591 visits/site. The project was run by the Association of Zoos and Aquariums, following standardized protocols that had been used since 1998. The project encouraged individuals and communities to learn about wetlands and help conserve amphibians by training volunteers to listen and report breeding calls of frogs and toads in local wetlands. Host institution coordinators were trained and then recruited, trained and supported local groups of volunteers. All resources and materials were available on the website. Data were used to help develop practical strategies for conservation.

A study in 2007–2009 of an online citizen science project, the Carolina Herp Atlas, in the USA (6) found that 698 volunteers registered and contributed 11,663 amphibian and reptile occurrence records. Numbers of records submitted by each volunteer varied from 1 to 4,452. Seventy-four people submitted 10 or more records. Distribution data were submitted for 32 frog and 51 salamander species, several of which were considered priority species in Carolina. Members of the public registered on the website to contribute data to the online database. Data were archived and distributed to a large online audience (registered and non-registered). Participants were initially recruited by contacting local and state wildlife managers, birding clubs, schools and other potentially interested groups. The project was advertised in wildlife-related publications and magazines and at scientific meetings in North and South Carolina.

A study in 2007–2012 of the National Amphibian and Reptile Recording Scheme in the UK (7, 8) found that 277 surveys in England, 106 in Scotland, 27 in Wales and 2 in Northern Ireland were undertaken by volunteers engaged in the project in 2007–2012. By 2009, 100 training events had been held across the UK, resulting in over 1,500 trained volunteers. In 2009, over 1,800 people were signed up to the project via the website. The project was coordinated by Amphibian and Reptile Conservation, the Amphibian and Reptile Groups of the UK and other partners including Statutory Agencies. Volunteers were trained in species identification, survey methodologies and bio-security. They then monitored ponds within randomly selected 1 km grid squares across the UK. Ponds were sampled during 1–4 annual visits in 2007–2012. Surveys involved visual and torch searches, netting and sometimes bottle trapping (only by experienced surveyors). Habitat characteristics were also recorded.

(1) Kendell K. (2003) Northern leopard frog reintroduction: year 4 (2002). Alberta Sustainable Resource Development & Fish and Wildlife Service Report. Alberta Species at Risk Report.
(2) Weir L.A. & Mossman M.J. (2005) North American Amphibian Monitoring Program (NAAMP). In M. Lannoo (ed) *Amphibian Declines: The Conservation Status of United States Species*, University of California Press, Berkeley, California. pp. 307–313.
(3) Cunningham A.A. & Minting P. (2008) National survey of *Batrachochytrium dendrobatidis* infection in UK amphibians. Institute of Zoology Report.

(4)   Genet K.S., Lepczyk C.A., Christoffel R.A., Sargent L.G. & Burton T.M. (2008) Using volunteer monitoring programs for anuran conservation along a rural-urban gradient in southern Michigan, USA. In J. C. Mitchell, R. E. Jung Brown & B. Bartholomew (eds) *Urban Herpetology*, SSAR, Salt Lake City.

(5)   FrogWatch USA (2010) 2010 Overview. Association of Zoos & Aquariums Report.

(6)   Price S.J. & Dorcas M.E. (2011) The Carolina Herp Atlas: an online, citizen-science approach to document amphibian and reptile occurrences. *Herpetological Conservation and Biology*, 6, 287–296.

(7)   Wilkinson J.W. & Arnell A.P. (2011) NARRS report 2007–2009: interim results of the UK National Amphibian and Reptile Recording Scheme Widespread Species Surveys. Amphibian and Reptile Conservation Report. Research Report 11/01.

(8)   Wilkinson J.W. & Arnell A.P. (2013) NARRS report 2007–2012: establishing the baseline. Amphibian and Reptile Conservation Report. Research Report 13/01.

Index